Darwin in Galápagos

Darwin in Galápagos

FOOTSTEPS TO A NEW WORLD

K. Thalia Grant and Gregory B. Estes

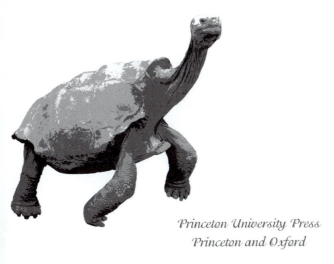

Princeton University Press
Princeton and Oxford

Copyright © 2009 by Princeton University Press

Published by Princeton University Press, 41 William Street,

Princeton, New Jersey 08540

In the United Kingdom: Princeton University Press, 6 Oxford Street,

Woodstock, Oxfordshire OX20 1TW

All Rights Reserved

ISBN: 978-0-691-14210-4

Library of Congress Control Number: 2009933541

British Library Cataloging-in-Publication Data is available

This book has been composed in Minion

Printed on acid-free paper ∞

press.princeton.edu

Printed in the United States of America

10 9 8 7 6 5 4 3 2 1

For Olivia and Devon

~~~~~~~~~~~~~~

"Whatever you can do, or dream you can do, begin it.
Boldness has genius, power and magic in it."
—*Goethe*

*"As you seem interested about the origin
of the "Origin" . . . I'll mention a few points . . ."*
—Charles Darwin, letter to Ernst Haeckel, 1864.

Contents

~~~~~~~~~~~~~~~~~~~~~~~~~~~~~~~~~~~

Part 2. Galápagos

Part 3. After Galápagos

Darwin in Galápagos

Introduction

~~~~~~~~~~~~~~~~~~~~~~~~~~~

K. Thalia Grant

The natural history of these islands is eminently curious, and well deserves attention.
—Charles Darwin, *Journal of Researches*[1]

Darwin's Islands

Galápagos is a group of 13 islands and numerous smaller islets and rocks that straddle the equator 1000 kilometers west of South America.[2] Like a cluster of forgotten crumbs on a well-swept floor they are isolated, arid, and barely visible on the world map. But insignificant they are not. Together they form a surprisingly heterogeneous world of their own, occupied by an astonishing diversity of endemic plants and animals. Their story, from their volcanic birth under the sea, to their colonization by organisms from distant lands, and the transformation of these colonists into diverse new species as the islands themselves change and diverge, is the story of evolution itself.

Viewed in this light it is no wonder that Galápagos played a pivotal role in the development of Charles Robert Darwin's famous theory of evolution by means of natural selection. For when Charles Darwin (referred to hereafter as Darwin) visited Galápagos for five short weeks in 1835 his experiences and observations there transformed his way of thinking about the natural world. They revolutionized our understanding of life on Earth.

Galápagos inspired Darwin, and both the place and the man have inspired us. It was Darwin and his theory of evolution that brought each of us to Galápagos in the first place. I arrived in 1973 with my parents, Drs. Peter and Rosemary Grant, at the start of their famous long-term study of evolution in Darwin's finches. Greg arrived nine years later as leader of the 1982 Cambridge Darwin Centenary Galápagos Expedition. Over the years we pursued university degrees in biology, conducted independent ecological research on various Galápagos species, and worked as naturalists in the islands. Our fascination with Galápagos and Darwin swelled. Then one day in early 1996, atop the most active and pristine volcano in the Galápagos archipelago, it took on a new form. The two of us were on the summit of Isla Fernandina (Narborough Island), making field observations on Darwin's finches and pledging a lifetime of adventure together. Wouldn't it be extraordinary, we agreed, to retrace Darwin's footsteps through Galápagos, to see the places he observed, to compare the wildlife he described with what can be found at the same sites today, and to investigate how else the archipelago has changed. We had little idea, at that point, of how closely we would be able to follow his path. How much physical detail had he recorded of the sites he had explored? Was it enough to reveal the treks he had taken, define the paths he had walked? Would we gain insights into the development of his ideas by following in his footsteps? The possibility was beguiling, the challenge irresistible.

We took our brainchild to England and nourished it with literature. Today Darwin's published works and transcriptions of some of his unpublished notes and manuscripts are accessible on the World Wide Web (www.darwin-online.org.uk and darwinlibrary.amnh. org), but this service did not exist at the time of our expedition. At the Cambridge University Library (CUL) we paged through volume after volume of published and unpublished Darwiniana. We feasted most heavily on Darwin's original manuscripts in the CUL Darwin Archive, for we guessed that Darwin's first-hand notes, and especially those pertaining to the geology of the islands, would contain the greatest number of clues to where he had landed and explored. We were astonished to find that Darwin's Galápagos geology notes had gone virtually unnoticed. A full transcription of those 115 manuscript pages became our first-priority assignment. We also transcribed the Galápagos portion of Captain Robert FitzRoy's log of the *Beagle* in the Public Records Office at Kew, and scoured his charts of Galápagos in the United Kingdom Hydrographic Office (UKHO) at Taunton. Back in Galápagos, at the Charles Darwin Research Station, we prepped and primed and plotted our course through the archipelago, based on all we had learned in England. Then, on October 19, 1996, one hundred and sixty-five years after HMS *Beagle* graced the waters of Galápagos, we found ourselves flying over a choppy ocean between our home island of Santa Cruz (Indefatigable) in the center of the archipelago and Darwin's starting point of Isla San Cristóbal (Chatham Island) on the eastern perimeter. Our dream was literally taking flight, with the two of us happily on board in search of what it meant to be Darwin in Galápagos.

Why Darwin?

Darwin is one of the most celebrated naturalists and influential persons in the history of mankind. Born a naturalist and an ambitious one, he spent his life seeking to explain the great diversity of life that exists in all its colors all around us, and to solve what contemporary luminary Sir John Herschel called the "mystery of mysteries"—how new species come into being.[3] Through his keen powers of observation, his interest in all aspects of the natural world, his ability to reason, and his rigorous approach to study, Darwin came up with answers that "shook the world."[4] His theory of evolution by means of natural selection not only restructured the entire science of biology, it revolutionized the way people perceive themselves. It created a whole new world of understanding.

Darwin will forever be credited with evolution, but no one pretends he was the first to think up the idea. Author Loren Eiseley, in his prize-winning book *Darwin's Century*, likens the discovery of evolution to "a new continent" that was glimpsed through lifting fogs by "master mariners" well before voyager–naturalist Darwin finally established its reality.[5] The first of these known "mariners" was

Anaximander, a 6th-century Greek philosopher who suggested that man had sprung from some form of aquatic animal.[6] The germ of an evolutionary idea may well have entered the minds of acute observers of nature even before this, in preliterate civilizations. What is known is that, in the western world, evolutionary speculations took off during the 18th century's Age of Enlightenment, when learning from nature began to replace unquestioning acceptance of dogma and myth. Darwin's own grandfather Erasmus contributed to this movement. There was even a theory of evolution when Darwin arrived on the scene, but the chief explanation for how it worked—Jean-Baptiste Lamarck's idea of the inheritance of acquired characteristics—failed to gain widespread acceptance at the time.

Darwin changed all that. By marshalling evidence in support of evolution, and coming up with a new and ultimately convincing explanation for how it works—natural selection—Darwin demonstrated the fact of evolution. He transformed evolution from a radical, somewhat illusory idea into a coherent scientific theory of consequence.

First published in 1859, Darwin's masterwork *On the Origin of Species by Means of Natural Selection, or the Preservation of Favoured Races in the Struggle for Life* (hereafter referred to as *The Origin of Species*) is Darwin's argument for evolution. In it he proclaims that all species are modified and diversified descendants of a single common ancestor and reasons that species change because of the following natural laws:[7]

1. Heritable variation exists among the individuals that comprise populations of species.
2. There is a struggle to survive and reproduce, in which individuals compete for limited resources.
3. More individuals are born than survive this struggle for existence.
4. Those individuals with variations that help them survive the conditions of their local environment are the ones most likely to reproduce and pass on their winning traits to the next generation.

As environments are neither uniform nor static it logically follows that as populations move into new environments, whether through time or space, they change. They form new species that diverge and diversify in an ever-branching tree of life.

As asserted by evolutionist Julian Huxley, Darwin's epochal book, "altered the substance and the direction of human thought more profoundly than any other publication of the age of print."[8] It transformed the prevailing view of the world as a static, biblical place inhabited by species created by God according to His will and design, into a dynamic landscape run by nature and subject to the rules of change. Ernst Mayr, another great evolutionist, goes as far as to say publication of *The Origin of Species* "almost single-handedly effected the secularization of science."[9]

This is not to say that Darwin's theory was accepted overnight. Darwin encountered major objections from scientists who doubted the validity of natural selection as an explanation for evolution. Over the 15 years following the first publication of *The Origin of Species*, Darwin revised his book five times to answer criticisms and clarify his arguments.[10] It took another 65 years for biologists to fully come to terms with natural selection as a cause of evolution.[11] Nonetheless, during his lifetime Darwin succeeded in bringing about a scientific movement that changed research programs and transformed the direction of science. In the words of science historian Peter Bowler, Darwin's "great achievement was to force the majority of contemporaries to reconsider their attitude towards the basic idea of evolution . . . despite the fact that many found natural selection unconvincing."[12]

The Origin of Species also made a huge, divisive impact outside the scientific community. The working classes, eager to "wrest control . . . from the old landed interests" hailed Darwinian evolution (or "Darwinism" as Darwin's theory and corrupted interpretations of it became known) as an endorsement for social progress and reform.[13] Members of the ruling class, those whose social and political position depended on maintaining the status quo, felt threatened. And because Darwin's theory offered secular answers for sacred questions—What is life? Who are we? Where do we come from?—it kicked up a storm of outrage among religious traditionalists. Even today, Darwin's theory of evolution remains contentious in society for its religious, social, and cultural implications.

Brilliant, ingenious, audacious, Darwin is loved and revered by many, hated by others. Above all he fascinates. As Ernst Mayr once said, "no biologist has been responsible for more—and for more drastic—modifications of the average person's worldview than Charles Darwin. . . . Almost every component in modern man's belief system is somehow affected by Darwinian principles."[14] Indeed, Darwin matters so much to human thought that he has been, and continues to be, the subject of intense scrutiny by scientists, historians of science, philosophers, psychologists, and theologians alike.

A substantial body of work has been written on Darwin and his life and seemingly no stone has been left unturned in an effort to understand the man who redefined the science of biology and the path he took to do so. In addition to a perennial supply of scientific and historical papers produced by members of this "Darwin Industry," several outstanding books have been written by some of these same scholars. They include Janet Browne's comprehensive two-volume biography[15] that describes Darwin's life, analyzes his achievements, and illuminates how his environment made him who he was; Adrian Desmond and James Moore's work of similar scope;[16] and the more concise, but insightful overviews of Richard Keynes,[17] Randal Keynes,[18] Niles Eldredge,[19] and David Quammen.[20] These books examine parts or all of Darwin's life—his childhood, his famous voyage round the world on HMS *Beagle*, his return to England, and the subsequent steps that led him to become the most influential scientist of the century—in the illuminating context of the Victorian era, the social circles in which he moved, and the scientists with whom he corresponded. Other authors adopt more oblique but equally revealing perspectives from which to explore the subject of Darwin and his intellectual journey, angles that reflect their own particular interests and expertise. For instance, Sandra Herbert investigates the key role that geology played in Darwin's thinking.[21] Edward Larson[22] and Peter Bowler[23] choose the history of evolutionary thought as the platform from which to examine Darwin and his theory.

By looking at Darwin's life from historical, philosophical, and scientific points of view, these authors paint a detailed picture of Darwin's development as he changed from an ordinary creationist to an extraordinary evolutionist. They vary on interpretation and the degree to which various moments in Darwin's life were important to his theory. Yet all agree on one thing; Galápagos was key to Darwin's conception of evolution.

Why Galápagos?

Darwin spent only 5 weeks in Galápagos, a minute fraction of the 248-week voyage of HMS *Beagle*, yet his experiences in the archipelago were of disproportionate importance to the development of his scientific thinking. Quite simply Galápagos convinced Darwin of evolution.[24] It did not happen overnight. Rather, it took several years for Darwin to fully recognize the significance of his Galápagos observations. Nonetheless, his appreciation of three fundamental features of Galápagos—its isolation, the geographical distribution of its organisms, and the affinities of these organisms to species on the South American continent and between the islands themselves—ultimately persuaded him that species could change. When he first began putting his ideas on transmutation to paper in 1837, Darwin declared the "S. American fossils —& species on Galapagos Archipelago . . . [are the] origin (especially latter) of all my views."[25] He maintained his emphasis on the importance of Galápagos, declaring to fellow evolutionist Alfred Russel Wallace 22 years later, while preparing for publication *The Origin of Species*, that the "Geographical Distrib[ution] & Geographical relations of extinct to recent inhabitants of S. America first led me to [the] subject [of evolution]. Especially case of Galapagos Isl[ds]."[26]

Galápagos was not the only factor in Darwin's conception of evolution. As Darwin pointed out in the quotes above, his recognition of the significance of Galápagos was influenced by his earlier observations on the mainland of South America and especially by his study of geographical barriers as they relate to species distributions. Nor was Galápagos central to the development of Darwin's theory of evolution. Three years after his visit to Galápagos Darwin hit upon the idea of natural selection as an explanation for evolution, not from contemplating Galápagos organisms but primarily from studying domestic breeds and reading Thomas Malthus's essay on human population theory.[27] He advanced his theory by spending decades conducting original research and gathering facts on various groups of organisms outside Galápagos, most notably on domestic pigeons, barnacles, orchids, bees, and worms. Nevertheless, Galápagos was the keystone of his conversion and the foundation for his understanding of evolution. It was the Galápagos organisms he observed and collected, and his recognition of their affinity to organisms found on the South American continent and their representation as similar but distinct species on the different islands of Galápagos, that persuaded Darwin of the mutability of species and against their miraculous creation. As one author avows, "It cannot be maintained that without Darwin the theory of evolution would not have come into being, but it can be insisted that had Darwin not taken the voyage of the *Beagle* to the Galápagos, it would have been seriously delayed."[28]

That Galápagos was important to Darwin is a well-accepted fact. The particulars—what about the islands stimulated Darwin, and how, when, and why they influenced his ideas on evolution—are generally not so well known. Indeed, they are the subject of ongoing study and exciting debate. Like a pendulum the question of whether Darwin was thinking in evolutionary terms while he was in Galápagos swings back and forth, each oscillation provoking new investigations and providing fresh insights into Darwin's Galápagos experience. In 1966 Julian Huxley wrote, "It was on the Galápagos in the early autumn of 1835 that Darwin took the first step out of the fairyland of creationism into the coherent and comprehensible world of modern biology, for it was here that he became fully convinced that species are not immutable—in other words, that evolution is a fact."[29]

While correct in essence, Huxley's words (and similar statements from other authors[30]) oversimplify the process of Darwin's conversion to evolution, and misleadingly suggest a eureka-like moment taking place in Galápagos. This has since been shown not to be the case. By using Darwin's spelling mistakes and changing orthography as a means of dating his notes, historian Frank Sulloway has mapped Darwin's evolving attitudes. He has shown that Darwin first wrote about the instability of species nine months after leaving Galápagos and has argued that Darwin did not become convinced of the mutability of species until he was back in England.[31] Sulloway has made great strides in elucidating the role Galápagos played in Darwin's thinking, and in debunking the myths of what Darwin did and did not do in the islands.[32]

Yet the pendulum swings. While it is no longer contended that Darwin reached an evolutionary standpoint while he was in Galápagos, just how close he got, and how important Galápagos was, is still under scrutiny. Darwin historian Sandra Herbert puts it this way: "[A]t first Darwin's experience in the Galápagos Islands was overemphasized as the turning point in his arrival at a transmutationist position. Now that most readers have learned that Darwin did not become an evolutionist until 1837, the Galápagos experience is possibly credited with too little."[33] After all, she reminds us, it was in Galápagos that Darwin "recorded patterns of variation among species on the islands . . . and that, ultimately, pushed him across the line to a transmutationist position."[34]

Darwin's Galápagos experience was multifaceted and the effects it had on his thinking multitiered. Thanks to the continued research of scientists and Darwin historians, there is a widening appreciation of what Galápagos meant to Darwin, and when it did so. The efforts of biologist Richard Keynes[35] and botanist Duncan Porter[36] have been particularly constructive. By examining Darwin's zoological and botanical notes and specimens they have identified the many Galápagos species that influenced Darwin's developing ideas and have shed light on how they did so. Paul Pearson[37] and Sandra Herbert[38] have helped expose one of Darwin's greatest geological discoveries in Galápagos, and have suggested that Darwin's developing ideas on evolution were closely tied to his concurrent theorizing on the origin and diversity of rocks. Darwin historian David Kohn and colleagues have recently elucidated the early origins of Darwin's interest in variation (a key element of evolution by means of natural selection) and have suggested that Darwin's collecting activities in Galápagos were influenced accordingly. They also suggest that although Darwin was not thinking in evolutionary terms while he was in Galápagos, he was operating within an intellectual framework to allow him to recognize varieties as incipient species, and to appreciate transmutation, soon after.[39]

In this book we take this rich cornerstone of Darwin's career as an evolutionist and put it under the magnifying glass. We examine Darwin's physical journey through Galápagos in unprecedented detail. By taking a step-by-step tour of his visit we demonstrate just how influential and inspirational it was. Using new facts drawn from Darwin's original unpublished notes, fresh insights from his sketches and publications, the studies of modern-day scientists, the analyses of Darwin historians, and our own intimate knowledge of the place, we explore how Galápagos shaped Darwin and his theory and how it defined the legacy he left behind. We show how Darwin's Galápagos experiences catalyzed his thoughts on evolution, how his Galápagos collections provided him with persuasive evidence to support his theory, and how his Galápagos observations fueled his speculations and motivated some of his later experiments. In doing so we shed light on the whole canvas of Darwin's life and work.

Into the Wild

> It has been said that the love of the chace [sic] is an inherent delight in man,—a relic of an instinctive passion. —if so, I am sure the pleasure of living in the open air, with the sky for a roof, and the ground for a table, is part of the same feeling. It is the savage returning to his wild and native habits. I always look back to our boat cruizes [sic] & my land journeys, when through unfrequented countries, with a kind of extreme delight, which no scenes of civilization could create. —I do not doubt every traveller [sic] must remember the glowing sense of happiness, from the simple consciousness of breathing in a foreign clime, where the civilized man has seldom or never trod.
> —Charles Darwin, *Beagle Diary*[40]

Our *Beagle* on the first day of our own voyage of discovery was a five-passenger Piper Aztec air ferry, but it metamorphosed over the following weeks into whatever means of transport we could find and afford: a fishing boat, the crews quarters of a cruise ship, a municipal launch, a dinghy, our legs for walking, our arms for swimming. At one point we really did travel on the *Beagle*. It was the current research vessel of the Charles Darwin Research Station, and the sixth vessel of that name to grace the waters of Galápagos since HMS *Beagle*. Darwin was our constant companion, speaking through his field notes,[41] geology notes,[42] zoology notes,[43] ornithology notes,[44] plant notes,[45] diary,[46] and specimen lists,[47] copies of which we had obtained and, in some cases (most notably the geology notes), transcribed in England. Captain FitzRoy was also a commanding presence, showing us, from the bearings and anchorages identified in his logs[48] and charts,[49] where to hit the shore and strike inland. Instead of spirit bottles and collecting bags we carried cameras, a GPS (Global Positioning System) receiver [50] and compass, and a stack of photocopies.

It was already known, from published accounts of the voyage, which four islands Darwin visited—San Cristóbal (Chatham), Floreana (Charles), Isabela (Albemarle), and Santiago (James)—and roughly where on each he explored. Frank Sulloway published a rough outline of Darwin's course through the islands but the details of where Darwin landed and explored remained vague.[51] Each island is topographically and ecologically heterogeneous, its habitats varied, and the distribution of its organisms uneven. Where Darwin walked determined what he saw and what he saw influenced what he thought. For the purposes of our study it was important to determine as closely as possible where Darwin stepped ashore and to define his movements inland and along the coast. Only in this way could we compare the natural history Darwin observed with what can be found in the same sites today. We could identify the sources of his geological insights, the exact features that triggered his understanding of the physical processes governing evolution in Galápagos.

By gleaning clues from Darwin's original notes, time and again we were able to figure out where Darwin walked, the land formations he examined, and the route he took to reach them. Indeed, on the very first day of the expedition, from clues in Darwin's forgotten geology notes and the *Beagle* logs, we identified the cove at Cerro Tijeretas

(Frigatebird Hill) as Darwin's first Galápagos landfall.[52] For 165 years this doormat of the most famous stopover on the voyage of the *Beagle* and in the history of evolutionary thought had remained unidentified. And now we had elucidated its location. Never again would Darwin's first landing spot be known only vaguely as somewhere on the southern end of Isla San Cristóbal (Chatham Island). It was a confident start to our expedition, and one that gave us no modest feeling of accomplishment. But not all the sites were so easy.

Despite traveling far faster than the *Beagle*, we took eight weeks to cover the area Darwin did in five. To determine the limits of Darwin's excursions we explored widely. As we pushed our way through thickets, tripped over knife-like lava, trod sun-scorched beaches, and clambered up crater after crater, we marveled constantly at Darwin's stamina. Not only were his hikes often long, some of the terrain was exceedingly rough. If he ever fell he never complained. We, on the other hand, went home sporting a few new scars. Of course, Darwin was only 26, and in fine form from his recent treks through South America!

One of the fascinating things about Darwin was his extraordinary power of observation and reasoning. He noticed "things which easily escape attention," questioned them, and endeavored to understand them.[53] Fortunately for us, he wrote down such thoughts, albeit in note form. Retracing his footsteps was like taking a guided walk through the countryside "to contemplate an entangled bank, clothed with many plants of many kinds, with birds singing on the bushes, with various insects flitting about, and with worms crawling through the damp earth."[54] Only instead of worms it was reptiles and instead of through damp earth, it was over arid lava.

In the prologue of his book *Fossils, Finches and Fuegians*, author Richard Keynes wrote, "When you have transcribed several hundred thousand words of [Darwin's] writings, concerned with places . . . not too greatly changed 160 years later, you may once in a while almost feel that you are talking to him."[55] Compared to Darwin's great-grandson we have transcribed but few words (and thank goodness, for Keynes's transcriptions made our footwork that much easier), but we have certainly read and relived many. How many times did I not look at Greg as he crouched to identify a plant, reach up to measure the thickness of an ancient stream of lava visible in a cliff face, or hike off toward a distant hill, and imagine, with allowances for costume, I was seeing Darwin himself. Nine years later I was awarded a sense of déjà vu when the British Broadcasting Corporation (BBC) filmed Greg acting as Darwin's double while we were working as on-site script consultants for their three-part television series, *Galápagos*. This time Greg was dressed for the part, with straw hat, waistcoat, and hob-nailed boots, but because of his dark hair, brown eyes, and beard—Darwin was fairer, blue-eyed, and possibly clean shaven in 1835—Greg was filmed from behind and at a distance, while another man played Darwin close up.

During the expedition, we often caught ourselves asking Darwin's ghost out loud which way he had gone, and admonishing (to put it mildly) the great man for not having made it clearer in his notes. For not all the sites were named in Darwin's day as they are today, and some were not named at all. We had to feel our way along by matching up landmarks with Darwin's imagery. While our task was helped enormously by the fact that Darwin was a geologist and generous in his descriptions of land formations, his footsteps became faint wherever outstanding geological features were lacking. Retracing Darwin's route, from hint to sometimes ambiguous or inconsistent hint, became a veritable treasure hunt, frustrating at times, but infinitely rewarding in the process. Nor did we limit ourselves to walking in Darwin's own footsteps; we readily branched out to explore formations that Darwin, having lacked the time and means to examine himself, had nonetheless described from the decks of the *Beagle*. Islote Tortuga (Brattle Islet) and Punta Cristóbal (Point Christopher) on Isla Isabela (Albemarle Island) are two such places.

In 1996, the same year as our own expedition, an attempt was made to follow Darwin's route through mainland South America on horseback. "All over the continent," the rider wrote, "I found that urban growth made following Darwin's precise routes dangerous and sometimes impossible, it being no joke to ride a fairly wild horse along main roads or through urban sprawls."[56] Other places had been "swallowed" up by resorts. Fortunately Galápagos has been largely spared this development. Several of the sites that Darwin visited now have a National Park trail running through or near them, but the only settlement from Darwin's day (on Isla Floreana/Charles Island) actually has fewer residents today than it did then. Unfortunately, this does not mean that Galápagos has been left untouched. Humans, and the plants and animals brought with them, have wreaked havoc on many of the islands, causing the local extinction of several endemic species. In many places we came across invasive plants choking out the native vegetation and saw the vandalistic signs of introduced insects and mammals. Nowhere was the destruction more apparent than in the highlands of Isla Santiago (James Island). Forests of endemic *Scalesia* trees (*Scalesia pedunculata*) that once covered the summits had been transformed into vast meadows by feral goats. We saw herds of hundreds of them running about the hills. In stark contrast, the native herbivores—the Galápagos tortoises that Darwin had reported swarming through the damp undergrowth of the highlands—were few and far between. Since our expedition, and over the first few years of the 21st century, the goats (and introduced pigs and donkeys) have been eradicated from the island, and the native vegetation has started to reinvade. Although there are now new alien contenders on Santiago—noticeably the invasive hill raspberry (*Rubus niveus*) and the aggressive paper wasp (*Polistes versicolor*)—there is much to rejoice about the natural state of Galápagos as a whole, for most of the wildlife described by Darwin is still there, albeit in reduced population numbers. Despite the increasing threats of a growing human population, "Darwin's Islands"[57] are still, for now, worthy of their epithet.

Darwin's Legacies

> *The mind is its own place, and in it self [sic]*
> *Can make a Heav'n of Hell, a Hell of Heav'n.*
> —John Milton, *Paradise Lost*[58]

Darwin lives on in Galápagos as nowhere else. The archipelago's unique plants, animals, and landscapes are responsible for its modern ranking as a Natural World Heritage Site, a Biosphere Reserve, and a National Park, but it is Darwin's association with this same fauna, flora, and geology that gives Galápagos its iconic status. Darwin's importance to the islands is reflected in the numerous species and places in Galápagos that bear his name (see appendix 2). It is also advertised in the titles of a local research institution (Charles Darwin Research Station), a road (Avenida Charles Darwin), a tour boat (M/V Darwin), and various other businesses (Darwin Hotel and Charles Darwin Travel Agency, for example) that operate in Galápagos. Indeed, the prominence of Darwin's name in Galápagos attests to the fact that the islands are "inexorably linked to the evolutionary views of Charles Darwin."[59]

Darwin revolutionized not only science and the way we perceive ourselves but also the way we view Galápagos. In the early days of its discovery the archipelago was commonly regarded (or at least portrayed) as a "monstrous" heap of islands upon which "God had showered stones"[60] in some places, "brimstone"[61] in others. They were inhabited by "creatures . . . the ugliest in Nature,"[62] "deformed fiends" and "devils"[63] and "imps of darkness,"[64] all readily sacrificed by the only men who dared frequent Galápagos waters—blood thirsty pirates, whalers, sealers, convicts, and men of war. American novelist and mariner Herman Melville summed up the gloomy impression of these early visitors when he wrote:

> Take five-and-twenty heaps of cinders dumped here and there in an outside city lot, imagine some of them magnified into mountains, and the vacant lot the sea, and you will have a fit idea of the general aspect of the Encantadas, or Enchanted Isles . . . Man and wolf alike disown them. Little but reptile life is here found: tortoises, lizards, immense spiders, snakes, and that strangest anomaly of outlandish nature, the *iguana*. No voice, no low, no howl is heard; the chief sound of life here is a hiss.[65]

Darwin began exploring the islands with some of the same prejudice. But as he traveled through the archipelago, then reflected on all he had seen, he realized that far from being dismal dumps of dust, the islands had a fascinating tale to tell. Using the language of science Darwin revealed the beauty of the archipelago's youthful landscapes and unique organisms. He lifted the veil of ignorance that had cursed Galápagos for the past three hundred years, and exhibited them in a new, secular light. In so doing he reclaimed the islands and their inhabitants from human condemnation and bequeathed them long-lasting fame and a life-saving future.

He did this in two ways. First, he revealed that the different islands of Galápagos are tenanted by different organisms of common ancestry adapted to different geographically isolated niches, and then issued a challenge for future scientists to "determine to what extent the fact holds good."[66] Numerous scientists picked up the gauntlet, expanding on Darwin's baseline collection and carrying out field studies to further understand the biodiversity of the archipelago.[67] Galápagos has now become a magnet for evolutionary biologists who, thanks to modern genetics, continue to reveal a greater amount of diversity in the islands than Darwin could ever have imagined. It is thanks to Darwin that Galápagos is now world famous and treasured as a "living laboratory."

Secondly, Darwin did the eminent service of inspiring the conservation of the islands. For he anticipated "what havoc the introduction of any new beast of prey must cause in [Galápagos], before the instincts of the aborigines become adapted to the stranger's craft or power."[68] His words were heeded by a handful of scientists who, on the centenary of Darwin's 1835 visit, initiated one of the first moves to protect the wildlife of Galápagos.[69] By 1959, one hundred years after the first publication of *The Origin of Species*, two organizations committed to conserving the Galápagos ecosystems were up and running. The Charles Darwin Foundation was founded to promote scientific research and environmental education about conservation and natural resource management in the archipelago. The Galápagos National Park Service dedicated itself to the protection by law of 97 percent of the landmass of the archipelago.

The efforts of these two institutions, with the support of various auxiliary conservation organizations,[70] have, to date, successfully managed to keep Galápagos one of the most pristine oceanic archipelagos in the world. It has not been easy. Tightly regulated nature-oriented tourism, introduced in 1967 with the hopes of it being a sustainable economy compatible with Darwin's legacy of science and conservation, now (2009) brings more than 150,000 visitors to the islands every year. The sheer volume threatens the wildlife that the scheme was designed to protect, for along with visitors and an increased awareness of the importance of Galápagos, comes development, immigration, invasive species and exploitation. Conservation measures struggle to keep the threats in check by restricting access to and activities on the uninhabited islands, prohibiting the harming and exportation of native wildlife, limiting the introduction of non-native plants and animals, conducting eradication programs to eliminate feral animals, and controlling fishing activities. However, the magnitude of the problem is such that on July 26, 2007, the United Nations Educational, Scientific and Cultural Organization (UNESCO) added Galápagos to their "World Heritage Site in danger" list. Now, with mounting pressures from a spiraling human population, vigilance and understanding are needed more than ever to ensure that Darwin's islands are not destroyed. It was through Darwin that people have come to appreciate Galápagos, and it is through intimate recollection of his visit—what he did, what he saw, what he interpreted—that its cherished worth can hope to be maintained. It is to this end that this book is dedicated.

Two Tales in One

In the year 2000, after making repeated field trips to fine-tune Darwin's paths in the islands, Greg and I published the results of our expedition as a scientific paper.[71] We lectured on the subject, continued with literature and field research, and designed an educational tour of the islands based on the *Beagle*'s route through the archipelago. We knew, however, that there was more to tell than where Darwin explored in Galápagos and a comparative analysis of what he saw there. The significance of Darwin's visit to Galápagos stretches beyond the geographical boundaries of the archipelago and the temporal limits of Darwin's time there. The repercussions of his visit encompass the world and exceed Darwin's lifetime. The story of Darwin in Galápagos was fabric for a book.

Darwin in Galápagos: Footsteps to a New World is primarily a tale of two expeditions woven into one. Darwin's visit to Galápagos on the voyage of the *Beagle* provides the warp and our revisit provides the weft. For without the second, and the research that made it possible, there would be little to write about the first beyond what has already been included in a chapter on Galápagos in a book about Darwin, or a chapter on Darwin in a book about Galápagos. We have used our unique and intimate knowledge of all the islands in Galápagos to interlace a scenic backdrop to blend with the natural history facts Darwin recorded. We have also applied the findings of modern scientists and our own insights accumulated from over 50 combined years of ecological research on various land birds, sea birds, reptiles, mammals, invertebrates, and plants in Galápagos to give a modern perspective to Darwin's observations.

The story of Darwin in Galápagos would have little meaning without some understanding of Darwin, the man. The maxim, "we are the sum of all the moments of our lives,"[72] cliché though it sounds, is key to understanding Darwin's visit to Galápagos. Everything Darwin did, observed, collected, and thought in Galápagos was directed by what he already knew, what he was in the process of discovering, and what he yearned to understand better. A look at Darwin's life before Galápagos, and especially at his education both before and during the voyage, is therefore needed to make sense of his Galápagos visit. It is important to know that Darwin attended two of the most respected universities in Great Britain, and was one of the most highly trained men of his age.[73] Before arriving in Galápagos he had been introduced to a wide range of scientific ideas and was aware that the scientific community, although heavily predisposed toward creationism, was also debating the concept of transmutation. Indeed, it can be said that many of the ideas Darwin developed were "lying fallow" in some form or another in England before he sailed on the voyage of the *Beagle*. It was the combination of Darwin's university training, his experiences during the voyage, and the "literary counsel" he received along the way that enabled him to pull it all together.[74]

This book does not pretend to be a biography; it purposefully emphasizes one stage in his life. But in order to put Darwin's Galápagos experience in perspective with the rest of his life and the development of his scientific thinking, we take a chronological pathway. Drawing principally from Darwin's autobiography, the Darwin biographies of Janet Browne and Adrian Desmond and James Moore, and the critical analyses of James Secord, Peter Bowler, and other historians of science, we provide glimpses of the experiences and ideas that helped shape Darwin and prepare him for Galápagos. We start with Darwin growing up as a child naturalist in England during the Industrial Revolution. We progress through his years at university, where he learned the scientific theories of the day. We then examine how Darwin matured on the voyage of the *Beagle*, and how his experiences in South America prepared him for what he would observe in Galápagos. We next take a step-by-step exploration of Darwin's journey through Galápagos and beyond, and demonstrate how the islands gradually changed his way of thinking. Finally, we look at Darwin after the voyage of the *Beagle* and show how Galápagos affected his life, and work, back home. The greatest portion naturally takes place in Galápagos for it is our principal aim to take the readers to the islands and have them put on Darwin's proverbial hiking boots and discover Galápagos as he did . . . as we did.

To make Darwin's visit to Galápagos a story rather than a fact sheet we have used some poetic license. It is impossible to determine what Darwin thought or how he felt at all times. He was a prolific writer[75] but a factual one; his personal journal, at least in the second half of the voyage, was often short on emotion, opinion, and the daily insignificances that one might expect to find in a diary. We have gleaned what we can of his psychological state from his letters home, though they were few and far between during the months surrounding his visit to Galápagos. Having said this, the book is nonfiction, and wherever we veer strongly away from known fact, we make clear our digression and our reasoning for it.

In the book we refer to both Darwin's diary and his journal. An explanatory note is in order to distinguish between the two, because Darwin confusingly used the word "journal" for both manuscripts. Darwin's diary is the "journal" of his activities and observations while on the voyage of the *Beagle*. This *Beagle* diary is one of the most illuminating volumes about the voyage. Darwin's granddaughter Nora Barlow published the first edited version in 1933 and Darwin's great-grandson Richard Keynes published a second version in 1988. Even though Darwin never published his diary himself, he used it and the letters he had written during the voyage to compose a more polished account of the *Beagle* voyage—his published *Journal*. This tome, originally entitled *Journal and Remarks*, was first published in 1839 as the third and last volume of Captain FitzRoy's *Narrative of the Beagle*. Later that same year Darwin's contribution was republished as a separate book under a new title: *Journal of Researches into the Geology and Natural History of the countries visited during the voyage of HMS Beagle round the world*. In 1845, after most of the *Beagle* specimens had been examined and described by taxonomists, Darwin rewrote his *Journal* to include what he had learned from them. The title remained the same, except the order of the words *Geology* and *Natural History* was switched to reflect a change in emphasis. Later editions of the same book were called *The Voyage of the Beagle*.[76] To avoid confusion, and for consistency's sake, whenever we quote from Darwin's *Journal* we revert by default to the 1839 edition, unless the material occurs only in later editions.

Except here in the Introduction we have chosen to use the original English names of the Galápagos Islands, because they were the ones used in Darwin's day. We identify the modern Spanish name of each island in the corresponding chapter headings, and in a table in appendix 1. Site names are also listed in this table. Species are identified by their modern nomenclature (common name and scientific name), unless otherwise stated.

For simplicity, throughout the book we use the terms "evolution" and "transmutation" synonymously, defined as the process by which populations of organisms change from one form into another.[77] Darwin first used "transmutation" and then "descent with modification" to express his developing ideas on biological change. While he used the word "evolve" once at the end of the first edition of *The Origin of Species* published in 1859, he did not adopt the term "evolution" until the 1870s.[78]

Many people were instrumental in producing this book and we have endeavored to give them full credit in a section devoted to their acknowledgment. Here, however, we wish to make it clear that there were really three authors to this book, just as there were three principal leaders on our 1996 expedition. Darwin "wrote" much of this book, and many of the quotations are his.

Part 1
Before Galápagos

Chapter I

~~~~~~~~~~~~~~~~~~~

Curious Beginnings

I was born a naturalist.
—Charles Darwin, *Life*[1]

Born to Science

Darwin entered the world in Shrewsbury, England on February 12, 1809, with all the makings for a bright future. His father, Robert Waring Darwin, was an esteemed medical doctor and a fellow of the Royal Society. His mother, Susannah (Sukey), came from the afflu- ent Wedgwood family, of pottery fame. Darwin had an older brother (Erasmus) and three older sisters (Marianne, Caroline, and Susan) to watch over him. He would soon have a younger sibling (Catherine) to play with. As for intellectual promise, his grandfathers on both sides of the family were famously intelligent. Erasmus Darwin was a physi- cian, inventor, poet, and natural philosopher. He was also somewhat of a dissident, a "hard-headed free-thinker"[2] who contributed some of the first speculations on evolution. Josiah Wedgwood was a cel- ebrated potter, inventor, and patron of the arts. Both were founding members of the elite Lunar Society, a philosophers club for politically liberal scientists and industrialists.

The house itself was called The Mount, an appropriate name for a supportive family at the apex of success and respectability. But for Darwin the name's active definition would have more significance, for *mount* means to begin a course of action and to climb, and it was here that Darwin started his life as a naturalist and embarked on his journey of discovery to the heights of scientific fame.

For an eminent-scientist-to-be Darwin was certainly born in a propitious place. He also grew up at a favorable time. These were the early years of the Industrial Revolution, a period that smiled on the sciences. In the quest for new technology to improve manufacture and transport, knowledge and brainpower were seen as valuable as any raw material. It was an age that nurtured scientific thought and encouraged the discussion of innovative ideas through philosophical meetings (like the Lunar Society), scientific publications, study tours, and lectures.

There was also a lively interest in the natural sciences and a fascination with nature. Just seven years before Darwin's birth, Christian apologist William Paley published a book called *Natural Theology, or Evidences of the Existence and Attributes of the Deity collected from the Appearances of Nature.*[3] It was an instant best seller. Amplifying and expanding upon the ideas of earlier natural theologians William Derham and John Ray,[4] Paley wrote that the intricate and varied adaptations of organisms are clearly designed and sustained by an intelligent being. Nature was verification of God's existence, and "[t]o study nature was to study the work of the Lord."[5] Paley evoked the now-famous watchmaker analogy to illustrate his point; behind the complex inner workings of a pocket watch is a watchmaker; therefore, behind the complex structures and adaptations of living things must be an intelligent designer. Paley's prose was so attractive and his arguments so persuasive that his book was instantly, hugely, and enduringly popular. His line of thinking became the dominant creed of the ruling classes for the first half of the 19th century and his book became standard university fare. Darwin would read it avidly when studying at the University of Cambridge.

Paley's view of life glorified nature and made it fashionable, but there was more motivating the study of nature in the early 19th century than just *Natural Theology*. There was the legacy and momentum of the previous century's Enlightenment.

The 18th century's Age of Enlightenment saw a huge expansion of natural history knowledge—the result of natural philosophers seeking rational answers for natural phenomena. Like their counterparts in previous centuries most of these 18th century "scientists"[6] looked at nature and the world as having been created by God. But instead of using the traditional methods of intuition, superstition, faith, and divine revelation to explain natural phenomena they sought natural history knowledge and ways of organizing the knowledge in rational ways.[7] The empirical methods of these men had the crucial effect of laying "a foundation for a science of biology."[8]

One of the most important naturalists of the Enlightenment was Swedish botanist Carl von Linné. Three-quarters of a century before Darwin's birth, Linné (referred to hereafter by his better known, adopted scientific binomial Carolus Linnaeus) opened the floodgates to systematic biology. He developed a hierarchical system of taxonomy, based on the physical similarities of organisms, to classify life. He also established binomial nomenclature, the formal, scientific convention for naming organisms. Although Linnaeus was in his grave when Darwin arrived on the scene, his influence was very much alive. His book *Systema Naturae*,[9] first published 74 years before Darwin's birth (and 100 years before Darwin's visit to Galápagos), inspired naturalists (including Darwin) to collect, describe, and systematically study all living species in a movement that swept through Europe, through the 18th century, and beyond.

As more was learned about the natural world in the 18th century, diverse ways of interpreting the facts arose. Some attempts were made to reconcile the new knowledge with the lore of the Bible, others to explain the phenomena in purely secular terms. Change was a common theme among the theorizers toward the end of the century for it became increasingly difficult to ignore the mounting fossil evidence showing the earth to have been populated by organisms that had changed over time.[10] Nor was it possible to overlook the geological facts that showed the earth to be much older than previously thought. The traditional view of the world was a literal interpretation of the story of Genesis—a static earth only a few thousand years old whose plants and animals had been miraculously created in six days.[11] God's creations were believed to form a perfect, unchanging, unbroken "ladder of life" or *Scala Naturae*; a hierarchy that started with the nonliving world and ascended "through [the] lower forms of life . . . to the highest, most spiritual of beings."[12] This worldview was now being challenged, and naturalists, religious thinkers, and secular philosophers alike had to adjust their views to accommodate the new facts.[13] It was during this period of free and open speculation that evolutionary thought began to emerge and develop.

The earliest 18th century proto-evolutionists were materialist thinkers Benoît de Maillet, James Burnett, Pierre Louis Moreau de Maupertuis, and several other European philosophers[14] who contemplated the origins of Earth's multifarious species and suggested, in vague and various terms that species adapt and change to survive.[15] French naturalist Georges-Louis Leclerc, Compte de Buffon, organized his ideas in a more scientific light. He proposed that similar species—all members of the cat family, for example—are "degenerated" forms of a single ancestral species, whose "internal mold" had changed in response to various environmental conditions. Darwin's grandfather Erasmus Darwin took the concept of organic change one step further by daring to imagine that everything now living had degenerated "from a single living filament."[16]

The progression of these pre-Darwinian evolutionary ideas can be pictured as streams of thought, flowing along similar lines, "without quite coalescing into an organized whole."[17] It was not until the 19th century that evolutionary thought became a cohesive idea. In 1809, the year Darwin was born, French naturalist Jean-Baptiste Pierre Antoine de Monet, Chevalier de Lamarck, published a book entitled *Philosophie Zoologique*,[18] in which he presented the idea of evolution as a theory. His causal explanation for how species change was that characteristics acquired by an individual during its lifetime are then passed down to its offspring. In a now oft-repeated example, he proposed that giraffes had transmuted from short-necked ancestors simply by stretching their necks to reach the leaves on high branches. Anticipating quarrel over the argument that animals with broken limbs do not beget offspring with broken limbs, Lamarck made it clear that only beneficial traits are inherited. Animals change through individual effort, so injuries are not passed on and traits that are not useful disappear through disuse.

Darwin would learn about Lamarck's ideas at the University of Edinburgh, and would later in life accept a limited view of the inheritance of acquired characteristics as a possible explanation for evolution supplementary to his own theory of natural selection.[19] But during Darwin's university years it was Lamarck's most severe critic, French Baron Georges Léopold Chrétien Frédéric Dagobert Cuvier, who had more concrete bearing on Darwin's scientific training. This was because Cuvier, a brilliant comparative anatomist and paleontologist, made numerous, important, and lasting contributions to the natural sciences. His career straddled the two centuries and many of his findings were made during Darwin's youth.

Two of Cuvier's most important contributions were to establish the fact of extinction and to fine tune Linnaeus's system of classification by comparing the internal structure of organisms. He proposed four distinct body types—vertebrates, invertebrates, radiates, and mollusks— and included extinct organisms as well as living in his classification of the animal kingdom.[20] He accomplished these feats through rigorous comparison of the anatomy of living animals and fossil remains. By further examination of the fossil record he showed that species become increasingly complex and more diverse as they are replaced through time. Such facts point to evolution, but Cuvier firmly believed in the fixity of species and propounded a different interpretation. God had simply made successive creations of species suited to increasingly favorable conditions. As for the alternative idea that the new species had mutated from old species, Cuvier argued that organisms could not possibly change because body parts are so highly correlated that any alteration in one part would sabotage the functioning of the whole. Cuvier's ruthless attack on Lamarck and all forms of evolutionary thought discouraged scientists from further speculation about the transmutation of species, and gave room for Paley's God-driven view of life to prevail. Nonetheless, and most importantly for Darwin, Cuvier's rigorous scientific methods and important findings helped lay the evidentiary foundation for evolution's eventual acceptance.

Darwin thus grew up during a period when there was glorious and noble purpose to studying nature but when exciting discoveries were being made to the tune of discordant theories. Great strides were being made in all branches of natural history, from astronomy and geology to botany and zoology. With encouragement from industrialists eager for scientists to improve economy, it was truly a bustling time for science.

As demonstrated by Janet Browne in her Darwin biography, Darwin was well "seasoned with the vigorous intellectual activity"[21] of this period during his school and university years. As a very young boy, however, Darwin was oblivious to all the noise the scientific community was making, much less to the reasons behind their racket. He was too busy creating his own hullabaloo at The Mount and causing his parents to fret over his future. For Darwin was "born a naturalist,"[22] but not in a way that reassured his father.

Shrewsbury

> [S]chool as a means of education to me was simply a blank.
> —Charles Darwin, *Autobiography*[23]

"Bobby," as Charles Darwin was affectionately called as a youngster, loved the outdoors. He was fond of climbing trees, examining flowers, taking long solitary walks, and collecting. He was also a mischievous little fellow, prone to fibbing, and "in many ways a naughty boy."[24] He made games of stealing fruit and professed the ability to turn primroses and crocuses different colors simply by watering them with colored fluids.[25] One of the few memories he retained of his mother, who died when he was eight years old, was her telling him that if she insisted he not do something, it was "solely for [his] own good."[26] Before he went to school his teenage sister Caroline was in charge of his education. So zealously did she try to improve her rebellious little brother that Darwin remembered habitually thinking, "What will she blame me for now?"[27] Soon after his eighth birthday, Darwin was sent to a local day school run by Unitarian minister Reverend George Case. There Darwin developed a passion for gardening and fishing for newts in the school pond, and after a close friend gave him a stone, he started rock collecting.

The following year Darwin was moved to a nearby boarding school (Shrewsbury School) but neither Darwin nor the headmaster, Reverend Samuel Butler, thought much of his attendance. Butler called him a "poco curante"[28] or trifler, while Darwin silently countered, "Nothing could . . . [be] worse for the development of my mind than Dr. Butler's school."[29] Fortunately, Darwin did not let school interfere with his education, for he "had the strongest desire to understand or explain whatever [he] observed"[30] in the natural world. When he wasn't exploring out of doors he spent hours paging through a school friend's copy of C.C. Clarke's *Wonders of the World*, vicariously traveling to the remote places described within. He also escaped into the "pleasure [of] poetry,"[31] delighting in the works of William Shakespeare, Lord Byron, Sir Walter Scott, and James Thomson.

Throughout his teens Darwin fostered his interest in natural history and cultivated his passion for collecting "all sorts of things,"[32] from shells and coins to birds' eggs and beetles. He remained a poor student but excelled at his hobbies, the pursuit of which would prove far more valuable to his later career than the classics he tediously studied at school. He loved dogs with a passion, and became "an adept in robbing their love from their masters."[33] He enjoyed hunting, and learned to ride at the age of 11, and to shoot at 15. He carefully recorded all the birds he shot, and in doing so initiated a lifetime habit of meticulous note taking.[34] During school term he often ran home in the evenings to spend the stolen hours conducting chemistry experiments with his older brother Erasmus (Ras) in the garden toolshed. The pastime earned him the unfortunate nickname "Gas,"[35] but Darwin proclaimed his hobby, "the best part of my education . . . for it showed me practically the meaning of experimental science."[36]

Darwin's nonacademic talents were lost on his father. Robert Darwin, "the kindest man [Charles Darwin] ever knew" but also a formidably large man, standing over 1.8 meters (6 feet) tall and weighing 136 kilos (300 pounds), was driven to despair, "You care for nothing but shooting, dogs, and rat-catching, and you will be a disgrace to yourself and all your family."[37]

Edinburgh

> *My father . . . declared that I should make a successful physician . . .*
> *but what he saw in me which convinced him . . . I know not.*
> —Charles Darwin, *Autobiography*[38]

In October 1825 Robert Darwin packed his son off to medical school at the University of Edinburgh. It was not a success. Not only were the lectures exceedingly dull, Darwin was horrified by the spectacle of surgery without anesthesia. Upon witnessing the amputation of a child's leg he ran from the operating theater before it was over. "Nor did I ever attend again," he wrote years later, "for hardly any inducement would have been strong enough to make me do so; this being long before the blessed days of chloroform."[39]

Darwin quit medical school after two years, but his time at Edinburgh was far from wasted. It was here that Darwin, through an assortment of lectures, discussions, and books on meteorology, hydrography, mineralogy, geology, botany, and zoology, was baptized in contemporary scientific thinking and introduced to a full range of scientific theories. Some were steeped with biblical implication; others were more secular. Serving as examples are the opposing ideas of 18th century geologists Abraham Werner and James Hutton. Werner, with his Neptunism theory, argued that all rocks are the precipitates and crystallized minerals of an ocean that once covered the earth in the not too distant past. Werner's ideas were attractive for their religious compatibility and aura of stability and order.

Hutton, on the other hand, believed that most rocks are formed below the ground. With his Plutonic theory he suggested that Earth is in a perpetual state of change, with volcanoes, earthquakes, and erosion constantly altering its aspect. He introduced the idea of deep time—time that stretches so far in to the past and future that he could imagine "no vestige of a beginning, no prospect of an end."[40]

Darwin was exposed to both Wernerian and Huttonian thinking by attending the lectures of professors Robert Jameson and Thomas Charles Hope, respectively. Jameson openly sneered at those who believed that rocks had been "injected from beneath in a molten condition."[41] Hope, on the other hand, promoted the idea that subterranean heat was the principal agent in forming the earth's surface. The dispute between these two professors was legendary within the university, and "relished" by students who experienced the controversy in their introductory courses. Darwin sided with Hope and his lectures, finding Jameson a bore.[42]

It was also at Edinburgh that Darwin was first exposed to the early concepts of evolution. He read *Zoonomia*,[43] written by his late grandfather Erasmus Darwin, and became well acquainted with Professor Robert Grant, an invertebrate anatomist, an atheist, and a great advocate of Lamarckian evolution.

Darwin enjoyed reading *Zoonomia* and was impressed by Grant's zeal for Lamarck, but he was not inclined to evolutionary thought himself. He later admitted that his exposure to early evolutionary thought in Edinburgh probably made a subconscious impression, but he could not remember it having any noticeable effect at the time.[44] He was more interested in the fascinating microscopic world of marine invertebrates and the pleasures of tide pooling to which Grant also introduced him. At the age of 17 Darwin made his first scientific discovery—the mode of generation of a marine leech and a sea mat (a bryozoan)—and gave his first scientific speech on the subject at a meeting of the Plinian Society, an undergraduate science club. It was also at Edinburgh that Darwin learned to skin and stuff birds. His instructor was John Edmonston, a freed black slave who had traveled with the famous ornithologist Charles Waterton in South America. Not only did he teach Darwin a skill that would be useful for his future career, Edmonston, with his first hand stories of the tropics, imbued Darwin with a burning desire to travel.

Cambridge

> *Considering how fiercely I have been attacked by the orthodox it seems ludicrous that I once intended to be a clergyman.*
> —Charles Darwin, *Autobiography*[45]

After Darwin dropped out of medical school, his father sent him to the University of Cambridge to study for the clergy. Darwin's early schooling had been in the Unitarian Church, once defined by Erasmus Darwin as "a feather-bed to catch a falling Christian."[46] However, he was also baptized in the Church of England, as appropriate for the gentrified society to which he belonged. The idea of becoming an Anglican minister appealed to Charles Darwin, for it was complementary to the life of a naturalist. After all, many respected clergymen were also leading naturalists. He stumbled through his theology and classics lectures and vigorously pursued his varied and growing outdoor interests on the side. He hunted, went horseback riding, and made many friends with whom he "sometimes drank too much, with jolly singing and playing at cards afterwards."[47] His greatest passion remained collecting beetles in the Linnaean tradition and sending them to entomologists and museums to be identified or described. So absolute was his zeal that he would go to any lengths to secure a new specimen. In later years he entertained his friends and his children with stories about these beetle days:

> [O]ne day, on tearing off some old bark, I saw two rare beetles and seized one in each hand; then I saw a third and new kind, which I could not bear to lose, so that I popped the one which I held in my right hand into my mouth. Alas it ejected some intensely acrid fluid, which burnt my tongue so that I was forced to spit the beetle out, which was lost, as well as the third one.[48]

After sending off a particularly rare species Darwin was rewarded with a great honor. As he reminisced in his autobiography, "No poet ever felt more delight at seeing his first poem published than I did at seeing in Stephen's *Illustrations of British Insects* the magic words, 'captured by C. Darwin, Esq.'"[49] Indeed, Stephen credited Darwin with collecting over 30 different insect species.[50] From then on it appeared that Darwin's collecting, of any object, was motivated as much by fame as by love of the sport. In 1836 when Darwin was at Ascension Island and received news that distinguished geologist Adam Sedgwick had taken an interest in the rock specimens he had sent home from South America, Darwin reacted with barely controlled

1. Charles Robert Darwin, age 31, four years after the voyage.

2A

2B

2A. Charles Darwin as a young boy, with his sister Catherine.
2B. "The Mount," in Shrewsbury, where Darwin grew up.

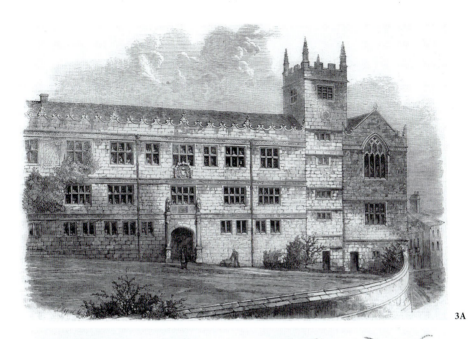

3A

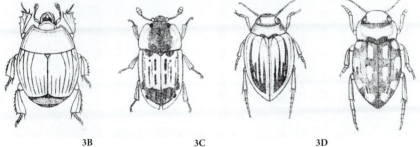

3B   3C   3D

3E

3A. Darwin attended the Royal Grammar School in Shrewsbury from 1818 to 1825.

3B–3D. Some of the British beetle genera collected by Darwin as a young man. *Hister* (3B), *Nitidula* (3C), and *Hydroporus* and *Hygrotus* (3D).

3E. Charles Darwin's father, Dr. Robert Darwin, circa 1826.

4B

4C

4D

4A

Influential men in Darwin's life.
4A. Adam Sedgwick taught Darwin the essentials of field geology.
4B. John Stevens Henslow was Darwin's mentor at Cambridge.
4C. Darwin derived great inspiration from the scientific ideas of geologist Charles Lyell.
4D. Joseph Dalton Hooker described Darwin's Galápagos plant specimens after the voyage.
    He was an invaluable friend to Darwin, and supporter of his ideas.

5A

5B

5A. Captain and commander of HMS *Beagle*, Robert FitzRoy.
5B. FitzRoy received detailed instructions for the voyage of the *Beagle* from the Lord Commissioners of the British Admiralty, pictured here at a meeting in 1843.

6A

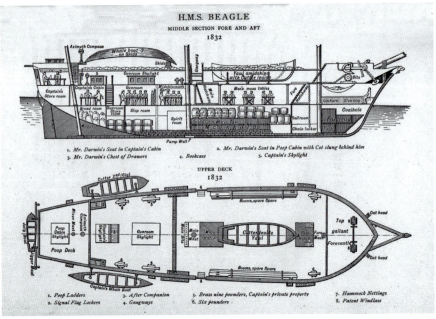

6B

6A. HMS *Beagle* in the Straits of Magellan.
6B. Diagram of HMS *Beagle*.

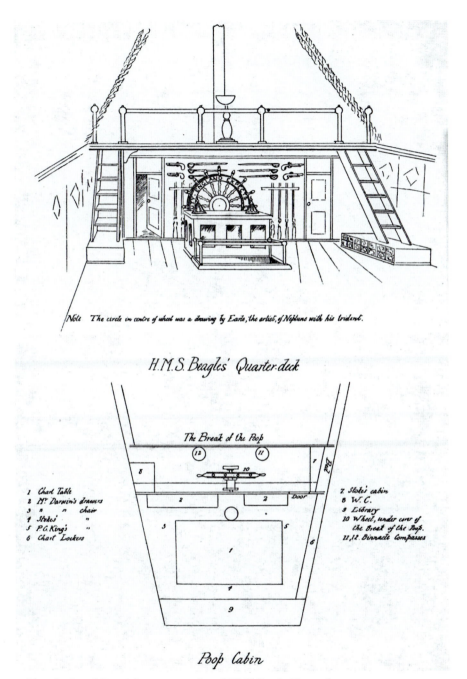

Note   The circle in centre of wheel was a drawing by Earle, the artist, of Neptune with his trident.

H.M.S. *Beagles'* Quarter-deck

The Break of the Poop

1 Chart Table
2 M:. Darwin's drawers
3 „   „   chair
4 Stokes'   „
5 P.G. King's   „
6 Chart Lockers

7 Stokes' cabin
8 W.C.
9 Library
10 Wheel, under cover of
   the Break of the Poop.
11,12 Binnacle Compasses

Poop Cabin

7. The wheel, and Darwin's quarters on board HMS *Beagle*. Drawn from memory, many years after the voyage, by Philip Gidley King.

8A

8B

8A. On December 27, 1831, the *Beagle* sailed from Plymouth Harbour, Devonshire.

8B. She returned to England four years and nine months later, arriving at Falmouth Harbour, Cornwall, on October 2, 1836.

9. "Among the scenes which are deeply impressed on my mind, none exceed in sublimity the primeval forests of Brazil, undefaced by the hand of man. . . ."
—Charles Darwin, *Beagle* Diary.

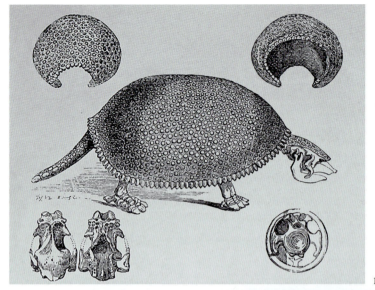

10A

10B

10C

10A. In South America Darwin discovered the fossil remains of several extinct mammals, including a glyptodont, an example of which is pictured here (10A). He was struck by its resemblance (except in size) to the extant armadillo (10B), several species of which were common in the region. It was one of the key observations that eventually led Darwin to be convinced of the mutability of species.

10B. A hairy armadillo. Darwin observed four species of armadillo in South America—the pichi (*Zaedyus pichiyi*), larger hairy armadillo (*Chaetophractus villosus*), southern three-banded armadillo (*Tolypeutes matacus*), and southern long-nosed armadillo (*Dasypus hybridus*).

10C. Darwin's fossil mammals were described by palaeontologist Richard Owen, assistant curator of the Hunterian Museum of the Royal College of Surgeons, London.

11A

11B

11A. Fuegian indians Jemmy Button (top left), Fuegia Basket (top right), and York Minster (bottom).

11B. Darwin was greatly struck by the difference in behavior between the "civilized" Fuegian indians on board the *Beagle*, and that of their "savage" counterparts in the wild.

12A

12B

12C

12A. Mount Sarmiento and glacier, from Warp Bay.
12B. Berkeley Sound, East Falkland Island.
12C. Port Louis, East Falkland Island.

13A

13B

13A. Darwin crossed the Uspallata range in April 1835. He was less impressed by the area's famous "Incas Bridge" (pictured here) than by his discovery, at an elevation of 7000 feet, of the remains of a silicified forest that had once grown at sea level.

13B. Travel through the Andes was often steep and hazardous.

14A

14B

14A. Coastal desert, Coquimbo, Chile.
14B. On September 7, 1835, 45 months after leaving England, the *Beagle* set sail for Galápagos from Peru.

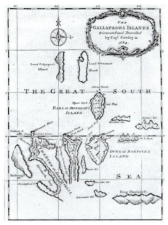

15A

15B

15A. One of the first maps of Galápagos was drawn after William Ambrose Cowley's visit to the islands in 1684.

15B. James Colnett's chart, published in 1798, was a vast improvement over Cowley's map, but still left much to be desired. Visits to the islands by British admiralty ships *Tagus* and *Indefatigable*, in 1814 and 1815, provided some additional detail for later editions of Colnett's map. However, an accurate chart of the archipelago remained lacking until Captain FitzRoy's five-week survey of Galápagos on HMS *Beagle* in 1835.

16A

16B

16A. September 20, 1835, at Fresh Water Bay (La Honda), Chatham Island. The *Beagle* returned to this site on October 11, after leaving Darwin camped on James Island, to take on fresh water for the upcoming passage to Tahiti.

16B. Leaving the "watering place" on Chatham Island. Although it appears this way, the island was not erupting.

pride. "I clambered over the mountains . . . with a bounding step and made the volcanic rocks resound under my geological hammer! All this shows how ambitious I was."[51]

Darwin looked back on his Cambridge days "with much pleasure"[52] and not only for the recreation. At Cambridge, Darwin met botany professor and clergyman John Stevens Henslow, "a circumstance," he declared, "which influenced my whole career more than any other."[53] Henslow was a kind man with a deep interest in "every branch of science."[54] He firmly believed in fixed species, created by God, but he saw that species could vary to a certain degree. He was interested in defining the limits of such variation and to this purpose practiced a unique method of collecting plants. He gathered specimens of a single species from various locations, and "collated" them on a single botanical sheet (normally devoted to just one specimen) in order to show how the species varied. Darwin took Henslow's botany course three times, learning his methods first hand and thoroughly. As demonstrated by Darwin historian David Kohn and botanists at the University of Cambridge, it was from Henslow that Darwin first learned to pay attention to variation, a crucial, influencing factor in the development of his theory of evolution by means of natural selection.[55]

Henslow held weekend meetings at his house for students interested in natural history and took them on instructive field excursions into the country. Darwin spent so much time with his mentor that he became known among the Cambridge dons as "the man who walks with Henslow."[56] Henslow encouraged Darwin to read Alexander Von Humboldt's *Personal Narrative of a Journey to the Equinoctial Regions of the New Continent* and John F. W. Herschel's *A Preliminary Discourse on the Study of Natural Philosophy*. From these works Darwin developed "a burning zeal to add even the most humble contribution to the noble structure of Natural Science." As he later asserted in his autobiography, "No one or a dozen other books influenced me nearly so much as these two."[57] Indeed, they spurred him to action.

Darwin began industriously planning his own collecting expedition—to Tenerife in the Canary Islands. Humboldt's descriptions of the island and particularly of a colossal tree with a circumference of

48 feet that grew there, were irresistible. "I never will be easy till I see the peak of Teneriffe [*sic*] and the great Dragon tree [*Dracaena draco*]," he wrote to his sister Caroline. "[S]andy, dazzling, plains, and gloomy silent forest are alternately uppermost in my mind . . . I have written myself into a Tropical glow."[58] Although Darwin's "Canary scheme"[59] was unexpectedly quashed with the sudden death of his traveling companion, Marmeduke Ramsay, the months of preparation, studying Spanish, and "working like a tiger"[60] were, in retrospect, invaluably spent.

Henslow, keenly supportive of Darwin's scientific ambitions, also encouraged Darwin to study geology. The science was well supported and fast growing and because it embraced theoretical work as well as empirical research, was one of the most stimulating branches of natural history to study in Britain in the 1830s.[61] Primarily concerned with defining the order of geological strata, it would lead the way in reconstructing the history of Earth and its inhabitants. For an upcoming naturalist wanting to make a contribution to science it was a promising field to study, and one that offered "opportunities for large views." It was also "ideally suited for travel in distant foreign lands."[62]

Henslow had a personal interest in geology for he had started his career as a mineralogist before turning to botany.[63] He introduced Darwin to geology professor Adam Sedgwick, and asked Sedgwick to take Darwin to North Wales to teach him "how to make out the geology of a country."[64] Sedgwick was a catastrophist; he believed a new and popular geological theory that claimed Earth had been subject to not just one great flood as previously taught in Wernerian geology, but several widespread, short-lived catastrophes. Catastrophism was loudly advocated by Cuvier, for it accommodated his ideas that God had created divine disasters to extinguish old species and make room for new ones.

Sedgwick was also a leader in the field of stratigraphy. He was interested in reconstructing local geological history by examining a district's strata, and was focused on tracing the oldest fossil-bearing sedimentary layers. Later he would be credited with proposing the Cambrian period of the geological timescale.

When in August 1831 Darwin accompanied Sedgwick to his study site in Wales, he learned an array of field skills that proved invaluable on the voyage of the *Beagle*. Historian Martin Rudwick goes as far as to say, "Darwin could never have made the kind of observations—or the kind of notes—that he did make in South America, if he had not first been initiated by Sedgwick into the tacit knowledge of field geologists and their routine practice of structural and stratigraphical geology."[65] But there was more to Sedgwick's training than methodology and note taking. During the two weeks that Darwin spent in the field with Sedgwick he discovered the process involved in constructing a scientific argument from empirical facts. "Nothing before had ever made me thoroughly realize," wrote Darwin years later in reference to his field trip with Sedgwick, "that science consists in grouping facts so that general laws or conclusions may be drawn from them."[66] Darwin had come of age as a scientist.

Darwin's experiences with Sedgwick were a far cry from Professor Jameson's dull and disparaging geology lectures that Darwin had attended in Edinburgh. They had had the effect of making Darwin vow "never as long as I lived to read [another] book on Geology or in any way to study the science."[67] Sedgwick had quite a different style. "I cannot promise to teach you all geology," he would say to his students. "I can only fire your imaginations."[68] It certainly worked for Darwin. Darwin became so "hot on the subject"[69] of geology that it became his "chief pursuit"[70] on a new expedition, one that would begin in just four months time.

When Darwin set off on the voyage of the *Beagle* he carried with him a broad understanding of contemporary scientific thinking, learned from the professors he had met and the books he had read. In Edinburgh Robert Grant had exposed Darwin to evolutionary thought, but Darwin's Cambridge mentors, John Henslow and Adam Sedgwick, had proved more immediately influential. Like them, Darwin believed in the fixity of species. He saw his role as a naturalist as adding to "the noble structure of Natural Science"[71] by collecting, observing, and describing God's creations. He had great dreams of finding something new for science, of contributing raw materials to the understanding of the history of the earth, and of earning recognition from the scientific community in the process.[72] However, his aspirations did not stop there. Darwin had been made aware that there were opposing theories, that scientists disagreed over the interpretation of facts, and "that important discoveries were waiting to be made and that he could help to make them."[73] This awareness gave Darwin the freedom to think for himself, and inspired him to question and theorize. The world was his to explain as much as it was to explore. He therefore set out on the voyage with an open mind. With his "enlarged curiosity,"[74] his interest in all facets of the natural world, and his keen powers of observation he had almost all the ingredients necessary to make a revolutionary contribution to science. All he lacked at this point was experience. And for that he had the voyage of the *Beagle*. It would take him around the world, filling him with facts, nourishing him with knowledge, stirring him to theorize. In South America, and especially in Galápagos, he would collect the makings for a grand new understanding of the world. He would discover evolution for himself, and then spend the rest of his days revolutionizing our perception of life on Earth.

Chapter II

~~~~~~~~~~~~~~~~~~~

Voyage of the Beagle

The voyage of the Beagle has been by far the most important event in my life and has determined my whole career. . . . I have always felt that I owe to the voyage the first real training or education of my mind.

—Charles Darwin, *Autobiography*[1]

Offer of a Lifetime

From a recommendation by Reverend John Stevens Henslow, Darwin was invited aboard HMS *Beagle* for a surveying voyage around the world. The opportunity was originally handed to Anglican clergyman Leonard Jenyns, a beetle-collecting friend of Darwin's and the man who would describe Darwin's fish collection from the voyage. However, as vicar of Swaffham Bulbeck, Jenyns was unwilling to abandon his parish, and turned the offer down. Henslow was tempted to accept the vacancy himself, but as a married man with a growing family he too was disinclined to leave home. Both men suggested Darwin go instead.[2]

The voyage was to be a scientific expedition, the principal goals of which were to[3]

1. Fix the longitude of Rio de Janeiro, Brazil, from which all distances would be measured.
2. Make a hydrographic survey of the coast of South America and other places visited on the voyage.

3. Make astronomical and tidal observations.

4. Report on the geology, climate, natural history, people, and cultures of the places visited during the voyage. It was to this objective that Darwin was expected to contribute.

During the voyage Captain FitzRoy would receive additional orders from the Admiralty, one of which was to negotiate the collection of a fine from the Queen of Tahiti. The penalty had been exacted for the unlawful capture of a British merchant ship in the Low Islands in 1831.[4]

Darwin was thrilled, "as happy as a king."[5] "[W]hat a stage it will make in my life," he projected.[6] He wrote an exultant note to Henslow. "Gloria in excelsis is the most moderate beginning I can think of. . . . [Until] today I was building castles in the air about hunting Foxes in Shropshire, now Lamas in S America. —There is indeed a tide in the affairs of men."[7] Before he could accept, however, there were two obstacles to overcome. First, his father rejected the offer outright. Darwin wrote out a list of his father's objections and handed it to his Uncle Jos (Josiah Wedgwood II) to contest. Fortunately, Josiah came to his nephew's rescue and convinced Robert Darwin that the voyage would be not only good for his son, but complementary to the life of a clergyman. The list of objections,[8] with Josiah's replies to Robert added in italics, read:

1. Disreputable to my character as a Clergyman hereafter
   I should not think that it would be in any degree disreputable to his character as a clergyman. I should on the contrary think the offer honorable to him, and the pursuit of Natural History, though certainly not professional, is very suitable to a Clergyman

2. A wild scheme
   I hardly know how to meet this objection, but he would have definite objects upon which to employ himself and might acquire and strengthen, habits of application, and I should think would be as likely to do so in any way in which he is likely to pass the next two years at home.

3. That they must have offered to many others before me, the place of Naturalist

The notion did not occur to me in reading the letters & on reading them again with that object in my mind I see no ground for it.

4. And from its not being accepted there must be some serious objection to the vessel or expedition

   I cannot conceive that the Admiralty would send out a bad vessel on such a service. As to objections to the expedition, they will differ in each mans case & nothing would, I think, be inferred in Charles's case if it were known that others had objected.

5. That I should never settle down to a steady life hereafter

   You are a much better judge of Charles's character than I can be. If, on comparing this mode of spending the next two years, with the way in which he will probably spend them if he does not accept this offer, you think him more likely to be rendered unsteady & unable to settle, it is undoubtedly a weighty objection—Is it not the case that sailors are prone to settle in domestic and quiet habits.

6. That my accommodations would be most uncomfortable

   I can form no opinion on this further than that, if appointed by the Admiralty, he will have a claim to be as well accommodated as the vessel will allow.

7. That you should consider it as again changing my profession

   If I saw Charles now absorbed in professional studies I should probably think it would not be advisable to interrupt them, but this is not, and I think will not be, the case with him. His present pursuit of knowledge is in the same track as he would have to follow in the expedition.

8. That it would be a useless undertaking

   The undertaking would be useless as regards his profession, but looking upon him as a man of enlarged curiosity, it affords him such an opportunity of seeing men and things as happens to few.

The second obstacle was raised by the captain of the *Beagle*, Robert FitzRoy, who, being a phrenologist, judged people's character by the shape of their face. Darwin's nose apparently displayed a lack of resolve to persevere with the whole voyage. However, Darwin's good breeding outranked his dubious physiognomy and FitzRoy soon became convinced of his suitability.

Darwin was not offered the official position of Ship's naturalist. That title went first to the ship's surgeon Robert McCormick, and after his release in April 1832 for "being disagreeable to the Captain,"[10] to assistant surgeon Benjamin Bynoe. Rather, FitzRoy wanted a "well-educated"[11] gentleman companion, with an active interest in the natural sciences (particularly geology), for meals and conversation during the long voyage. Such a privileged position gave Darwin carte blanche to pursue his interests in geology, natural history, and collecting; activities that were not only welcome, but helped justify his station as "supernumerary" on board the *Beagle*. Only once the voyage was well underway did Darwin earn the right to be considered, if never officially, principal naturalist aboard the ship.

The *Beagle* was scheduled to sail from England in November 1831, and Darwin was bursting with anticipation. "What a glorious day the 4th of November will be to me," he exclaimed in a letter to FitzRoy. "My second life will then commence, and it shall be as a birthday for the rest of my life."[12] But the 4th came and passed and the *Beagle* remained firmly moored to the docks of Devonport. There were many delays before HMS *Beagle* finally sailed from England; the quarters needed redesigning, the weather turned foul, the crew got drunk on Christmas Day. With mounting impatience Darwin was soon grumbling to Henslow, "I look forward even to sea sickness with something like satisfaction, anything must be better than this state of anxiety."[13] For truly, as he was "waiting in suspense"[14] for the voyage to begin he became increasingly nervous at the prospect of leaving his family and friends for so long. He had only left the British Isles once before—on a summer trip to Paris in 1827—and had been back home within the month.[15] The voyage of the *Beagle* was expected to last at least three years, possibly four. "Time, which no one can alter, is the only serious inconvenience," he wrote to his cousin, William Fox. Conquering his nerves, he joked:

> Why, I shall be an old man, by the time I return, far too old to look out for a little wife. What a number of changes will have happened; I suppose you will be married & have at least six small children . . . & I shall sit by the fire & tell such wondrous tales, as no man will believe. —When I think of all that I am going to see &

undergo, it really requires an effort of reasoning to persuade myself, that all is true.[16]

At the forefront of his mind was the current object of his affections,
the "belle"[17] Fanny Owen. He clearly suspected that she, "the pretti-
est, plumpest, [most] Charming personage that Shropshire posseses
[*sic*],"[18] would not wait for his return. But he could only have guessed
that by the time he received his first mail packet, just three and a half
months into the voyage, Fanny would be engaged to another man and
married soon after. He began to suffer heart palpitations but would not
see a doctor, fearful that he should be prevented from traveling.[19]

The *Beagle* finally set sail from England on December 27, 1831,
when Darwin was only 22 years old. He would not see England again
for almost five years.

The Beagle

> *To say that the* Beagle *was extremely cramped, even given the
> expectations of the time, would be a supreme understatement. The
> ship was, after all, no longer than the distance between bases on a
> baseball field. If you imagine both baseball teams, plus umpires and
> trainers standing along the base line, that would still be fewer than
> the number of people on board.*
> —Keith Thompson, HMS Beagle, The Story of Darwin's Ship[20]

The *Beagle* was originally built as a 90-foot-long, 25-foot-wide, 10-
gun brig. She was launched in 1820 and then lay at mooring at the
Woolwich dockyards for the next five years. In 1825 she was re-rigged
as a three-masted bark (barque), and in May 1826 set sail on her first
commission, a hydrographic surveying voyage to Patagonia and
Tierra del Fuego that lasted almost four and a half years. Lieutenant
Robert FitzRoy was placed in command of the ship halfway through
that first voyage, when Captain Pringle Stokes shot himself, unable
to handle the stress of surveying the treacherous coasts of Tierra del
Fuego. The *Beagle* was rebuilt once again for her second surveying

voyage—the one Darwin made so famous—and the worthy Robert FitzRoy was retained as commander.[21] Under FitzRoy's guiding hand the *Beagle* was transformed into "the most perfect vessel ever turned out of the Dock yard." As Darwin explained in a letter to Henslow, "no vessel has been fitted out so expensively & with so much care.— Everything that can be made so is of Mahogany, & nothing can exceed the neatness & beauty of all the accommodations."[22]

The *Beagle* was the length of a typical 16-passenger, 6-crew tourist yacht operating in Galápagos today but with a ship's complement over three times as great. Sometimes she carried more, for FitzRoy was obliged to offer assistance and transport to any stranded or shipwrecked sailors they came across. And he was not averse to acquiring the occasional and sometimes permanent mascot. Among such animals, a coatimundi from the Andes[23] and several young Galápagos tortoises were allowed free range of the ship from their acquisition until the end of the voyage. Not that there was much room for wandering the decks; the ship was cluttered with no less than seven boats –four whalers (each between 7 and 8½ meters in length), a 7-meter-long cutter, an 8-meter yawl, and a small jolly boat (dinghy).[24] Superfluous though these boats may have appeared, they were indispensable tools for the hydrographic survey. They were used for charting multiple coastlines simultaneously, and for accessing harbors too shallow for the *Beagle* to enter.

Darwin's own space on the *Beagle* was a "most wofully [sic] small"[25] corner of the chart room in the stern of the ship. This poop cabin, approximately 3 meters by 3½ meters, accommodated the ship's library of about 250 books, a wash stand, a chest of drawers, an instrument cabinet and two hammocks all squeezed around (and over) a large central chart table.[26] Darwin shared the room with two men; Midshipman Philip Gidley King occupied the other hammock and Ship's Mate John Lort Stokes slept in a small annex to the cabin. Fortunately, Darwin was provided an additional locker for his specimens. At 1.813 meters (almost 6 feet) tall—the height of the cabin—Darwin was forced to stoop when moving below decks. Yet despite the cramped conditions he nobly declared the ship "a very comfortable house."[27]

There were 74 men on board the *Beagle* when she left England, and a dozen less when she returned.[28] Some sailors simply left the ship and were replaced by new recruits, or their positions were filled by crew members already on board. Five men died. In June 1832 three seamen (Morgan, Boy Jones, and Charles Musters) "fell victim to a fever"[29] (most likely malaria) caught while snipe hunting in Macúcu near Rio de Janeiro, Brazil. Darwin had tried to join this side trip, but, as he expressed in his diary, "[M]y good star presided over me when I failed."[30] In March 1833 ship's clerk Edward Hellyer "drowned at the Falkland Islands, in attempting to get a bird he had shot," and in June 1834 George Rowlett "died, at sea, of a complaint under which he had laboured for years."[31] When the *Beagle* was in Galápagos there were roughly 65 men on board (see appendix 3).

Captain and Crew

> *Can one wonder at pride in the Captain, when he knows that all*
> *& everything bends to his will?*
> —Charles Darwin, *Beagle Diary*[32]

Descended from royalty, Captain Robert FitzRoy was an imposing man with impeccable qualifications. Just four years older than Darwin, he was an outstanding naval officer, hydrographer, and pioneer meteorologist. He was also a handsome and "uncommonly agreeable open sort of fellow."[33] Soon after meeting FitzRoy, Darwin wrote to his sister Caroline, "It is ridiculous to see how popular he is. Ladies can hardly splutter out big enough words to express their big feelings."[34] And it was not only the fair sex that admired FitzRoy. "Everybody praises him," insisted Darwin to his sister Susan, "& indeed, judging from the little I have seen of him, he well deserves it." To tease his sisters, Darwin sometimes referred to FitzRoy as "my beau ideal of a Captain."[35] When Darwin became panicked at leaving England on such a small ship for such a long time it was Captain FitzRoy who assuaged his worries. "[He] is such an effectual & goodnatured [*sic*] contriver," Darwin effused in his diary, "that the very drawers enlarge on his appearance & all difficulties smooth away."[36]

Like Darwin, FitzRoy was interested in the natural sciences, and as a housewarming gift of sorts, he presented Darwin with the newly published first volume of Charles Lyell's *Principles of Geology*.[37] Lyell's book introduced Darwin to a new geological theory called uniformitarianism. The theory was a revision of James Hutton's Plutonic theory that claimed the earth's crust was constantly changing through volcanic activity, and a direct attack on the dominant geological theory of the day, catastrophism. Reverend Henslow had already recommended Darwin read Lyell's book "but on no account to accept the views therein advocated."[38] Now that he had a copy of his own, Darwin studied it attentively and, contrary to his friend's warning, became fully convinced of uniformitarianism. Indeed, during the course of the voyage Darwin found much evidence in support of Lyell's conviction that the shape of the earth's crust is the result of gradual and ongoing processes rather than past catastrophic events as interpreted through the Bible. Lyell's book (together with Volumes II and III acquired later in the voyage) also inspired Darwin to question the geographical distribution of past and present species in the context of a dynamic physical world. For in his latter volumes Lyell openly speculated on the extinction and creation of species, and believed that new species were periodically created to fit the physical conditions of their place in time. And although Lyell fell short of advocating organic transmutation, he certainly saw the physical world in evolutionary (that is, gradually changing) terms. It was by entering Lyell's framework of thinking, looking for patterns of temporal and spatial change in the organic world and relating these patterns to geological and geographical features in the inorganic world, that Darwin was eventually able to recognize the mutability of species. Had FitzRoy, a devout Christian, realized that his gift would be so influential to the development of Darwin's "heretical" theory of evolution, he would undoubtedly have chosen a less controversial welcome-aboard present.

Darwin suspected that the "violent admiration" he felt upon first greeting Captain FitzRoy would not last throughout the voyage. "No man is a hero to his valet," he wrote to his sister Susan, "& I certainly shall be in much the same predicament as one."[39] Sure enough Darwin soon discovered that FitzRoy had a quick temper and could be excessively harsh in his treatment of wayward crewmen. Darwin admitted that he could well "fancy [FitzRoy] being a Napoleon or a Nelson."[40] This, however, did little to dampen his esteem of the man. "The Captain keeps all smooth by rowing every one [*sic*] in turn, which of course he has as much right to do, as a gamekeeper to shoot Partridges on the first of September,"[41] he wrote rather defensively to his sister Catherine in 1834.

Thanks to mutual respect the two men got along well. They did, however, have the occasional argument. Their worst spat occurred early on, in Brazil, over polar views on the slave trade. Darwin, decidedly antislavery, found the practice despicable. FitzRoy "defended and praised"[42] slavery and in a fit of contempt threw Darwin out of his cabin as they were dining. The incident ended quickly with a gentleman's apology from FitzRoy and by the end of the voyage the captain acknowledged that slavery was an "evil" and the slave trade an "abominable . . . traffic."[43]

Contrary to popular myth, Darwin and FitzRoy did not bicker incessantly over religious matters.[44] They were both believers for most of the voyage, Darwin only beginning to question the role of the "Creator" toward the end. In fact, on the *Beagle* Darwin remembered "being heartily laughed at by several of the officers (though themselves orthodox) for quoting the Bible as an unanswerable authority on some point of morality."[45] Several months after leaving Galápagos Darwin joined forces with FitzRoy in writing an impassioned defense of the missionary work being conducted in Tahiti and New Zealand. Their letter was published in the *South African Christian Recorder*, and became, strictly speaking, Darwin's first publication from the voyage.[46] Only once Darwin and FitzRoy were back in England did their relationship change. Darwin became convinced of evolution. FitzRoy became increasingly religious. Darwin was lauded by the scientific community for his work in geology and natural history, and with the arrogance of a rising star, became less tolerant of FitzRoy's erratic moods and close mindedness. The rift widened when in 1837 Darwin, albeit unintentionally, offended FitzRoy "almost beyond mutual reconciliation" by insufficiently acknowledging the officers of the *Beagle* for his accomplishments on the voyage.[47] It became an unbridgeable gulf when, in 1859, Darwin published "so unorthodox a book . . . as the *Origin of Species*."[48] But all that animosity came later. During the voyage they were allies.

FitzRoy ruled that for safety reasons no one could leave the ship on his own and as a result Darwin never lacked for company. While Darwin enjoyed his solitude, he found the system more of a boon than a burden, for the officers of the *Beagle* assisted him greatly with the logistics of exploration and the acquisition of specimens. The officers were a youthful lot, some still in their teens. Ship's purser George Rowlett was the oldest officer on board—only 35 years of age at the time of sailing.[49] Darwin got along with them all. He charmed them with his "genial smile," his even temper (a welcome contrast to the captain's), and "his energy and ability."[50] With a natural abhorrence of idleness[51] and time "fritter[ed] away,"[52] Darwin was always ready to offer a helping hand. He enjoyed a privileged position at the captain's table, and was initially addressed by some of the officers as "Sir,"[53] but he never made anyone feel he was their better. Lieutenant Bartholomew Sulivan certainly felt comfortable enough to play an April fool's joke on Darwin early in the voyage, having him rush to the decks at midnight with the false sirens call, "Darwin, did you ever see a Grampus [i.e., Orca]: Bear a hand then!"[54] As Darwin secured the affections of everyone on board he acquired two new nicknames: "Flycatcher" for his passion for insect collecting and, FitzRoy's favorite, "Philos" or "Philosopher,"[55] short for "the dear old Philosopher."[56]

Darwin's opinion of the ship's officers was equally magnanimous. "We all jog on very well together, there is no quarrelling on board, which is something to say," he penned to his sister Catherine halfway through the voyage. [57] Of course, Darwin had his favorites and, of those, Lieutenant John Wickham topped the list. Wickham was 11 years Darwin's senior, and Darwin revered him like an exemplary older brother. "[T]here is not another [in] the ship worth half of him," Darwin told his sisters.[58] "He is far the most conversible [*sic*] being on board."[59] His feelings of respect were apparently reciprocated. On the third surveying voyage of the *Beagle*, Wickham (having been promoted to commander after FitzRoy stepped down to become Governor-General of New Zealand) and John Lort Stokes named the Australian port of Darwin in honor of their old shipmate and friend.

Sulivan was a chatterbox. Darwin wrote to his sister Catherine, "in that respect [he] quite bears away the palm."[60] Sulivan's garrulity, however, did nothing to stop Darwin from considering Sulivan "one of his best and truest friends"[61] throughout his life.

Darwin also thought highly of midshipman Philip Gidley King. "[He] is the most perfect, pleasant boy I ever met with & is my chief companion," Darwin wrote to his sister Caroline in April 1832.[62] King took on the role of ship's artist when painter Conrad Martens left the *Beagle* in 1834.

A year and a half into the voyage Darwin hired Syms Covington, "fiddler & boy to Poop-cabin,"[63] to help him collect and preserve specimens. Surprisingly, Darwin was not overly fond of Covington. In a letter to Catherine in 1834 he wrote: "My servant is an odd sort of person; I do not very much like him; but he is, perhaps from his very oddity, very well adapted to all my purposes."[64] As the voyage progressed Darwin was relieved to leave much of the shooting to Covington. He discovered "that the pleasure of observing and reasoning was a much higher one than that of skill and sport."[65] He was able to dedicate more time to geology and collecting plants and invertebrates, while Covington shot most of the vertebrates. Covington kept a private journal for most of the voyage, but apparently, and somewhat mysteriously, wrote nothing of his time in Galápagos. Nevertheless, it is clear from Darwin's writings that Covington accompanied Darwin on several shore excursions in Galápagos, including all his overnight trips. Darwin was clearly satisfied with his servant's work and Covington remained in Darwin's service for a total of six years, continuing as his personal secretary after the voyage.

Life on Board

I [am] very sick & miserable. —This second attack of sea-sickness has not brought quite so much wretchedness as the former one. But

*yet what it wants in degree is made up by the indignation which is
felt at finding all ones efforts to do anything paralysed [sic].*
—Charles Darwin, *Beagle Diary*[66]

Because Darwin's role was to examine the land while the surveying
officers of the *Beagle* took care of the hydrography, Darwin ended up
spending most of the voyage on shore.[67] This amounted to over two-
thirds of the voyage, or 39 months. His extensive exploration of the
South American continent on foot and horseback helped keep him in
top form. The 18 months he spent on board the *Beagle*, however, was
another matter.[68] Darwin suffered dreadfully from seasickness. His
duty as dinner companion to Captain FitzRoy was probably rarely
put into practice, for the slightest swell made Darwin "unspeakably
miserable."[69] When unable to face the standard meal of fresh fish or
meat, rice, peas, bread, pickles, dried apples, and lemon juice (for
scurvy), he confined himself to his hammock where he picked at bis-
cuits and raisins, the only foods he could stomach.[70] It may help explain why, when he returned home, Darwin was "looking very thin,"
sporting just 67 kilos (148 pounds) on his 1.813 meter ("5 ft, 11⅜ in.")
frame.[71] He gained 8 kilos (18 pounds) after just two months back
home, presumably returning to his normal weight.

Darwin's worst experience at sea fell on Sunday, January 13, 1833,
when the *Beagle* was "sorely tried"[72] while doubling Cape Horn in a
storm. It was an appalling day for Darwin and almost a catastrophic
end to the *Beagle*. Darwin wrote to his sister Caroline, "The Captain
considers it the most severe [gale] he was ever in. —We have already
heard of two vessels which were wrecked at the very same period . . .
[O]ne of our boats was knocked to pieces & was immediately cut
away: the water being deep on the deck, it did me an infinity of harm,
as it wetted a great deal of paper & dried plants . . . —I suffered also
much from sea-sickness."[73] FitzRoy knew they were lucky to be alive,
for their "little diving duck"[74] had been hit by "three huge rollers"
in deadly succession. "For a moment, our position was critical," he
remembered after the voyage, "but, like a cask, she rolled back again,
though with some feet of water over the whole deck. Had another sea

then struck her, the little ship might have been numbered among the many of her class which have disappeared."[75]

Sailing the great blue open was not all misery. Darwin acknowledged that some aspects of ocean travel were delightful, especially the quiet and "the beauty of the sky & brilliancy of the ocean."[76] He even looked forward to some of the longer passages, for he enjoyed the simplicity of "bare living on blue water" as long as the sea remained calm.[77] By the end of the voyage he could reminisce of "a moonlight night, with the clear heavens, the dark glittering sea, the white sails filled by the soft air of a gently blowing trade wind, a dead calm, the heaving surface polished like a mirror, and all quite still excepting the occasional flapping of the sails."[78] Furthermore, the oceans gave generously to his labors, as revealed in the fact that over half of his zoology notes are concerned with marine invertebrates that he harvested from the sea.[79] Still, rough seas rendered Darwin useless and apathetic, and during these frequent periods Darwin reverted to "the Arabian" description of the "illimitable ocean [as a] tedious waste, a desert of water."[80] After four and a half years he was thoroughly sick of it. On August 4, 1836, safely on the homestretch to England, and writing his last letter on the voyage, he exclaimed to his sister Susan, "I loathe, I abhor the sea, & all ships which sail on it!"[81]

Chapter III

~~~~~~~~~~~~~~~~~~~~~~~~~~~~~

Gearing for Galápagos

It appears to me that nothing can be more improving to a young naturalist, than a journey in distant countries.
—Charles Darwin, *Beagle Diary*[1]

It took three years and nine months for the *Beagle* to reach Galápagos. During that time Darwin matured from a wide-eyed student of the natural sciences to a skillful collector, well-honed observer, and shrewd theorizer.[2] All he did and saw in South America primed and conditioned him for Galápagos. By the time he reached the islands he had forged a framework of thinking for contemplating its mysteries as he could never have done had he arrived any earlier. To understand how requires a look at the route of the *Beagle*, and Darwin's most significant stops along the way.

The first two months of the voyage of the *Beagle* were spent sailing to Brazil and stopping at the islands en route. For the next 27 months the *Beagle* moved gradually up and down the East Coast of South America, fixing the longitude of Rio de Janeiro from which all subsequent distances were measured, and charting the coastline. Another year and a quarter were spent moving up the West Coast to Peru. The Galápagos Islands were next on the route, followed by a final year of island hopping to Australia, South Africa, back to South America, and home. This slow progress gave Darwin ample time to make observations and collect specimens of rocks, fossils, terrestrial plants and animals, and marine life. He spent much of the time exploring inland, riding roughly 5000 kilometers across the pampas of Argentina and over the Andes on horseback, and staying with locals or sleeping under the stars.[3] He shipped numerous specimens back to Professor Henslow, sent scores of letters to friends and family, filled 18 field notebooks, four volumes of zoological observations, and 13 volumes of geological observations, composed several essays, kept a descriptive specimen catalog, and maintained a diary from beginning to end. In all, the voyage took him to three continents, 16 countries (as defined today), and over three dozen islands, filling him with such wild and varied experiences, not to mention revolutionary thoughts, that upon returning to England, his father wittily exclaimed, "Why, the shape of his head is quite altered."[4]

First Stop, Paradise

The first stop on the route of the *Beagle* was to be Tenerife, the Holy Grail of Darwin's first expedition plans. But Darwin was never destined to land on the island of his dreams. Upon arrival Captain Fitz-Roy was informed that no one from the *Beagle* would be allowed on shore for fear they might introduce cholera. There had been a recent outbreak of the illness in England. "Oh misery, misery," sobbed Darwin into his diary, "we were just preparing to drop our anchor within ½ a mile of Santa Cruz when a boat came alongside bringing with it our death-warrant."[5] Rather than undergo a lengthy quarantine, Fitz-Roy shouted "Up Jib"[6] —and gave orders for the *Beagle* to continue on. Darwin was devastated. "[I]t is like parting from a friend,"[7] he lamented, as the island's peak faded from view.

On January 16, 1832, 20 days after leaving England, the *Beagle* arrived at the Cape Verde Islands, and Darwin's melancholy evaporated with his first landfall. His diary fairly bubbles with the awe and excitement he felt upon exploring the interior of St. Jago (Santiago), his first tropical island:

> I returned to the shore, treading on Volcanic rocks, hearing the notes of unknown birds, & seeing new insects fluttering about still

newer flowers. —It has been for me a glorious day, like giving to a blind man eyes. —he is overwhelmed with what he sees & cannot justly comprehend it. —Such are my feelings, & such may they remain.[8]

The officers of the *Beagle* were highly amused by Darwin's enthusiasm, and FitzRoy joked in a letter to a friend at the Admiralty that "a child with a new toy could not have been more delighted than [Darwin] was with St Jago."[9] After all, much of the volcanic island was "utterly barren . . . nothing meets the eye but plains strewed over with black & burnt rocks rising one above the other." Yet Darwin found "a grandeur in such scenery & . . . the unspeakable pleasure of walking under a tropical sun on a wild & desert island."[10]

Upon examining his first volcanic rocks, Darwin's "latent passion for theorizing" was unleashed.[11] Finding a layer of "triturated" shell and coral fragments embedded within the lava, he hypothesized that the lava had initially "flowed over the bed of the sea," and then been "upheaved" above sea level. By following the line of white rock he also concluded that "there had been afterwards subsidence round the craters," and that these craters had then poured forth yet more lava.[12] He was inspired to dare that he "might perhaps write a book on the geology of the various countries visited."[13] He did indeed write that book. In fact, he got five books (with multiple volumes and editions) out of his experiences on the voyage: *Journal of Researches,*[14] *The Structure and Distribution of Coral Reefs,*[15] *Geological Observations on the Volcanic Islands visited during the Voyage of H.M.S. Beagle,*[16] *Geological Observations on South America,*[17] and *The Zoology of the Voyage of H.M.S. Beagle,* in five parts.[18]

Darwin's geological triumph on St. Jago also encouraged him to keep geology top priority for the remainder of the voyage. In total he wrote four times more geology notes (1383 pages) than zoology notes (368 pages)[19] and collected almost 2000 rock specimens.[20]

With its stark volcanic landscape the Cape Verde archipelago offered Darwin a taste of what he would find in Galápagos. But while the physical environments of the two archipelagos proved similar in climate, and height and size of their islands, the productions of each were surprisingly disparate. "[W]hat an entire and absolute difference in their inhabitants!" he would exclaim after visiting Galápagos.[21] The organisms of Galápagos, far from resembling those of the Cape Verde Islands, were similar to species occupying more temperate habitats on the mainland of South America. The plants and animals of the Cape Verde Islands, in turn, resembled the flora and fauna of neighboring Africa. To Darwin these dual facts were "perhaps more striking than almost any others"[22] for they suggested that oceanic islands are "indebted to neighbouring lands for their . . . productions."[23] And while they did not negate "the ordinary view of independent creation,"[24] which evoked divine "centers of creation" to explain geographical patterns in species distribution, they favored an entirely different explanation. When, after the voyage, Darwin worked out the details, he argued that oceanic islands are colonized by organisms from the nearest mainland, and the descendents of these colonists evolve into new species as they adapt to their new environments.

Darwin's initial exhilaration at entering the Tropics took a long time to wane. Six weeks after leaving Cape Verde he beheld his first tropical rainforest in Brazil. The scene was like "a view in the Arabian Nights" and his mind "a chaos of delight."[25] "It is not only the gracefulness of their forms or the novel richness of their colors," he wrote of the varied trees, "it is the numberless & confusing associations that rush together on the mind, & produce the effect."[26] Darwin had always dreamed of seeing tropical vegetation and experiencing his first tropical rainforest was one of his most memorable spectacles of the voyage. Indeed, tropical vegetation became one of his greatest and most enduring loves, second only to geology. As the voyage progressed he saw a variety of stunning landscapes, yet, like a faithful lover, often contrasted them unfavorably with the splendor of his Brazilian forests. And any scene that remotely recalled to mind the lush, "primeval forests"[27] of the Tropics was instantly revered.

Darwin's style of writing at the beginning of the voyage was clearly influenced by his early love of poetry and by the vivid, often emotionally charged writings of his hero Alexander von Humboldt. Humboldt had explored South America between 1799 and 1804 and had written a glorious *Personal Narrative* of his travels. After reading and rereading the account, Darwin felt that Humboldt, "alone gives any notion, of the feelings which are raised in the mind on first entering the Tropics."[28] For the months Darwin remained in Brazil he maintained that Humboldt "like another Sun illumines everything I behold."[29] Caroline thought that Darwin went overboard in emulating Humboldt's "flowery" language; it sounded "unnatural" coming from her younger brother. So in October 1833 she wrote Darwin a letter beseeching him to temper the prose and to stick to his "own simple straight forward & far more agreeable style."[30]

As the voyage progressed Darwin expressed himself with less emotion and more scientific intensity. Compared to the early entries of his diary, his chapters written in Galápagos seem somewhat dry in tone, and more factual. This is not to say that he did not lapse poetic on occasion, when so inspired, but Darwin himself admitted to a growing change in outlook. When he returned to Bahía toward the end of the voyage he lamented in a letter to his sister Susan, "It has been almost painful to find how much, good enthusiasm has been evaporated during the last four years. I can now walk soberly through a Brazilian forest; not but what it is exquisitely beautiful, but now, instead of seeking for splendid contrasts; I compare the stately Mango trees with the Horse Chesnuts [*sic*] of England."[31] And after the voyage he adopted such an analytical way of thinking that he "wholly lost, to [his] great regret, all pleasure from poetry of any kind."[32] His mind became "a kind of machine for grinding general laws out of large collections of facts."[33] But what a brilliant machine it was, for it allowed Darwin to make sense of the diversity of life that had captured his imagination so vividly during the voyage.

A Massive Discovery

Ten months into the voyage, on September 22, 1832, near Bahía Blanca, Argentina, Darwin made his first important discovery. In some low cliffs he found the fossil remains of some extinct giant mammals. "Immediately I saw them" he wrote to Henslow, "I thought they must belong to an enormous Armadillo, living species of which genus are so abundant here."[34] There were many more fossils. Here at Punta Alta and later (January 1834) at Port St. Julian (Puerto San Julián), Patagonia, over 1200 kilometers to the south, he found several different species of extinct Pleistocene mammals and all appeared to be giant varieties of present-day animals found in the same areas. Most were new to science, and back in England comparative anatomist Richard Owen ascribed them names. Among them were the giant armadillo-like glyptodont (*Hoplophorus*),[35] a massive ant-eater (*Scelidotherium*), an enormous capybara-like mammal (*Toxodon*), an oversized llama-like creature (*Macrauchenia*), and a huge ground sloth (*Megatherium*).[36] (In October 1833 he also found mastodon bones and a prehistoric horse tooth near Santa Fé, Argentina.) The megatherium had already been studied by Cuvier in the late 1700s and used to introduce the principal of extinction. But when Darwin sent home a skull that he purchased from some "Gauchos" in Monte Video (Montevideo, Uruguay) in 1833 it caused great excitement among paleontologists because it included bones missing from Cuvier's specimen. Unknown to Darwin at the time, some of his fossil mammals and letters to Henslow about his fossil finds were presented at a meeting of the Cambridge Philosophical Society on November 16, 1835. With these fossils Darwin made his name as a collector before he even reached home. More importantly, they were, in retrospect, Darwin's first evolutionary clues. As proclaimed by Richard Keynes, September 22, 1832, "was truly a red-letter day for biology, marking the discovery of the first of the lines of evidence that eventually led [Darwin] to question and ultimately to reject the doctrine of the fixity of species."[37] Along with the geographical distribution of the unique organisms of Galápagos, these extinct mammals and their resemblance to living species in the same area stimulated Darwin's initial conception of the mutability of species.

Punta Alta was important to Darwin for yet another reason. It was here that Darwin earned his position as chief collector on board the *Beagle*. Many of the officers, most notably FitzRoy, Bynoe, and Sulivan, assembled collections of their own, but it was hereupon agreed that Darwin, being the most serious of the collectors, would have first dibs over the important, rare specimens.[38] FitzRoy was amused by Darwin's enthusiasm and often watched with a smile as Darwin hauled "cargoes of apparent rubbish" aboard the ship.[39] However, not all the officers regarded Darwin's activities with equal magnanimity. "Wickham always was growling at my bringing more dirt on board than any ten men,"[40] Darwin cheekily remarked in a letter to his youngest sister in July 1834.

A Change of Perspective

A year into the voyage, the *Beagle* arrived at the southern tip of South America and in January 1833 Captain FitzRoy set about fulfilling a personal mission. On board were three native Indians from Tierra del Fuego: York Minster, Fuegia Basket, and Jemmy Button. Captain FitzRoy had captured them in 1830 during the first voyage of the *Beagle* and had taken them to Britain to be "civilized." A fourth captive, Boat Memory, had died of small pox soon after arriving in Britain. The three survivors were now being returned home in the hope that they would spread the Christian religion among their tribe.

Upon arrival at the Beagle Channel, Darwin saw several men running along the shore. They were "absolutely naked & with long streaming hair; springing from the ground & waving their arms around their heads, they sent forth hideous yells."[41] Darwin was astounded at the difference in behavior between these "savage" Fuegians in their natural state and the "tame" Fuegians he'd come to know on board the *Beagle*.[42] He found the contrast "greater than between a wild and domesticated animal."[43] The fact that the captives had been transformed from such a state of savagery to one of civilization provoked Darwin to think of humans as animals, and "as an integral part of the natural world."[44] It altered his whole way of seeing himself and the men around him. For the remainder of the voyage Darwin paid particular attention to the various races of people he encountered. By doing so, and in the words of historian Janet Browne, he came to recognize that culture was "nothing but an outer garment for humanity" and that there was no "group of human beings on earth exempt from the most basic of impulses."[45]

As for the Fuegian captives, they had a difficult time returning to their original ways. Their tribal existence was a hostile one; family ties were weak, and thievery and warfare were commonplace. Missionary Richard Matthews, sent by the Church Missionary Society in England to live with the Fuegians, had to be rescued within the month. He was invited to rejoin the *Beagle* until the ship reached New Zealand. The three anglicized Indians, however, stayed in Tierra del Fuego. At first Jemmy Button was "quite disconsolate" about the prospect of returning to his native land, for he had "quite forgotten his language." "It was pitiable, but laughable," Darwin wrote in his diary, "to hear [Jemmy] talk to his brother in English & ask him in Spanish whether he understood it."[46] "I am afraid," Darwin added, after saying his first good-bye to the reluctant repatriates, "whatever other ends their excursion to England produces, it will not be conducive to their happiness."[47] But Darwin was in for a surprise. A year later when the *Beagle* returned to the area, Darwin saw that things had changed. Jemmy had found a "wife" and now "had not the least wish to return to England." "I hope & have little doubt", Darwin penned into his diary, "he will be as happy as if he had never left his country; which is much more than I formerly thought."[48]

At the same time the officers of the *Beagle* were releasing their Fuegian passengers back into the wild, Britain was taking possession of a new captive, the Falkland Islands. When the *Beagle* arrived at East Falkland Island in March 1833, Darwin was astonished to learn that Argentina was no longer sovereign and "that the [British] Flag was now flying."[49]

As Darwin explored East Falkland Island he found that far more significant forces than mere politics had overwhelmed the Falklands in the past; forces that had spanned vast periods of time, transformed the environment, and shaken the earth's crust. Darwin found the present day Falklands "dreary,"[50] barren, and desolate, yet the rocks were full of fossil Paleozoic shells and corals, proving the area had long ago teemed with life under a tropical climate. He also discovered a wide "stream of stones" remarkable for their enormous size ("from that of a man's chest to ten or twenty times as large"), angular shape, and even distribution on the ground.[51] Today it is believed the stones were formed little by little, over eons, by the weathering process of freeze and thaw. Darwin, however, thought that some monumentally powerful tremor of the earth had cracked the massive boulders into their present shape and vibrated them into an even plane. He wrote, "never did any scene, like the 'stream of stones,' so forcibly convey to my mind the idea of a convulsion."[52] Although it was an incorrect interpretation, it left an enduring and significant impression. Dramatic changes in the physical landscape as a result of violent movements of the earth's crust, via earthquakes and volcanic eruptions, played a fundamental part in Darwin's conception of the evolution of organisms as they adapt, over time, to new conditions fashioned from such dynamic changes.

The Falkland Islands saw a change of pace in the surveying efforts, a move that backfired and almost ended the voyage prematurely. In March 1833, shortly after arriving at the Falklands, FitzRoy purchased a schooner (the *Unicorn*) from a "notorious"[53] but "enterprizing [*sic*] and intelligent sealer"[54] named Captain William Low.[55] FitzRoy renamed the ship *Adventure*, hired Low as her pilot, and placed Wickham in command. For the next 18 months the *Beagle*'s daunting task of charting the southern reaches of South America with its multitude of islets and passages, was eased with their support.

Low, "the son of a respectable land-agent in Scotland . . . brought up as a sailor"[56] was an invaluable addition to the crew. He knew the shores of Tierra del Fuego and Patagonia intimately, and was an "excellent practical observer, long acquainted . . . with the productions of th[ose] seas."[57] He provided Darwin with valuable natural history observations, and furnished FitzRoy with useful information about the places en route. Much of what Low imparted was originally learned from the local Indians, for Low had been in the habit of inviting, and sometimes forcing, young Indians aboard his ship, in order to use their "sagacity and extensive local knowledge"[58] to help with seal hunts. On one occasion, while working on a ship called the *Adeona*, he had captured a 10-year-old Chonos Indian boy (nicknamed Bob) and taken him all the way to Galápagos and back.[59]

FitzRoy was severely rebuked by the Admiralty for purchasing the *Adventure*. He had to use his own funds to keep her going, and ultimately was forced to sell the ship, and say goodbye to Low. In a fit of depression, a recurring condition that would one day (on April 30, 1865, at the age of 60) cause him to take his life, FitzRoy resigned from his post. Acting against all personal ambitions of becoming captain himself, the gallant Lieutenant Wickham convinced FitzRoy to resume command. Had he not, Darwin would never have reached Galápagos, for the Admiralty had made it quite clear that without FitzRoy at the helm the voyage would be cut short.[60] As it turned out, FitzRoy bounced so readily back from his gloom that he had the audacity to purchase another surveying vessel, the schooner *Constitución*, in Chile. Again the Admiralty disapproved, but this time FitzRoy was defiant. He placed surveying officer Alexander Usborne and Charles Forsythe in command of the tender, and ordered them to survey the coast of Peru, while he took the *Beagle* on to Galápagos.

Throughout the voyage Darwin was interested in global questions, and especially in the "laws which regulate the creation and distribution of species."[61] By the time Darwin reached the Falkland Islands in March 1833, he was already scrutinizing the relationship between the geography of a place and the distribution of its organisms.[62] In the Falklands he was particularly interested in geographical barriers such as those presented by oceanic channels between islands and a continent. Fresh in his mind were the observations he had made on the fauna and flora of Uruguay and Patagonia, and, more recently still, of the island of Tierra del Fuego. While exploring East Falkland Island he questioned whether certain groups of organisms, such as reptiles, mosses, and aphids, were present or not and noted the relative abundance of those that were. He stressed a desire to compare "differences of species & proportionate Numbers" between locations and to define the "characters of [the] different habitations"[63] in which they occurred. He also became interested in migration, and plant and animal associations, as factors influencing the distribution of organisms on and between islands and the continent. He would ask these same types of questions when exploring Galápagos.

One animal that grabbed his attention was the famously fearless and now-extinct Falkland Island fox (*Dusicyon australis*). The "Gauchos & Indians" and sealers who collectively knew "all parts of [the]

Southern part of S. America" assured Darwin it was not found on the mainland. Darwin thought it "very curious, thus having a quadruped peculiar to so small a tract of country." His interest was piqued further when Captain Low asserted that the animals differed in size and color between the two principal islands of the Falklands.[64] Darwin only visited East Falkland Island but the surveying officers of the *Adventure* collected a fox specimen for FitzRoy from the West Island to compare with three collected from the East Island.[65] Darwin viewed the West Island fox and the East Island fox as varieties of the same species, as indeed they were. Later he would take the concept of geographical distribution to daring new levels, ultimately examining how physical isolation gives rise to the evolution of new species from such varieties. Near the end of the voyage when Darwin was mulling over his Galápagos specimens and first questioning the stability of species, the Falkland Island fox was his only example outside Galápagos of an organism that exists as distinct varieties on separate islands.[66]

The Proof of the Pudding

As Darwin explored both sides of the South American continent his interest in the geographical distribution of organisms grew accordingly. He saw the Andes, the backbone of the continent, as potentially "a great barrier" dividing eastern species from western. He kept "Birds" and "Animals" lists, and noted whether each species occurred on the East, on the West or on both sides of the Andes.[67] Through observations such as these Darwin was able to record a "marked difference between the vegetation of these eastern valleys and that of the opposite side . . . the same . . . with the quadrupeds, and in a lesser degree with the birds and insects."[68] Another important pattern that Darwin noticed was "the manner in which closely allied animals replaced one another in proceeding southwards over the Continent."[69] Darwin's most famous evidence for this pattern was obtained in a rather peculiar way.

Darwin had heard from the locals about a small, rare rhea in Patagonia that differed significantly from the large common rhea

(*Rhea americana*) found north of the Rio Negro. In vain Darwin attempted to glimpse the bird, let alone collect a specimen. Then in January 1834, Conrad Martens, the *Beagle* artist, shot one for supper. It was cooked and eaten before Darwin realized what it was, but fortunately the "head, neck, legs, wings, many of the larger feathers, and a large part of the skin, had been preserved."[70] Darwin was able to put together a nearly perfect specimen. He was especially proud of his "Avestruz petise"[71] (little ostrich), for he was convinced that it was a new species, new to science. When Darwin returned home to England in 1836, ornithologist John Gould examined Darwin's little ostrich, declared it distinct from the common rhea, and dutifully honored it's collector by naming it *Rhea Darwinii*. However, the bird was not strictly new to science. The French naturalist Alcide d'Orbigny, who had been journeying through parts of South America ahead of Darwin and causing him to fret that he "will get the cream of all the good things, before me,"[72] had already seen the bird, described (but not collected) it, and by 1834 proposed giving it the name *Rhea pennata*.[73] Today both men are honored for its discovery; the bird is commonly known as Darwin's rhea but the modern scientific name, *Rhea pennata*, retains d'Orbigny's specific descriptor.

By now, the third year of the voyage of the *Beagle*, Darwin was preoccupied with patterns of species distribution and how they related to a varied and changing physical world. He had truly become a master theorizer, weaving together the observations he collected along the way, threading them into his working theories, always striving for the big picture yet never afraid to discard a fraying idea. As revealed by Sandra Herbert, his skillful, inclusive, bio-geological way of thinking shines through in a rough, unpublished essay he wrote in early 1834, called "Reflections on reading my Geological notes."[74] In it Darwin theorized on the formation of different parts of South America and discussed the distribution of specific past and present organisms in context with his geological findings. He asked questions about the home ranges of the extinct animals whose bones he had come across, puzzling over how the "most sterile plains [where they were discovered, could have] support[ed] such large animals." He also hypothesized on the distribution of modern species in relation to the geological history of the area. One train of thought began: "As Patagonia has risen from the waters in so late a period [in the history of the world] it may be interesting to consider whence came its organized being[s]." He believed the area should have few or no indigenous species "owing to no Creation having taken place" subsequent to the deposition of its top strata and latest elevation. To check his idea he listed the region's mammals (skunks, otters, jaguars, etc.) and found, with two notable exceptions (a guinea pig and an armadillo), that they "nearly are all characterized by large Geographical range[s] —& therefore may easily have traveled from their Northern original homes."

Through her analysis of Darwin's words, Herbert highlights their implication to Darwin's later conversion to evolution. His essay, she stresses, shows that in 1834 he was already thinking of both the geology of the earth and of the movement of its organisms through time and space in highly sequential terms.[75] He had yet to think that species were mutable and still believed the common assumption that all species came from "centers of creation," but with his focus on change, movement, and series of organisms he was certainly pondering the threads of a dangerous new tapestry of thought. By following Lyell's belief that God periodically created new species, which then spread out from their points of origin, by seeking to identify these "centers of creation" and paths of distribution, he was, unknowingly, priming his mind to think in evolutionary terms and collecting the data to support it.

A Bump in the Road

After Darwin returned to England at the end of the voyage he struggled with a chronic illness that lasted the rest of his life. During the voyage, however, Darwin was superbly fit. He suffered from seasickness and the occasional malady but was otherwise healthy. His physical health not only determined the success of his collecting excursions, it also helped with the surveying efforts. His constitution was put to the test in Tierra del Fuego where for the chance to explore on shore he endured prolonged and dangerous trips in the *Beagle*'s small, loaded surveying boats. On January 29, 1833, during one of these excursions, a glacier suddenly calved off while the men were lunching on shore. The impact of the ice hitting the water kicked up a tremendous swell with enormous rollers that threatened to sweep the boats from the beach, or dash them to pieces. Darwin leapt to the rescue in the nick of time, seizing the ropes and helping the seamen haul the boats to safety. The incident inspired a grateful FitzRoy to name Darwin Sound after his brave and stoical "messmate."[76] FitzRoy also named one of the high peaks in the area Mount Darwin, an act that made Darwin excessively proud when it was measured and calculated (although erroneously) to be taller than Mount Sarmiento, the highest known peak in Tierra del Fuego.[77] A year later, in Argentina, another shore party began suffering severe exhaustion and dehydration during a long and arduous hike. Darwin alone among the officers had the stamina to strike out in search of badly needed drinking water.[78]

Darwin weakened, however, on one occasion in Chile. In September 1834, a year before reaching Galápagos, he drank some homemade wine after which he became bedridden for over a month, "a grievous loss of time"[79] for his collecting ambitions. His father and sisters wrote urging him "to think of leaving the *Beagle*, and returning home, and to take warning by this one serious illness."[80] The thought of quitting the voyage had crossed Darwin's mind.[81] By now, almost three years into the voyage, he longed for home and was "beginning to plan the very coaches by which [he would] be able to reach Shrewsbury in the shortest time."[82] He was also continuing, "to suffer so much from sea-sickness, that nothing . . . [could] make up for the misery & vexation of spirit."[83] But the prospect of exploring Peru ("where the . . . country [is] hideously sterile but abounding with the highest interest to a Geologist") and then "crossing the Pacific," were too alluring to abandon. As he explained in a letter to Catherine, "I could not give up all the geological castles in the air, which I had been building for the last two years."[84] He rallied himself together and wrote to his family with "determination of remaining in the *Beagle* till the expedition is over."[85] FitzRoy had once labeled Darwin a quitter, simply because of the shape of his nose. Perhaps the memory of this unreasonable criticism also strengthened Darwin's resolve to persevere with the voyage.

Indeed, determination was clearly the hallmark of Darwin's character, the force that forged his success as an explorer and theorist. It was determination that gave Darwin the resilience to endure miserable sea journeys throughout the voyage, a crippling illness after the voyage, and the prejudicial fury unleashed upon him with the publication of his theory of evolution. Best of all, it was determination that got Darwin to the Galápagos Islands.

Uplifting Andes

Darwin ventured into the Andes on several occasions. In doing so he accumulated a mountain of evidence in support of Lyell's view that the earth's topography is the result of ongoing processes of tectonic activity and erosion. On March 21, 1835, upon climbing to an elevation of over 3700 meters to the Portillo Pass between Chile and Argentina, he discovered "fossil [marine] shells on the highest ridge."[86] A couple of weeks later, on April 2, 1835, he crossed the Uspallata range, and at an altitude of over 2000 meters, found the silicified remains of a coastal forest. The sight of petrified trees, which had once graced the shore, now lying at angles among grossly tilted strata, having been "tossed about like the crust of a broken pie"[87] filled him with awe. In a letter to Henslow he wrote, "I cannot imagine any part of the world presenting a more extraordinary scene of the breaking up of the crust of the globe than the very central peaks of the Andes."[88] Everything pointed to the Andes having been pushed up into its present state, bit-by-bit, rising and crumpling and rising again, over vast periods of time.

Darwin's confirmation that the physical world is one in constant flux was based as much on immediate experience as on historical clues. In January 1835, from the safe distance of the Chiloe Islands, he witnessed the sky-illuminating eruption of Chile's Osorno Volcano, 100 kilometers to the north. Then a month later, on February 20, he experienced a severe earthquake in Valdivia. "I was on shore & lying down in the wood to rest myself. It came on suddenly & lasted two minutes (but appeared much longer). The rocking was most sensible . . . there was no difficulty in standing upright; but the motion made me giddy. —I can compare it to skating on very thin ice or to the motion of a ship in a little cross ripple."[89]

Upon arriving at the town of Concepción two weeks later Darwin observed just how devastating the earthquake had been to the entire area. It was "the most awful but interesting spectacle I ever beheld," he wrote in his diary. Seventy villages had been razed, the majority of their citizens having been saved only by "the *constant* habit of these people of running out of their houses *instantly* on perceiving the *first* trembling."[90] A tidal wave had ravaged the shore and deposited large slabs of marine deposits high up on the beach. Fissures split the ground in all directions and the entire offshore island of Santa Maria "was upheaved nine feet."[91] Here was proof in excess that seismic violence was not just a thing of the past but a recurring force that kept the world's surface perpetually changing. Experiencing this force in the flesh made a huge impact on Darwin. The earth's surface is "a mere crust over a fluid melted mass of rock,"[92] he proclaimed in his diary. "The Earthquake & Volcano are parts of one of the greatest phenomena to which this world is subject."[93]

The more Darwin saw, the more he appreciated just how much the geography of a place determined the distribution of its organisms, and how much it all could, and did, change over time. Every subsequent step he took in South America was in search of further evidence of the internal workings of the planet and of fossils to reveal its dynamic past.

Bound for "Novel Ground"[94]

The *Beagle's* last stop on the South American continent was Lima, Peru, a town Darwin enjoyed for the ladies and the "chilimoya" (cherimoya, or Peruvian custard apple). The former, he wrote in his diary, are "certainly . . . better worth looking at than all the churches & buildings" in the city, and the latter "is a very good & large fruit & that is all I have to say about it."[95] There was no time to visit the mainland of

Ecuador, the country to which Galápagos belonged politically then as it does now. On September 7, 1835, after having hugged the South American continent for almost four years, the *Beagle* broke away from the coast and "steered direct towards the Galapagos Islands . . . [and on to] novel ground."[96] Darwin was now 26 years old and as excited as a child on Christmas Eve.

Before leaving England Darwin had consulted naturalist Charles Stokes for some specific suggestions on what to do during the voyage. Collect "every thing [*sic*] especially from Gallipago [*sic*, Galápagos]"[97] had come Stokes' reply. Few naturalists had explored the islands and fewer still had brought home specimens. Alcide d'Orbigny, the most recent collector to have explored South America, had not included Galápagos in his travels, and nor would he, for his voyage ended in 1833. Here was Darwin's biggest chance to make an important collection, find something new for science, and really promote his name as a collector.

Darwin already knew something about the islands. Galápagos was a well-known whaling area and sealing ground that had been frequented by British mariners for the past 40 years. Earlier in the voyage, at the Falkland Islands, Darwin and FitzRoy had met one of these mariners, sealer William Low. Darwin thought that Low, in "manners habits &c . . . strikingly resembled the old Buccaneers."[98] On the other hand, he and FitzRoy had been impressed by Low's "readiness at description, and . . . extraordinary local memory." FitzRoy had milked him of some general information about the places they were to visit, including the Galápagos Islands.[99] It was Low who told FitzRoy where to find fresh water on Chatham Island, vital knowledge as it turned out to be, for all those on board.[100]

Darwin learned even more about Galápagos from reading. In the *Beagle* library were upward of 245 natural history and travel books, dictionaries, and reference volumes, the known titles of which are listed in the first volume of *The Correspondence of Charles Darwin*.[101] Among the tomes were four books containing extracts about the Galápagos archipelago written by previous British visitors to the islands: James Colnett, Lord George Anson Byron, Basil Hall, and William Dampier. From references made in Darwin's notes and diary it is clear that Darwin read at least the first three accounts before arriving in the islands, or during his time there.

James Colnett had been sent to Galápagos by the British whaling firm Enderby & Sons to assess the commercial worth of the archipelago as a potential whaling station. He scouted the islands in HMS *Rattler* from June 22 to July 2, 1793, and again from March 12 to May 17, 1794, and in the process gathered a wealth of information about the islands and their wildlife. His book, *A voyage to the South Atlantic and round Cape Horn*,[102] includes a rough map of the archipelago with many of the islands named (or renamed) by him (see black and white plate 15b). It was the most comprehensive treatise of Galápagos of all the books aboard the *Beagle*.

The purpose of Vice-Admiral George Anson Byron's voyage had been to return home the bodies of the Sandwich Islands' (Hawaiian) king and queen. They had died of measles during an official visit to England. On the way to the Sandwich Islands HMS *Blonde* stopped at Galápagos between March 25 and April 2, 1825, and visited the islands of Albemarle and Narborough. In the resulting book, *Voyage of H.M.S. Blonde to the Sandwich Islands, in the years 1824–25*,[103] the natural history of the places Byron saw is described in poetic detail, hinting at a talent reminiscent of that of his cousin, the famous poet Lord George Gordon Byron. However, the account was not written exclusively (or perhaps at all) by Byron, but is an edited compilation of the journals of several of the men on board the *Blonde*.[104] From references in FitzRoy's and Darwin's writings this book seems to have been a favorite on board the *Beagle*.

Also containing valuable information about Galápagos were *Extracts from a journal written on the coasts of Chili, Peru and Mexico for the years 1820, 1821, 1822* by Basil Hall[105] and *A New Voyage round the world* by William Dampier.[106] Captain Basil Hall sailed HMS *Conway* to Galápagos in January 1822, to "see whether any assistance was required by [the South Seas whaling ships] . . . that important branch of the British shipping interests."[107] Finding only two whalers in the islands, Hall took advantage of his allotted time in Galápagos to conduct a physics experiment. His aim was to measure the acceleration of gravity at the equator with an invariable

pendulum of Captain Kater's construction, and "determine the figure of the earth."[108] Strong currents forced him somewhat north
of the line, to Abingdon (Pinta) Island, where he laid up for nine
days constructing a field lab and performing his experiment. Hall
restricted his comments on the wildlife of Galápagos to the tortoise,
for which he quoted heavily from American merchantman Amasa
Delano's published account of his visits to Galápagos in 1800 and
1801.[109] Hall also collected one of the very first scientific specimens
of Galápagos tortoise, a young individual that he preserved in a cask
of spirits and gave to the Museum of the College at Edinburgh (now
the Royal Scottish Museum). Darwin spent a considerable amount
of time at the museum during the two years (1825–1827) he spent
studying medicine at the University of Edinburgh.[110] Although there
is no record of Darwin seeing the specimen, one cannot help but
wonder if he did and what he thought about it (see chapter 5). The
Abingdon population of Galápagos tortoise (*Geochelone nigra abing-
doni*) is now virtually extinct, with one old and famous male (Lonesome George) surviving in captivity at the Charles Darwin Research
Station on Indefatigable (Santa Cruz) Island.[111]

William Dampier was a "gentleman pirate" and an amateur naturalist. He and his buccaneering companions used the Galápagos Islands as a hideout, while awaiting the arrival of Spanish prizes along
the South American coast. They arrived on May 31, 1684, on board
the *Nicholas* and the *Bachelor's Delight*. The latter was a Danish slaving
vessel captured off the coast of Guinea, and presumably renamed to
reflect the pirates' devil-may-care exploit. The ships had three Spanish
prizes in tow, full of rather disappointing cargoes of flour, timber, and
marmalade. In the two weeks he spent in the islands, Dampier took
careful notice of his physical, botanical, and zoological surroundings.
He is often described as the first naturalist to visit Galápagos. His book
offered the world its first peek at the Galápagos Islands—an evocative
view of their physical appearance and exotic fauna and flora. William
Ambrose Cowley, one of Dampier's companions, gave the islands
their first names, in honor of British royalty and aristocracy. One of
the earliest and most famous maps of Galápagos has also been attributed to Cowley (see black and white plate 15A).

From these literary accounts on board the *Beagle* Darwin learned that the islands harbored enormous tortoises, swimming iguanas, and strangely fearless birds. He could barely wait so see it all for himself. The "Zoology cannot fail to be very interesting,"[112] he penned excitedly to his sister Caroline.

Even more tantalizing was the geology of the islands. Fresh in Darwin's mind were his paleontological discoveries in the mountain chains of South America, his distant view of the eruption of Mount Osorno, the powerful earthquake at Valparaiso, its effects on Concepción, and the resulting rise of the isle of Santa Maria off Chile. In Galápagos he anticipated finding more fossils, more volcanoes, more uplift. They could only feed his understanding of the physical laws governing the planet Earth and of the patterns of distribution of extinct and living species upon it. "I look forward to the Galapagos, with more interest than any other part of the voyage," he wrote to his cousin William Fox. "They abound with active Volcanoes & I should hope contain Tertiary strata."[113] Officer Alexander Usborne was to continue surveying the northern coast of Peru while the *Beagle* sailed to Galápagos. Before parting Darwin asked him to keep an eye out for "beds of shells lying on the surface of the land" as proof of "recent elevation of the land above the level of the ocean."[114] Darwin planned to do exactly the same in Galápagos.

Most of all Darwin wanted to see an active volcano up close. Before leaving Lima he wrote to Henslow, "In a few days time the *Beagle* will sail for the Galápagos Is[ds]. —I look forward with joy & interest to this, both as being somewhat nearer to England, & for the sake of having a good look at an active Volcano. —Although we have seen Lava in abundance, I have never yet beheld the Crater."[115]

Darwin did not find any tertiary fossils nor witness an eruption in Galápagos but he was delighted to discover "that there must be, in all the islands of the archipelago, at least two thousand craters."[116] The geology of "that land of Craters"[117] kept Darwin preoccupied for the entire five weeks the *Beagle* remained in Galápagos. In Galápagos and on the way home to England he wrote over 100 manuscript pages of notes on the geology of Galápagos,[118] compared to just 37 pages of notes on their zoology.[119] He also came up with several fundamental geological concepts from his observations in Galápagos that are considered axiomatic to the science today. Although the revolutionary impact of *The Origin of Species* overshadows his geological accomplishments, it was Darwin's thinking as a geologist in Galápagos that helped him recognize evolution in its species. It was also his work as a geologist on the voyage that first earned him recognition among the scientific establishment when he returned home to England. It was as a geologist that Darwin entered the Galápagos archipelago, geology that determined his footsteps within the islands, geology that framed his contemplation of the living world, and geology that set the stage for his illustrious future.

Part 2
Galápagos

One night I slept on shore, on a part of the island where some black cones—the former chimneys of the subterranean heated fluids—were extraordinarily numerous. . . . [T]hey gave the country a workshop appearance . . .
—Charles Darwin, *Journal and Remarks*[1]

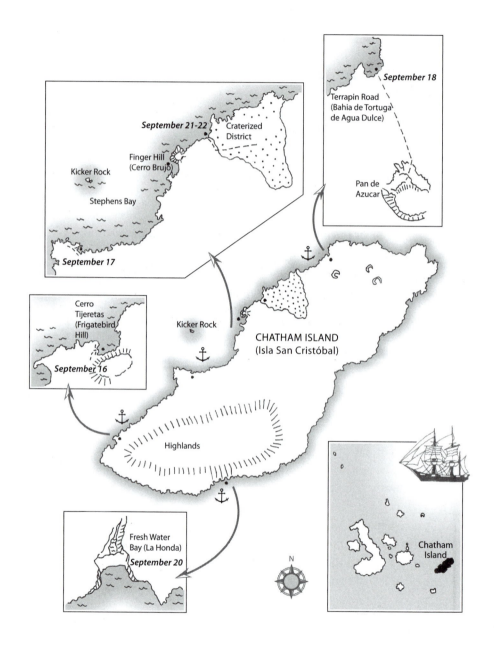

September 18

Terrapin Road
(Bahia de Tortuga
de Agua Dulce)

Pan de
Azucar

September 21-22

Craterized
District

Finger Hill
(Cerro Brujo)

Kicker Rock

Stephens Bay

September 17

Cerro
Tijeretas
(Frigatebird
Hill)

Kicker Rock

September 16

CHATHAM ISLAND
(Isla San Cristóbal)

Highlands

Fresh Water
Bay (La Honda)

September 20

N

Chatham
Island

Chapter IV

Chatham Island
(Isla San Cristóbal)

On Tuesday, September 15, 1835, nine days after leaving the coast of Peru, and 45 months after leaving England, HMS *Beagle* sailed into Galápagos waters. At 10 o'clock in the morning Darwin was rolled from his hammock by the long anticipated call, Land Ho! He and FitzRoy peered through their telescopes and made out what looked like a small islet on the northwestern horizon. The "islet" turned out to be "the summit of Mount Pitt, a remarkable hill at the north-east end of Chatham,"[1] the easternmost island of the archipelago. Most of the island's mass remained invisible in the hazy atmosphere and it was close to evening by the time they could perceive its entire 48-kilometer length. In some places the land was "quite barren —in others covered with a stunted and sun-dried brushwood —and . . . the heights, on which the clouds hung, were thickly clothed with green wood."[2] The variegated appearance of the island beckoned to Darwin with the promise of rich discovery. Little did he know what a Pandora's box Galápagos would prove to be.

During the night currents bore the *Beagle* south toward Hood (Española) Island, and the next morning, without ceremony, the *Beagle* began what it had come to do—chart the islands. Two of the surveying officers (Chaffers and Mellersh) were sent off in one of the whaleboats "to examine this island [Hood] and the anchorages about it."[3] Then at noon a larger surveying party of 13, "victualled [*sic*] for three weeks"[4] and headed by Lieutenant Sulivan, left in the 26-foot yawl "to examine the central islands of the archipelago."[5] The *Beagle* then steered toward Chatham where it would spend the week circumnavigating the island and allowing Darwin his first explorations on shore in Galápagos.

Surveying Galápagos

No people have such cause anxiously to watch the state of weather —as Surveyors. —Their very duty leads them into the places which all other ships avoid & their safety depends on being prepared for the worst.
—Charles Darwin, *Beagle Diary*[1]

The *Beagle* went to Galápagos to make a map of the archipelago for the British Admiralty. Ever since 1794, when James Colnett established the reputation of Galápagos as a whaling ground, the archipelago had received a rising influx of British whalers, sealers, and merchantmen, all eager to profit from her wealth of marine mammals and tortoises. And ever since the Napoleonic Wars, when American commander David Porter of the USS *Essex* crippled over half the British whaling fleet in the South Seas before his capture in 1814, the British Admiralty became increasingly aware of the need for a comprehensive chart of the archipelago to aid British shipping interests. Colnett's rough map of the archipelago was the best there was. While Captain Philip Pipon of HMS *Tagus*, who visited Galápagos in 1814, and Captain John Fyffe of HMS *Indefatigable,* who passed through the archipelago in 1815, provided "Additions & Corrections" to Colnett's map,[2] several of the islands remained depicted as uncertain blobs, and the others were grossly misshapen or misplaced. Basil Hall, captain of HMS *Conway,* who visited the islands in 1822, drew attention to the problem in his published *Journal*, by writing, "We had no time to survey these islands, a service much required, since few if any of them are yet properly laid down on our charts." He incited action by further writing, "It is to be regretted, that the true geographical position of these islands is still uncertain, and the hydrographical knowledge respecting them so exceedingly scanty."[3] So it was that the British Admiralty picked up the hints and ordered Captain FitzRoy of HMS *Beagle* to include Galápagos in the hydrographic survey, "if the season permits."[4]

The *Beagle* was superbly equipped as a surveying vessel. On board were sextants and reflecting circles, bearing compasses, theodolites, micrometers, artificial horizons and "boards" for calculating latitude and triangulating distance, leads for sounding depths, mechanical logs for measuring current speed, and no less than 24 chronometers for fixing longitude by comparing local time with that of Greenwich, England. The reason for so many timepieces was their potential unreliability. To compensate for "irregularities and errors of individual watches" accurate chronometrical measurements were calculated from "the mean of the whole."[5] Most were permanent fixtures on board the *Beagle*, suspended in gimbals, cushioned in sawdust and housed in a special room in the center of the ship to minimize vibration.[6] Others were pocket timepieces to be taken on shore after being calibrated to the best of the clocks on board the *Beagle*. Of all the instruments, the chronometers were most important for the expedition's success. They were tended to religiously by one specialist, George James Stebbing, who wound them precisely at 9 a.m. every day.

Captain FitzRoy was a meticulous hydrographer and he trained his surveying officers well. Darwin wrote of an air of intense concentration that reigned whenever the ship was depth sounding. Everything was in "a state of preparation; Sails all shortened & snug: anchor ready to let fall: no voice or noise to be heard, excepting the alternate cry of the leadsmen in the chains."[7]

With up-to-date tools and "intelligent assistants," FitzRoy succeeded in making the first practical chart of the Galápagos Islands (see color plate 28). His measurements of latitude

and longitude are slightly skewed to the north and east, respectively, and Tower Island lacks its distinctive horseshoe-
shaped bay. Otherwise, FitzRoy's map was so well drawn that
it was used effectively for navigation, with just a few altera-
tions, for the next 100 years.

Considering that the *Beagle* stopped at only five islands
(Chatham, Hood, Charles, Albemarle, and James), how was
it that FitzRoy was able to produce such a comprehensive and
detailed chart of the entire archipelago with its 13 principle
islands and many additional islets and rocks? The answer lies
in the small boats aboard the *Beagle*. The surveying officers
were sent out in these boats, spending days, even weeks at a
time surveying the coastlines of the various islands. Lieutenant Bartholomew Sulivan, the principal surveyor, mapped
the central islands, which were the most poorly represented
on earlier maps. FitzRoy honored his achievement with two
place names—Sulivan Bay (today commonly misspelled Sul-
livan Bay) on James Island, and nearby Bartholomew Islet.
William Ambrose Cowley had already named James Island
in 1684, in honor of King James II.[8] The sites were an appropriate choice, for James was Bartholomew Sulivan's middle
name. Somewhat surprisingly, Sulivan was the only member
of the *Beagle* to be awarded a place name in Galápagos, other
than Darwin and the *Beagle*, both honored posthumously
(see appendix 2). The authors have since given some of the
other men aboard the *Beagle* fair due by naming the islets in
the lake of Albemarle's Beagle Crater after Chaffers, Bynoe,
King, Covington, and Fuller, in honor of their services to
both Darwin and the surveying efforts.[9]

It should be noted that the Admiralty, in its instructions
for the voyage, had discouraged FitzRoy from naming newly
discovered places after private friends or public figures. It pre-
ferred that the names "convey some idea of the nature of the
place."[10] This was more a suggestion than a rule, for the Admiralty also recognized that the honor of naming new places that
naturally falls on the officers and crew of ships, "helps excite

an interest in the voyage."[11] FitzRoy certainly appears to have interpreted the instructions this way, for he named several landmarks around the world in honor of people he knew.[12]

In addition to Sulivan, the *Beagle* surveyors who helped map Galápagos were Edward Main Chaffers, John Lort Stokes, Arthur Mellersh, and Philip Gidley King. Chaffers charted Hood Island (with Mellersh), the satellite islets of Charles, and the northern islands of Tower, Bindloe, and Abingdon. Stokes surveyed Charles Island, and Mellersh and King surveyed the western coast of Albemarle Island, including Elizabeth Bay, and Narborough Island. Their explorations were as useful to Darwin as they were to the mapping efforts, for Darwin "saw . . . specimens & received notices respecting the [other islands], from the various officers employed in the survey,"[13] thus expanding his horizons well beyond the four islands he himself visited. The men frequently landed, for a trick of the trade was to climb "the summits of the highest hills near the harbour"[14] to fix bearings and triangulate distances. The view from such heights also afforded them "eye sketches of the coast line"[15] and showed them the location of dangerous rocks and shallows. Their grueling surveying schedule left little time for collecting, but Chaffers managed to pick up several rock specimens for Darwin.

One of the advantages of living in the *Beagle* chart room was that Darwin was able to watch the map of Galápagos take form. By studying the charts and consulting his copy of Aaron Arrowsmith's *A New General Atlas*, he calculated that the Galápagos Islands "are scattered over a space equaling in extent the whole of Scicily [*sic*] & the Ionian Isles."[16] He also saw that "the principal Volcanic vents & . . . the islands themselves" are aligned along "a NW & SE line"[17] and was the first to notice how the islands are placed in a rectilinear grid, rather than an arc, as are the vast majority of archipelagos in the world.[18] He offered no explanation for this phenomenon (today known as the "Darwinian trends") other than pointing out a directional link between the American continents and the archipelago. The islands, he wrote, are "parallel to the whole coast of N. America, & . . . nearly in the same line with . . . part of the shores of S. America."[19] Why, he could not say. Even today the geometry of the archipelago remains poorly understood. The Galápagos Islands are the result of a slow-moving tectonic plate passing over a stationary hot spot, or mantle plume, which erupts periodically to form new volcanic land.[20] This Nazca plate moves eastward at the rate of about 7 centimeters per year, such that when Darwin was in Galápagos the islands were about 10 meters further west from South America than they are at the writing of this book. However, the Darwinian lines are not directionally related to this movement. Nor do they correspond to the north–south direction of seafloor spreading that occurs between the Nazca Plate and neighboring Cocos Plate. The best explanation is that they are the result of a complex combination of stresses on the earth's crust resulting from the close proximity of the Nazca, Cocos, and Pacific plates.

Fit for Pandemonium

At 4:15 p.m. on Wednesday, September 16, 1835, HMS *Beagle* anchored off the southwestern end of Chatham Island and Darwin and FitzRoy landed for an hour. There are several sandy beaches exposed along this section of coastline but the men headed for a more protected area—a small rocky cove (00°53'17.36"S, 89°36'29.54"W). They found it a most unwelcoming spot. The cove is shadowed by a looming cliff and bound on either side by "dismal-looking heaps of broken lava"[6] covered with stunted trees and cacti. Darwin and FitzRoy arrived during the dry season when many of the plants in the arid zone are leafless and the trees look decidedly skeletal. The men imagined they had stepped into John Milton's *Paradise Lost* (a favorite book of Darwin's and one he often carried on his excursions away from the *Beagle*[7]) and were entering Satan's "high Capital," itself.[8] FitzRoy called it "a shore fit for Pandemonium."[9] Darwin wrote, "the country may be compared to what we might imagine the cultivated parts of the Infernal regions to be."[10]

Darwin and FitzRoy examined the coast on which "innumerable crabs and hideous iguanas started in every direction as [they] scrambled from rock to rock."[11] They then bushwhacked inland so that FitzRoy could climb to the top of the unnamed cliff and get his bearings.

Today Frigatebird Hill or Cerro Tijeretas, as the site is now called, is a tourist attraction. It is a popular site for snorkeling, for observing frigatebirds in flight and boobies and pelicans feeding in the cove, and for the panorama afforded from the top of the hill. The site is reached by boat, or by hiking a National Park trail from the nearby town of Puerto Baquerizo Moreno. In 1835, however, the island was uninhabited. The only sign of man was a "hand-barrow . . . lying at the landing-place, which showed that terrapin were to be got near [by], though [they] did not then see any." The barrows, FitzRoy explained, were used by "men from whalers and sealing vessels [to] carry the large terrapin, or land-tortoises [to shore]."[12] On each of the four islands they visited, the *Beagle* men found evidence of human visitation. This is hardly surprising considering that FitzRoy directed the *Beagle* to places he knew were most important for shipping interests.

Golden Age of Whaling

I am disposed to believe that we were now at the general rendezvous of the spermaceti whales from the coasts of Mexico, Peru, and the Gulf of Panama, who come here to calve. . . . There is great plenty of every kind of fish. . . . But all the luxuries of the sea, yielded to that which the island[s] afforded us in the land tortoise, which in whatever way it was dressed, was considered by all of us as the most delicious food we had ever tasted.
—James Colnett, A Voyage to the South Atlantic[1]

Darwin arrived in the Galápagos Islands in the middle of the "Golden Age of Whaling," when untold numbers of whales (primarily sperm whales) were hunted for their spermaceti oil. Captain FitzRoy recorded seeing half a dozen whaling ships, one of which, the *Science* of Portland, Maine, he particularly admired for the "nine whale-boats!" she carried.[2] The

Beagle was overloaded with just seven boats (four of which were whaleboats). American whaling ships dominated the archipelago at this time, but it had not always been so. The British had started grand-scale whaling in Galápagos at the end of the 18th century, and had ruled the seas for the first two decades of the 19th.

In 1789, 20 years before Darwin's birth, the British whaling ship *Emilia* made history by rounding the horn on a pioneering passage to the coast of Peru, thereby inaugurating British whaling in the Pacific Ocean.[3] Her feat immediately opened the floodgates to whalers of many countries. Those who made it as far as Galápagos sent back reports of an abundance of whales, a bonus of fur seals and sea lions, and an endless supply of gourmet tortoise meat to be had for the taking. The Spaniards occupying the west coast of South America were already taking advantage of such resources, and Britain schemed to take over. Enderby & Sons, the whaling company that had sent the *Emilia* to scout out the Pacific, commissioned HMS *Rattler* and naval officer James Colnett to evaluate Galápagos as a potential whaling station for the British South Seas fishery. Between June 1793 and May 1794, Colnett spent almost two and a half months cruising Galápagos waters. With his sharp observational skills and keen scientific curiosity, he reaped a wealth of information about the islands and their wildlife. His book, *A Voyage to the South Atlantic and Round Cape Horn into the Pacific Ocean*, became the most comprehensive treatise there was on the geography, zoology, and botany of Galápagos. Four decades on, it was still considered the essential guide to the islands, and a well-thumbed copy was to be found in the *Beagle* library.

Colnett educated the world about Galápagos, but his words beckoned exploitation. Not only were the Galápagos seas rich in marine life, but "the place was, in every respect, calculated for refreshment or relief for crews after a long and tedious voyage."[4] Whalers and sealers from all over the world soon flocked to the archipelago to reap the islands' riches.

For the first decade, sealing was as rampant as whaling. As much as it was lucrative, it was a cutthroat profession. It was often said that "a Sealer, Slaver & Pirate are all of a trade"[5] and Darwin had to agree. In the Falkland Islands he had met the type in sealing captain William Low, "a notorious & singular man who ha[d] frequented these seas for many years & been the terror to all small vessels."[6] It must have come as no surprise when, on James Island in Galápagos, Darwin heard tell of another sealing captain who had been murdered by his crew, and saw with his own eyes the skull lying testament in the bushes.[7]

The sealers killed so many fur seals in Galápagos (in 1817, one ship alone took 8000 pelts and another, in 1825, took 5000) that by the time Darwin arrived on the scene the practice of sealing had virtually exhausted itself.[8] It is no wonder that Darwin made no mention of seeing fur seals in Galápagos; there were probably few left in the archipelago. They were not, however, completely wiped out. Sealing was renewed in the late 1800s and the logbooks show that at least 9485, and undoubtedly many more, were taken in the last 30 years of the century, leading once again to their virtual extinction.[9] Scientists from the California Academy of Sciences found only one fur seal in the entire archipelago during their year-long survey of the islands in 1905–1906.[10] Fortunately, the fur seals recovered a second time, and they are now protected under Galápagos National Park legislation. The latest fur seal census, conducted by scientists from the Charles Darwin Research Station in 2001, revealed there to be between 6000 and 8000 individuals in the archipelago.[11]

Sea lions were also killed during the 1800s, for their oil. Darwin never mentioned seeing a sea lion in Galápagos, but Captain FitzRoy reported that "hair seals" were "abundant" on Barrington (Santa Fé) Island.[12] FitzRoy did not visit Barrington himself, but presumably got the information either from Sulivan, who surveyed the island in the Beagle's yawl, or perhaps from a visiting whaler. The current population of sea lions in Galápagos is estimated to be roughly 18,000 individuals.[13]

During the US–Anglo war of 1812, American naval captain David Porter of the frigate USS *Essex* cruised Galápagos for seven months, capturing British ships and crippling the British South Seas whaling industry. By the time he was captured off Valparaiso, Peru on May 28, 1814, he had destroyed over half Britain's fleet in the Pacific. While the British whalers were quick to return to Galápagos, from then on American whalers dominated the scene.[14]

Whaling was never so industrious as during the first half of the 19th century. Between 1811 and 1844 there were more than 700 American whaling ships operating in the islands, and perhaps over a thousand including other nationalities.[15] Whalers are to blame for the slaughter of countless whales in Galápagos. They are also responsible for killing hundreds of thousands of giant tortoises. Tortoises were considered an ideal food source, for they could be kept fresh on board, without consuming any food or water, for months on end. In the early 20th century, historian Charles Townsend examined a sample of log books from American whaling ships that operated in Galápagos. He found that prior to 1835 tortoises were primarily taken from the southern islands of Charles and Hood.[16] Because Hood Island is an old eroded island, low in elevation, its relatively small saddlebacked tortoises could be acquired with little difficulty. The race probably survived only because Hood Island does not have any fresh water, making it a less desirable stopping point for mariners. Charles Island, on the other hand, was heavily visited for its freshwater supply and its saddlebacked tortoises were hunted until there were literally none left. As population numbers declined on Hood and Charles islands in the 1830s, James Island became heavily visited, so that by 1845 that island too had lost its status as a prime hunting ground. Chatham became increasingly popular in the 1840s, when it became settled, but again, the tortoise population soon dwindled. A report of HMS *Pandora*'s visit to Galápagos in 1847 states that the Chatham Island settlers took "as many as 10,000 from this Island alone, so that they are now far from numerous —the wild dogs & the supply of food they afford to Whalers & Settlers will soon destroy all that remain, indeed Mr. Gurney [Mayor of Chatham] thought three months work would pretty nearly clear off those that are available for Food or oil."[17]

As predicted, the tortoises disappeared from the southern end of the island. However, a small number survived in the northern part of the island, and there is now a healthy population of tortoises living there today.

Albemarle was the last destination to receive the bulk of the tortoise hunters. It became the best place to find tortoises in the 1860s, and remained so until the end of the century.

Prior to the 1830s, catches of up to 350 tortoises per vessel were taken, with an average catch of 210 tortoises per ship. The average number of tortoises taken progressively declined to 122 tortoises per vessel between 1831 and 1868.[18] It is not known just how many vessels traversed the islands during those years, but Governor Villamil recorded 27 ships anchoring off his settlement on Charles Island in a six-month period in 1835,[19] and American commodore John Downes of the US Frigate *Potomac* recorded 31 ships in a 10-month period between 1832 and 1833.[20] With an extrapolated average of 45 ships per year, and an estimate of 200 tortoises per ship, an annual catch could have tallied over 9000 tortoises. At this rate over 180,000 tortoises would have been slaughtered in the time it takes a hatchling tortoise to reach sexual maturity—roughly 20 years. Populations had little chance of recuperating from such an onslaught and it is no wonder that the Charles Island tortoise was driven to the brink of extinction by the time Darwin arrived, and was pushed over the edge within the following few years.[21] Today there are estimated to be between 20,000 and 30,000 tortoises in the entire archipelago, from an original population of perhaps 250,000 or more.[22]

Nineteenth-century whalers sometimes marked their presence in Galápagos by carving the name of their ship in a suitable cliff face. At Tagus Cove on Albemarle Island, various ship names can be seen on the tuff cliffs that circumscribe the oval bay. Most are 20th-century ship names that have been painted onto the rock, but a few older names can be found etched in the tuff. The oldest legible engraving—*Phoenix, 1836*—can be viewed from the modern national park trail at this visiting site. Protected by an overhang it has withstood excessive erosion by wind and rain. The *Phoenix* was an American whaler from Nantucket that cruised Galápagos waters in both 1835 and 1836. Her name was carved at Tagus when "all hands were on shore" on Sunday, January 10, 1836.[23] This was just three months after the *Beagle* anchored in the very same cove. Excerpts from the logbook of the *Phoenix* are worth paraphrasing for they reveal the typical activities of a whaling ship from Darwin's time. The crew reported taking tortoises from two of the islands: 40 from Hood Island, where they were "hard to get," and "8 boat loads" (perhaps 100) from James Island where they were still common. At one point the ship was smoked to rid the hold of "25–30" rats. While the task was performed at open sea, between Abingdon and Albemarle islands, it is easy to imagine how rats were introduced to Galápagos by ships performing this chore closer to shore (see Alien Invaders-The Human Factor).

The *Phoenix* encountered several whaling vessels, at least two of which—the *Science* of Portland, Maine carrying 400 barrels of whale oil and the *America* of New Bedford, Massachusetts carrying 450 barrels—were also seen by the *Beagle*.

Darwin came to Galápagos looking for evidence of land elevation, and here, at his very first landing spot, he found it. At the base of Frigatebird Hill were (and still are) some "huge fragments of vesicular & compact Basalt" cemented together "at a height of several ft above high water mark [with] . . . a hard calcareous sandstone." To Darwin this was undeniable "proof of elevation to a small degree within recent times."[13] Small scale though it was, Darwin was disproportionately excited. Chipping away at the pinkish conglomerates with his geologic hammer at eye level he extracted his first Galápagos specimen—a chunk of rock studded with the plates of chitons and the broken shells of limpets and bivalves that had once lived in the littoral at his feet. Encouraged, Darwin searched for more signs of uplift. Over the next few weeks, whenever he came across "calcareous" matter or wave-rounded boulders situated above the highest tidemark he asked, "Has this been a horizontal upheaval[?]." At the time he was quick to conclude, "Everything shows that in place of these Islands being formed by pile of poured out matter, there has been upheaval extending over these different Is^ds."[14] It was a premature conclusion but an interesting one. It showed that he was conjuring to mind an archipelago that had been pushed up from beneath the ocean into separate islands of different heights and it indicated just how influenced Darwin was by his recent study of the orogeny (the process of mountain formation) of the Andes. Ultimately, Darwin downplayed the role of uplift in the formation of the islands and declared, "Proofs of the rising of the land are scanty and imperfect"[15] in Galápagos. This change of mind was typical of Darwin. Reflecting on his methodology later in life Darwin admitted, "I cannot remember a single first-formed hypothesis which had not after a time to be given up or greatly modified."[16]

There are several areas of uplift in Galápagos, for instance on the two Plaza islets, on the two Seymour isles (Baltra and North Seymour), and on Indefatigable Island (Santa Cruz). Here large-scale tectonic events have caused major faulting, with accompanying uplifts (horsts) and down drops (grabens) on the order of tens of meters. In other places, such as the modern visiting site of Punta Espinosa on Narborough Island (Fernandina), comparatively mild uplift on the order of a couple of meters has occurred as a result of sills of magma intruding into the flanks of a volcano. Darwin, however, visited none of these spots. Had he been around to witness the great uplift of Urvina Bay on Albemarle Island in 1954 when one and a half square kilometers of marine reef were raised four meters above sea level he might have been loath to give up his initial hypothesis. As it was, his final conclusion was correct. Uplift in Galápagos is of local significance only; the Galápagos Islands have been shaped primarily by lava exuded or exploded from eruptions and only partially by tectonic uplift and subsidence and fluctuations in sea level.[17]

Darwin was distracted from his geologizing by the marine iguanas (*Amblyrhynchus cristatus*) that draped the rocks. Marine iguanas are an uncommon sight on Chatham Island today, the population having been decimated by introduced dogs in the 20th century. In 1835, however, the rocks were covered with the animals. In color as in contour they closely resembled the jagged, "black [and] porous rocks over which they crawl[ed]"[18] and even they seemed hard put to tell the difference. They moved over each other with utter indifference, treading on head and tail as if they were but lumps in the lava.

Darwin found them to be "most disgusting, clumsy Lizards."[19] On land they were "sluggish" and easy to catch by the tail. Despite their ferocious countenance they did not attempt to bite but merely "squirt[ed] a drop of fluid from each nostril"[20] when frightened.[21] Over the next few days Darwin and Bynoe examined the stomach contents of several, and found that they fed, not on fish as Darwin had at first supposed from reading James Colnett's account, but on seaweed.[22]

Darwin was a practical scientist as well as a collector and theoretician, and to indulge his boundless curiosity he experimented readily, if somewhat recklessly. As there was little scope for rigorous experimentation while traveling, Darwin's experiments on the voyage were crude, improvised affairs, but nonetheless informative. In Argentina, on October 15, 1833, he "exposed [his] hand for five minutes" to the night air and watched as it turned black with at least 50 mosquitoes "all busy with sucking."[23] At Iquique (then part of Peru), in July 1835, he observed a Vinchuca bug (*Triatoma infestans*) suck blood from a finger (not his own this time) until it was replete "in less than ten minutes."[24] The animal was then kept to see how long it could last between meals—four months.

In Galápagos he experimented with iguanas. Wishing to observe the feeding behavior of the marine iguana Darwin "carried one to a deep pool left by the retiring tide, and threw it in several times."[25] Instead of feeding, the iguana "invariably returned in a direct line to the spot where [Darwin] stood."[26] Darwin admired the "perfect ease and quickness" of the animal as it moved through the water with "a serpentine movement of its body and flattened tail," and marveled at its ability to climb out onto slick and vertical lava, with its "admirably adapted" limbs and long claws.[27] He chased it back toward the sea but "nothing would induce it to enter the water."[28] The poor iguana was merely trying to absorb the warming rays of the fading afternoon sun, but Darwin, unaware of its thermoregulatory needs, attributed its apparent preference for land solely to having "no enemy whatever on shore, whereas at sea it must often fall a prey to the numerous sharks."[29] As he arrived in Galápagos several months before the marine iguanas' breeding season, he could not know that for a hatchling iguana the Galápagos hawk is a far worse threat than any aquatic animal.[30] Nonetheless, it was clear that the marine iguana was well adapted to an aquatic existence. One of the seamen sank an iguana "with a weight for nearly an hour," after which it was still "very . . . active in its motions."[31]

Darwin found marine iguanas everywhere he landed, and on Albemarle he noticed that they "appear to grow much larger than in any other place."[32] Darwin bothered collecting only one young specimen (and that for dissection), perhaps because the species had already been described. There was also some question as to whether the species was even unique to Galápagos. Byron had written that the type "Amblyrhynchus Cristatus . . . [was first] described . . . from a specimen brought to Europe by Mr. Bullock among his Mexican curiosities . . . [and] was found . . . probably on the Pacific shore."[33] Darwin, however, was skeptical. Sometime after leaving Galápagos he thought back to the marine and land iguanas he had seen, and wrote, "Is any other genus amongst the Saurians Herbivorous? I cannot help suspecting that this genus, the species of which are so well adapted to their respective localities, is peculiar to this group of Is^{ds}."[34]

When back in England he concluded that Bullock's specimen must have come "through some whaling ship from these [Galápagos] islands."[35] He also became convinced by "M. Bibron—one of the best authorities in Europe on reptiles"— that there were two species of marine iguana in the archipelago.[36] Today just one species of endemic marine iguana (*Amblyrhynchus cristatus*) is recognized in the islands. Depending on how it is classified, it is divided into seven morphologically distinct subspecies or three genetically similar clades (a clade being a monophylectic group consisting of a common ancestor and all its descendants). There is limited compatibility between the two types of differentiation, but either way it can be said that Darwin saw three different marine iguanas on the four islands he visited.[37] Although most of the Galápagos islands retain healthy populations of marine iguanas, introduced predators (cats, dogs, and rats) and contamination from oil spills pose a serious threat to them, especially on the inhabited islands.[38] Alarmingly there appears to be almost no recruitment of iguanas at all on Chatham and northern Albemarle islands.[39]

As it was the wrong season Darwin saw no sign of breeding in the marine iguanas, and left the islands frustrated that he could learn nothing more about their behavior. Even after talking to the Charles Island settlers, some of whom were camped on James Island when Darwin was there, he was left in the dark. "I asked several of the inhabitants if they knew where [the marine iguana] laid its eggs," he wrote in his journal. "[T]hey said, that although well acquainted with the eggs of the other kind [i.e., the land iguana], they had not the least knowledge of the manner in which this species is propagated;— a fact, considering how common an animal this lizard is, not a little extraordinary."[40]

Many years later Darwin begged scientist Osbert Salvin, who was planning an expedition to the islands, to find out whatever he could about marine iguana reproduction.[41] But Salvin never made it to Galápagos. Darwin never learned that during the hot, rainy season (the first months of the calendar year) the male iguanas become brightly pigmented and aggressively territorial, that the females typically lay one to three leathery eggs in a sandy burrow dug at an oblique angle beyond the influence of the tide,[42] and that the hatchlings emerge after three to four months under ground.[43]

Other than some tantalizing geology and his curious encounters with the marine iguanas Darwin was not impressed with his surroundings on that first day ashore in Galápagos. The jumbled terrain offered no satisfactory means of stretching ones legs after 10 days on a cramped ship. Just when Darwin had felt most like striding he could only stumble. "Nothing could be less inviting than the first appearance," he wrote in his diary. "A broken field of black basaltic lava is every where covered by astunted [*sic*, a stunted] brushwood, which shows little signs of life. The dry and parched surface, having been heated by the noonday sun, gave the air a close and sultry feeling, like that from a stove: we fancied even the bushes smelt unpleasantly."[44] FitzRoy mirrored Darwin's sentiment. "This first excursion had no tendency to raise our ideas of the Galapagos Islands,"[45] he grumbled. Fortunately their rather bleak impression of Galápagos did not last.

Strangers to Man

The next day (Thursday, September 17) spirits rose as the *Beagle* sailed up the western coast of Chatham and into the southern reaches of "St Stephens harbor," a generous bay alive with "Fish, Shark & Turtles . . . popping their heads up in all parts."[46] By late morning the anchor was down and the ship's company was reveling in a "sport [that] makes all hands very merry"—fishing. The sailors caught great numbers of large groupers to the sound of "loud laughter & the heavy flapping of the fish . . . on every side."[47] Before long a flourish of frigatebirds (*Fregata* sp.) converged on the scene "from a great height," to pick fish scraps "from the surface of the water without wetting even [their] feet."[48] Brown pelicans (*Pelecanus occidentalis urinator*) paddled round the ship like oversized ducks, lunging at the spills with their huge bills.

Here, and at Charles Island, Darwin acquired all of his Galápagos fish specimens, 15 in total, "*all . . . new*, without exception,"[49] and five of them endemic to Galápagos.[50] Among the catch was a bizarre looking Pacific red sheephead (*Semicossyphus darwini*) which was named *Cossyphus darwini* by Leonard Jenyns in 1842, "in honour of Mr. Darwin, whose researches in the Galapagos Archipelago, where he obtained it, have been so productive in bringing to light new forms."[51] A scorpionfish (*Scorpaena histrio*) was brought up—carefully, so as not to be cut by the sharp dorsal spines. Later a lentil moray eel (*Muraena lentiginosa*) was pulled from a tide pool close to shore, undoubtedly twisting itself into knots as it climbed the line that hooked it.[52] Darwin was particularly taken by the bullseye puffers (*Sphoeroides annulatus*) swimming idly about the ship. They made "a loud grating noise"[53] when captured. He attempted to sketch their dorsal pattern of concentric circles, despite professing a life-long "incapacity to draw."[54] It was one of several simple sketches that Darwin made in Galápagos.

Buoyed by their success at fishing, a party of men went on shore to try their luck "turpining," but returned empty handed. Darwin would have to wait another day to see his first giant tortoise.

In the afternoon Captain FitzRoy took a boat to the far north side of Stephens Bay. His goals were to look for fresh water, as indicated on James Colnett's map of the island, and to climb Finger Hill (now called Cerro Brujo—00°45′36.76″S, 89°27′35.45″W) for an elevated view of the coastline and any hidden reefs or shallows.[55] It was a long way and, lacking sufficient daylight hours to accomplish these tasks and return to the ship in one day, FitzRoy decided to spend the night on shore. It is unlikely that Darwin joined FitzRoy on this excursion. He certainly described Finger Hill in detail in his geology notes, and 5 of his 13 rock specimens from Chatham Island come from this location. However, it is most likely that Darwin explored Finger Hill on September 21 or 22 in conjunction with his explorations of the Craterized District.[56]

While FitzRoy was off at Finger Hill, another shore party (including, most likely, Darwin) landed closer to the ship (00°49′24.57″S, 89°31′25.52″W). Here, on the flat "south-west side"[57] of Stephen's Bay, some of the *Beagle* men set to work measuring the altitude of the sun and determining the local time. They compared this with Greenwich meantime registering on 12 of the chronometers on board the *Beagle*. The data were then compared with the last observations taken on shore, 13 days previously at Callao, Peru, and the difference was converted to longitudinal distance, one hour equaling 15 degrees. Back on board the *Beagle*, in the chart room that was also Darwin's cabin, the first scratch was made to position Galápagos on the map.

When on shore Darwin "proceeded to botanize" and as his collections grew so did his appreciation of his surroundings. Although it was the dry season and most of the plants appeared lifeless, on closer examination Darwin found "¾ of Plants in flower."[58] Most were white or yellow and so unassuming that Darwin dismissed them as "insignificant, ugly little flowers, as would better become an Arctic, than a Tropical country."[59] Among the low-growing herbs were gray mat-plants (*Tiquilia darwini*), appearing quite dead but harboring myriad white flowers among their diminutive leaves. There were also blue-leafed heliotropes (*Helitropium curassavicum*), with scorpioid spikes of flowers, growing amid patches of rust-colored carpetweed (*Sesuvium edmonstonei*) with star-shaped flowers. A little further inland Darwin found clumps of chaff flower (*Alternanthera filifolia*), looking as dry as their common name suggests, and pearl berry shrubs (*Vallesia glabra*), with pretty white drupes but insipid green flowers.

The terrain behind the coast was a stunted forest of leafless "Balsamic"[60] incense trees (*Bursera graveolens*) and spindly croton trees (*Croton scouleri*) under whose gray and shadeless scaffolding grew a variety of shrubs. An endemic composite (*Scalesia incisa*) was "the commonest bush,"[61] followed by a "wild cotton"[62] that would later bear Darwin's name, *Gossypium darwinii*. (Interestingly, Darwin reported cotton only from Chatham and James Islands. He did not report cotton growing on Charles Island even though it now grows there in abundance. Conversely, both he[63] and Colnett[64] recorded cotton growing on James Island where it is not known to occur today.[65]) Most of the plants were truly leafless but a few individuals of each species could be found with leaves and flowers still clinging to their branches. Lone bushes of leafy cordia (*Cordia lutea*), waltheria (*Waltheria ovata*), and leather leaf (*Matenus obovatus*) provided restful spots of green in an otherwise stark gray landscape.

In another 10 years, when his plant specimens were finally described by plant taxonomist Joseph Dalton Hooker, Darwin would look at his "Galapageian" plants[66] in a much more positive light. His poor opinion of the plants had not stopped him from collecting "every plant, which [he] could see in flower"[67] and as a result Darwin ended up with more botanical specimens from Galápagos than any other group of organisms. They would greatly support his theory of evolution. Numerous, separated by island and consisting of representatives from all the islands Darwin visited, they would show species diversification to an unexpected and unparalleled extent.[68]

The manner in which Darwin made such a valuable plant collection was a continual source of amusement for him. As he was oft to repeat, far from being purposeful in his collections, he had "not attempt[ed] to make complete series, but just took every thing in flower blindly."[69] He collected widely, most likely to please his mentor, Professor Henslow, who had a special interest in plant variety and island floras, but also, perhaps, as a way of compensating for his inexperience (and lack of confidence) in plant taxonomy and identification. Darwin had a keen interest in plants stemming from his childhood.[70] But despite once, at the age of eight, asserting to a teacher that he could find the name of any plant simply by looking inside its flower, despite taking Professor Henslow's botany course three times as a university student, and despite spending much of his later life researching plant variation and behavior, Darwin always claimed he was no botanist. Shortly after returning to England at the end of the voyage of the *Beagle*, Darwin found himself cornered by botany Professor David Don in the library of the Linnean Society. Don asked about the "habitation" of a certain plant "with an astoundingly long name" that Darwin had collected. To which Darwin replied to the dumbstruck professor, " I [know] no more about the plants, which I . . . collected, than the Man in the Moon."[71] Ten years later Darwin still retained his self-deprecating attitude when it came to botany. He called himself a "non-botanist" and "a man who hardly knows a daisy from a Dandelion."[72]

The happy result of Darwin's collecting "from [his self-professed] ignorance of botany, . . . more blindly in this department of natural history than in any other" was that he ended up, "not *intentionally*,"[73] with by far the largest and most diverse Galápagos plant collection that had ever been accrued. With a handful of specimens collected by other visitors to the islands, Darwin's collection formed the basis of botanist Joseph Dalton Hooker's *An Enumeration of the plants of the Galapagos Archipelago*.[74] Not only were three-quarters of the specimens new records for Galápagos, over a third were new species, including a new endemic genus that Hooker named after Darwin, *Darwiniothamnus*.

More importantly, many of the genera (most noticeably *Scalesia, Borreria,* and *Chamaesyce*) were represented by different species that were "*wonderfully* peculiar" to the separate islands.[75] They were hugely important to Darwin's ideas that species could change, for they showed over and over again that "whilst the genus is common to two or three islands, the species are often different in the different islands."[76] They bolstered Darwin's post-Galápagos views on transmutation of species "in a *glorious* Manner."[77]

Darwin was as struck by the boldness and curiosity of the land birds as Byron and Dampier had been before him. Finches and mockingbirds "quietly hopped about the Bushes"[78] and flycatchers and yellow warblers peered inquisitively at Darwin from eye-level branches. "So tame and unsuspecting were they, that they did not even understand what was meant by stones being thrown at them."[79] The collecting guns remained unfired. Darwin easily caught the small birds with his hat, and with the muzzle of his superfluous weapon "pushed a hawk off the branch of a tree."[80] Later, at Charles Island, Darwin watched "a boy sitting by a well with a switch in his hand, with which he killed the doves and finches as they came to drink."[81] Darwin had observed the same fearlessness among the land birds of the Falkland Islands, and attributed the unique condition in both places to their relatively recent human history. "The birds [in Galápagos] are Strangers to Man," he wrote in his diary.[82] They have not "yet learnt that man is a more dangerous animal than the tortoise, or the amblyrhyncus [iguana]."[83]

Darwin quickly saw the fatal flaw of this otherwise charming behavior, foretelling, "what havoc the introduction of any new beast of prey must cause in a country, before the instincts of the aborigines become adapted to the stranger's craft or power."[84] How prophetic his words. Among the Chatham Island land birds alone, the vermilion flycatcher (*Pyrocephalus rubinus dubius*), large ground finch (*Geospiza magnirostris*), and Galápagos hawk (*Buteo galapagoensis*) have since disappeared from this island as the direct result of humans and the exotic plants and animals they introduced. On other islands and among other groups of organisms the casualties have been even worse (see Alien Invaders: The Human Factor).

Fearlessness was just one aspect of the land birds that Darwin found noteworthy; he was equally impressed by the "American character"[85] of the birds. He jotted in his field notebook, "I certainly recognize S. America in Ornithology. Would a botanist?"[86] The mockingbird was the most striking case; in appearance, song, and habits it seemed "closely allied to the Thenca [*Mimus thenca*] of Chili [*sic*]"[87] yet different enough to be a distinct species. How remarkable that a land bird in Galápagos should resemble a species found over 3000 kilometers away in quite a different sort of habitat, and climate. It opened Darwin's eyes to similar affinities among Galápagos organisms to their continental counterparts. As he would later write, "It was most striking to be surrounded by new birds, new reptiles, new shells, new insects, new plants, and yet by innumerable trifling details of structure, and even by the tones of voice and plumage of the birds, to have the temperate plains of Patagonia, or the hot dry deserts of Northern Chile, vividly brought before my eyes."[88] Here was evidence staring him in the eyes that organisms from the South American mainland had colonized Galápagos and over time these organisms had evolved into new species. He may not have recognized this principle until the conclusion of the voyage, but he was beginning to notice the supporting facts.

A League from the Coast

On the morning of Friday, September 18, Captain FitzRoy was picked up from his bivouac at Finger Hill. He had been unsuccessful in his search for fresh water, and had failed to obtain a good view from Finger Hill. Indeed, it does not appear from FitzRoy's writings that he reached the summit at all. The authors have found only one, difficult, way up the steep slopes themselves.

The *Beagle* then sailed north to Terrapin Road (now known as Bahía Tortuga de Agua Dulce). As anchor was dropped at noon, Darwin had one object in mind—to explore the "amorphous"[89] mound of an eroded crater he could see rising from the even, interior plain (00°43′23.41″S, 89°21′20.37″W). Darwin, Officer Stokes, and several crewmen were landed on a beach glistening with olivine crystals and bordered with a welcome mat of pink morning glory (*Ipomoea pres-caprae*). Behind a pretty portico of green saltbush (*Cryptocarpus pyri-formis*) and crimson saltweed (*Sesuvium edmonstonei*) they found a natural road leading inland. It took Darwin to his first tuff crater (a volcanic cone now called Pan de Azucar) and the beginning of a fascination with how palagonite tuff (or "Volcanic sandstone"[90] as Darwin called it) is formed.

The "road" was in reality a shallow gully, meandering through the ubiquitous dry and scrubby vegetation. It was naturally paved with broken slabs of basalt messily caulked with white "calcareous tuff, apparently of submarine origin."[91] Much of the "calcareous tuff" that Darwin found throughout the area ("Calcareous Tufa,"[92] as he originally called it in his notes) is caliche, a form of calcium carbonate that is produced by the alkaline weathering of substrate exposed to the air in an arid environment. Despite its resemblance to marine limestone it is not an indication of marine deposition. However, near the shore, some uplifted shell-containing marine carbonate can be found.

Signs of giant tortoises were everywhere. Wide, flattened, and "well chosen [tortoise] tracks" meandered through the shrub land.[93] Sun-bleached carapaces lay overturned in the bushes, surrounded by scatterings of broken clay tobacco pipes and black glass bottles that had been tossed aside by "turpining" whalers. Great elliptical lumps of tortoise dung, "resembl[ing] that of the S. American Ostrich,"[94] and as large as Darwin's boot, clogged the path.

The *Beagle* crew spread out in search of the leviathans themselves, but Darwin remained bent on reaching his tuff cone. He and Stokes pressed further inland and after "about a league" came to "some small hills in part detached, in others joined"[95] to the "low but broard [*sic*] crater"[96] that had drawn Darwin from the ship. Here he scrambled up the crumbling yellow slopes to the "800 ft"[97] (244-meter) summit, measuring angles of strata and collecting geological specimens (six in all) along the way. Upon reaching the peak he was surprised to find that the bowl of the crater, which should have opened before him, no longer existed but was merely a gentle slope melting into the surrounding terrain. The central mass and external slope of the southern side had been completely "wasted away."[98] He hypothesized, from the presence of more "Calcareous Tufa" and the strange yellow-brown, "semi-resinous"[99] texture of the palagonite tuff that the crater had formed "while standing immersed in the sea."[100] The volcano had ejected ash and "mud"[101] that had collected around the base and finally filled up the center of the crater. The sides had later been eroded by the action of waves "at the same time that the lower lavas were smoothed over with Calcareous Tufa."[102] All that was left was the northern rim he stood upon, much like the remains of a buttressed sandcastle worn by the lapping tide. That the tuff cone was now isolated from the coast by "a league" (actually closer to half a league, or 2.4 km) could be explained by uplift. "[T]he idea of the submarine origin of much of this sandstone," he wrote in his geology notes, "necessarily presupposes an horizontal elevation to account for their present position."[103] Darwin was certainly correct about one thing; Pan de Azucar had formed in contact with water, but whether with groundwater or open ocean is still not clear. If Pan de Azucar had formed in open water as Darwin supposed, the surrounding plains were not necessarily uplifted, but could have been filled in by lavas produced after the formation of the tuff cone.

Darwin returned to the ship a satisfied man. He would declare the Galápagos tuff craters "the most striking feature in the geology of this Archipelago."[104] From his study of a variety of tuff cones over the next few weeks he concluded that seawater is the crucial element in tuff formation and that much of tuff is formed by the "trituration"[105] or pulverization of hot magma interacting explosively with the ocean. Darwin's hypothesis for the origin of palagonite and tuff cones was original, and has stood the test of time. It was proved with the eruption of Surtsey Island off Iceland in the 1960s (as described later, in Land of Craters).

The hunting party had an equally successful afternoon searching the lower environs of Pan de Azucar. Darwin wrote that the men "brought back 15 Tortoises: most of them very heavy & large."[106] Captain FitzRoy remembered things a little differently: "Our party brought eighteen terrapin on board," he wrote. "In size they were not remarkable, none exceeding eighty pounds."[107] Whatever their number and size, they were the first Galápagos tortoises the men had ever seen, and the first that they had tasted. Galápagos tortoise meat had a reputation of being a wholesome, delicious food. Colnett had declared it the best meat he had ever eaten. Darwin, however, was not overly impressed with the way the tortoises were cooked that night, describing the meat as, "to my taste, indifferent food."[108]

Chatham was the only island from which the *Beagle* crew collected tortoises for food. On October 12 the men took an additional "thirty large terrapin on board," from Fresh Water Bay on the southern end of Chatham, where the animals were "very large, deserving the name of elephant-tortoises."[109] Tortoises can still be found on Chatham Island, and the visiting site La Galapaguera Natural, located approximately five kilometers southeast of Pan de Azucar, is considered one of the best places to see large numbers in the wild.[110] They have, however, completely disappeared from the wild from the southern part of the island, where once they "wallow[ed] like hogs, and [could] be found by [the] dozens."[111]

"Waterfalls!"

September 19 and 20 were spent rounding the northeastern extreme of Chatham, and then tacking against the current, through "gloomy" weather to the southern end of the island.[112] At 1:30 p.m. on the 20th, the *Beagle* arrived at Fresh Water Bay (now spelled Freshwater Bay, and in Spanish called Bahía Agua Dulce), where "somewhat brighter green" valleys descend from the highlands to the sea.[113] Here Darwin was delighted to behold real "waterfalls of Water!" streaming from the cliffs.[114] How incongruous the silver streams looked in the arid landscape. For FitzRoy these "little rills of water, & one small cascade"[115] were no less than a godsend. He knew the *Beagle* would run out of drinking water before they reached their next stop on the voyage (Tahiti), and he desperately needed to find a place in Galápagos where he could take on more. Colnett's map indicated a source of fresh water at Finger Hill but FitzRoy had found no such thing when he searched the area on September 17. Fresh Water Bay was not mentioned as a watering place on Colnett's map, nor named in any of the books on board the *Beagle*, but FitzRoy knew of its existence from conversations he had had with sealing captain William Low in the Falklands. It was a fortuitous piece of information for nowhere else in the dry desert isles of Galápagos would FitzRoy find fresh water in sufficient quantity to fill the ship.

To determine the feasibility of collecting water for the *Beagle*, Captain FitzRoy landed (evidently without Darwin who had little to say about the site) "on a stony beach in the cove, and found a fine stream of excellent water" (00°56′25.19″S, 89°29′26.53″W).[116] The landing at this cove (now called La Honda) was "bad, sometimes impracticable"[117] due to the frequency of heavy swells on the beach.[118] But as FitzRoy later reported to the Hydrographer of the Admiralty, "the abundance, and goodness of the water, and the quantity of Terrapin (Land Tortoise) which may be collected at the same time, make up amply for the additional trouble."[119] FitzRoy did not take on water this day but returned on October 11, after leaving Darwin camped on James Island, to spend two days filling the *Beagle* for the next leg of their journey round the world.

Today no one but the occasional fisherman collects water at La Honda because of the same adverse conditions that frustrated Fitz-Roy. The little waterfalls are left to empty, untapped, into the surf. Another highland source supplies the modern coastal town of Baquerizo Moreno, located about 20 kilometers west of La Honda.

On the evening of September 20 the *Beagle* continued round to Stephens Bay, anchoring in the same place as before. There they met up with Chaffers and Mellersh who had "returned in the Whale Boat"[120] from Hood Island. They brought with them a souvenir—a mister "David Walton who had been left on Hoods Island (without water or provisions) from the American Whaler, Hydaspy [*sic*, Hydaspe]—of New Bedford."[121] At some point two young tortoises from Hood Island were also brought on board and presented to Captain FitzRoy.[122] The little tortoises may have been acquired at this time, or possibly on October 14 when the *Beagle* returned to Hood Island and spent two hours sounding the depths of Gardner Bay. Walton remained FitzRoy's guest for four days, until the *Hydaspe* was sighted near Charles Island on September 24, and the marooned sailor was returned to his ship. The tortoises stayed on board for the duration of the voyage.

Land of Craters

By September 21 Darwin was eager to explore on land again, having been shipbound for the previous two days. Not only would Fitz-Roy grant him shore leave, he would drop him off in a geologist's paradise—an undulating field of raw lava that Darwin dubbed the "Craterized District."[123]

The *Beagle* was to stay at anchor in Stephens Bay for two days so that the crew could take on wood and employ themselves "variously on Ships duty."[124] The sail maker was ordered to make a tent, perhaps the one Darwin would use the following month on James Island. The captain sent a whaleboat to scout the northeastern part of the island, for according to Colnett's map the northern tip of the island was another place where fresh water could be found. Darwin and his servant went along.

On the way (or on the return journey the following day) Darwin asked that they stop at Finger Hill, the northern headland of Stephens Bay where FitzRoy had spent a night the week before. Ever a geologist, Darwin jumped briefly ashore to chip a sample from one of the black volcanic dikes that striate the plummeting yellow cliffs of this lofty cape on its western side. It was the only means of reaching the dikes as the vertical slopes of Finger Hill make access from the beach impossible. Darwin expected the black dikes to be basalt and was greatly surprised to find that his hammer struck not basalt, but a hard form of tuff. Some of the dikes at Finger Hill are intrusions of basalt leaked from a large pool of basalt in the center of Finger Hill, but others consist of a dark, compact form of tuff. Darwin clearly tapped into one of the latter kinds. From his position on the coast Darwin could not have seen any sign of the central basalt pool, but once within the Craterized District and looking back toward Stephen's Bay he had a grand view of Finger Hill's funnel-like core of basalt exposed to the southeast (see black and white plate 20). From there the scene reminded Darwin of a picture he had seen in George Scrope's *Considerations on Volcanos*,[125] a copy of which was on board the *Beagle*. The drawing is of "the craters in central France,"[126] and consists of one volcano, looking superficially like Finger Hill in consisting of two types of lava, set against a backdrop of several steep-sided craters like those of the Craterized District. Darwin was clearly excited to see this phenomenon himself, for he drew an uncharacteristically detailed sketch of Finger Hill in his geology notes (see black and white plate 21b and c).[127] He also, at some point, obtained a specimen of the basalt plug (specimen #3235) and several samples of the tuff surrounding it (see endnote 56). As for the dikes, he concluded that they were fissures that had filled up with tuff before the eruption of basalt had pooled in the interior of the crater.

Finger Hill, or Cerro Brujo as the area is known today, is a National Park visitor site. Provided that sea conditions are favorable, visitors may take a dinghy or kayak ride past Darwin's landing spot where a series of dykes meet the sea. They can also land on a nearby beach to look at sea lions, and observe shorebirds in a saltwater lagoon.

Finger Hill, with its unusual basalt plug, may have formed somewhat like Iceland's island of Surtsey, the birth and growth of which were well documented in the 1960s. Surtsey started as an island of pillow lava developing on the seafloor. Between 1963 and 1967 it then erupted in a series of phreatomagmatic explosions as the reduced water pressure over the growing volcano caused the magma to interact explosively with the sea. These ash eruptions eventually formed two tuff rings above the surface of the sea and in doing so sealed the vents from all further contact with water. Subsequent eruptions produced flows of basalt, which, after filling the rings, poured over the lips of the craters and covered the bulk of the island. Both tuff and basalt were thus produced from the same vent.

As the men continued up the coast to the Craterized District, "gannet[s] [Nazca boobies, *Sula granti*], . . . beautifully white & black"[128] skimmed the waves as they glided out to sea. Closer to shore, blue-footed boobies (*Sula nebouxii*) plunge-dived for fish in the shallows. Darwin thought they were the same species as "the common gannet [i.e., Peruvian booby, *Sula variegata*] as at Callao"[129] and took little further notice. Instead his gaze rested on the singular, flat-topped spire of Kicker Rock that lies in the center of Stephens Bay roughly six kilometers offshore. It looked to be formed of the same substance as Finger Hill and was roughly the same height. He wondered if Kicker Rock and Finger Hill had been lifted to their present height by "horizontal upheaval."[130] Perhaps they had been joined by an enormous plain that had since eroded away. He jotted these ideas in his geology notes, but later scribbled them out. Finger Hill and Kicker Rock were the remains of tuff craters that had formed in the ocean as separate islets and then been eroded by the waves. Whereas "a recent, great stream of lava"[131] emanating from the Craterized District had subsequently grasped Finger Hill firmly to the main island of Chatham, and protected it from further erosion, Kicker Rock remained exposed as an offshore islet.

The whaleboat continued north to Mount Pitt (Punta Pitt) where Philip Gidley King landed and collected some geological specimens similar in constitution to the tuff at Finger Hill.[132] But first, and at a distance of "6-miles [9.7 kilometers] from the ship"[133] Darwin and his servant were dropped off on a small white beach bordering the southern edge of "a strange black district, bare of all vegetation & studded over with small Craters, so as to resemble those parts of Staffordshire & Shropshire where Iron Foundries are most common" (00°44′48.34″S, 89°26′18.85″W).[134] Here a raw lava flow, as rough as "a sea petrified in its most tempestuous moments" overlaps an older, smoother flow "partially clothed with a stunted vegetation."[135] Both flows had clearly originated from the same inland area, a district of craters so densely packed that Darwin estimated there were "in [the] space of [a] few miles little less than 100."[136] The area reminded Darwin of a sketch of the Phlegraean fields of Italy he had seen in Lyell's *Principles of Geology*[137] back on board the *Beagle*. Darwin, rather prosaically, named the area the "Craterized District" but in light of his reference to Lyell's sketch, and the fact that he used the word "Galapageian"[138] in referring to the plants of Galápagos, perhaps a better name would have been the Galapageian fields.

Darwin was eager to explore the area, for, as he wrote in his diary, "it is always delightful to behold anything which has been long familiar, but only by description."[139] Darwin was clearly referring to

Lyell and Scrope's respective descriptions of similar volcanic formations elsewhere in the world, but he also recognized the area from
having read James Colnett's account of Galápagos. James Colnett had
walked through this same area during his visit to Chatham Island in
1793. Although Colnett had not named the site, he had provided a
telling, and fearsome, description of the terrain: "[T]he earth is . . .
frequently rent in cracks that run irregularly from East to West, and
are many fathoms deep: there were also large caves, and on the tops
of every hill which we ascended was the mouth of a pit, whose depth
must be immense, from the length of time during which a stone, that
was thrown into it, was heard."[140]

Darwin may also have been referring to his familiarity with an entirely different kind of book, for the Craterized District is a Miltonian
landscape if there ever was one. With *Paradise Lost* surely tucked in
his pocket alongside his field notebook Darwin jumped ashore and
"immediately started to examine . . . [the] district."[141]

Darwin and Covington explored both flows, at first walking with
ease across the wobbly broken pavement of the older flow while following its border with the younger flow. Colnett had described the
lava as "a kind of iron clinker, in flakes of several feet in circumference, and from one to three inches thick: in passing over them they
sound like plates of iron."[142] Darwin had to agree. Annotating Colnett's description in his field notebook he wrote, "*Clinking* plates of
"iron Lava["].[143] Later he enlarged on the words in his geology notes.
He wrote, "[The] loose cakes . . . sound, when a person walks over
them, like plates of Iron."[144]

During the walk, Darwin and Covington stumbled upon two giant
tortoises. Darwin later wrote, "One was eating a piece of cactus, and
when I approached, it looked at me, and then quietly walked away:
the other gave a deep hiss and drew in its head."[145] Darwin had already seen giant tortoises on board the *Beagle*, but these were the
first he had seen in the wild. He estimated they were about "7 ft"[146]
(2 meters) in circumference and weighed "at least two hundred
pounds"[147] (91 kilos) each. He tried moving one and found it was as
difficult as shifting a small boulder. How well they "match the rugged
lava,"[148] he penciled into his field notebook.

Near the coast the lava of the younger flow had piled so deep that the surface of the overlay was considerably higher and rougher than that of the older flow. Surmounting the newer flow here would have necessitated climbing an abrupt and jagged wall and of course neither Darwin nor Covington carried a ladder. Several hundred meters inland, however, the levels of the two flows gradually even and here Darwin and Covington were able to cross onto the newer lava with little difficulty.

The two men then struck off toward the center of the craters, picking their way from one chimney cone to the next through a veritable mine field of volcanic pitfalls. As Colnett had warned, the area was traversed with "deep & long chasms," and undermined with pit craters and thinly roofed chambers, "from 30 to 80 ft [9 to 25 meters] deep."[149] Today National Park wardens consider the terrain to be one of the most hazardous in Galápagos, and needless to say the area is not a visitor site. The way was perilous and the day "glowing hot"[150] but Darwin, in his element, found the walk delightful and his exertion well repaid by the novel scene. For here he was, on a vast, undulating sheet of lava, pitted and pimpled by a profusion of craters, apparently so recently brought into existence by the perpetual churnings of the earth, that the drippings of molten rock appeared to yet ooze from their subterranean ovens. The metallic resonance of the ground spoke eerily of an underground workhouse and one could almost smell the sulfurous belchings of the extinct furnaces. It is easy to imagine Darwin surreptitiously sliding Milton from his pocket and reciting the following passage:

There stood a Hill not far whose griesly top
Belch'd fire and rowling smoak; the rest entire
Shon with a glossie scurff, undoubted sign
That in his womb was hid metallic Ore,
The work of Sulphur.[151]

Under Milton's influence Darwin might have been tempted to believe that the heat haze shimmering across the surface emitted from below rather than from the sun above. But Darwin was no fool. He recognized that the lava had not formed yesterday; it had merely aged well. In the arid air the crisp lava had resisted decomposition to a remarkable degree. It would have been a different story had the area been in contact with the scouring forces of the ocean, he thought. When back on board the *Beagle* he wrote, "Looking at this district I was much struck by . . . how easily the sea would entirely remove these rings of scoria, and . . . how difficult it would then be to distinguish the different streams: the whole would appear as the mouth of one great eruption from one point: instead [of] from very many point[s] and *at least* at two epochs."[152] He thought back to when he had seen such weathered and undistinguishable flows, 44 months earlier in the Cape Verde Islands. "I now understand St. Jago," he wrote in his notebook, —"50 years in the sea —would remove the Crater & the upper surface of Lava. Who could tell points of upheaval."[153]

When in the late afternoon Darwin and Covington returned to the coast, without accident, they surely celebrated with a rousing swim. The sea was a nippy 68.5° F (20.3 °C) but would have felt decidedly polar compared to the heat on land.[154] The contrast to the senses would have been shocking, to the mind disturbing. Darwin may well have sensed they were in an entirely different world, where nothing was quite how it seemed. Old, new, hot, cold—the words here took on new meanings. It was as if Galápagos was telling him a story, but in a language he could not yet decipher.

That night the two explorers slept on one of the sandy beaches banking the Craterized district, under a moonless sky as dark, vast, and silent as the petrified subterranean workshop behind them. Only the scratchings of fearless nocturnal rice rats scurrying around the tent gave life to the desolate night. One can only imagine the mysterious dreams (perhaps nightmares) Darwin may have had during those hours of darkness, inspired by the "Cyclopian" landscape behind him, with its giant tortoises looking like "most old-fashioned antediluvian animals; or rather inhabitants of some other planet."[155]

The following morning (Tuesday, September 22) Darwin and Covington set to work collecting the "plants, birds, shells & insects," and rice rats, that had understandably been ignored the previous day.[156]

The Chatham Island rice rat (*Oryzomys galapagoensis*) stands out as being the only endemic mammal species that Darwin collected in

Galápagos. Because of its small size, soft brown fur, and "large thin ears"[157] Darwin recognized that this "mouse or rat" was not merely a ship's rat. When Darwin returned to England his specimen was examined by George Waterhouse and named as a new species, *Mus galapagoensis.*[158]

While it was then "very numerous,"[159] this rice rat has since become extinct on Chatham, losing out to the introduced ship (black) rat (*Rattus rattus*), which arrived on the island with 19th-century whalers. It was one of twelve species of Oryzomyini rats endemic to Galápagos, eight of which are now extinct. Of the four species still in existence, *Oryzomys bauri* is found on Barrington (Santa Fé), *Nesoryzomys fernandinae* and *N. narboroughi* occur on Narborough (Fernandina), and *N. swarthi* inhabits northern James (Santiago) Island.[160] *N. swarthi* is the only species that has managed to co-exist with introduced rodents, namely the black rat (*Rattus rattus*) and the house mouse (*Mus musculus*).[161]

The Chatham Island rice rat is just one of several organisms that have disappeared from Chatham as a result of human occupation. When the island was settled in the 1840s the immigrants brought domestic plants and animals that were destructive to the native wildlife. Today (in 2009) about 10,000 people live on Chatham, and while most of the island is a National Park reserve, the park boundaries do little to prevent the spread of invasive species over the entire island. Chatham Island has become one of the most altered islands in Galápagos, agriculture having obliterated much of the original highland habitat in the southern third of the island, and goats, with their voracious appetite for almost all greenery, having munched their way over the entire island. Incredibly, in terms of species, the majority of the plants and animals that Darwin collected or noted from Chatham still survive on the island, though some are currently endangered. Species survival has been helped by the topography of the island. It is partially thanks to the inhospitable nature of the Craterized district that some of the endemic plants Darwin found on Chatham exist in the area today. Mature candelabra cacti (*Jasminocereus thouarsii*) cling to the precipitous walls of the chimney cones, where elsewhere they are invariably chewed and toppled by goats.

Rare herbs, such as the Galápagos tomato plant (*Lycopersicon chees-manii*) and cut leaf daisy (*Lecocarpus darwinii*), can sometimes be found growing in patches of sunlight at the bottom of pit craters, out of reach of the indefatigable herbivores whose droppings litter the plain like hailstones.[162]

Unfortunately, the "prickly pear" cactus (*Opuntia megasperma*) that Darwin recorded being eaten by tortoises in the Craterized district has not withstood the onslaught of goats.[163] Small patches of *Opuntia* do exist elsewhere on the island but its effective disappearance from the island is significant because the especially large seeds were almost certainly the main food of the now extinct, extra-large form of large ground finch (*Geospiza magnirostris magnirostris*).[164] Darwin, FitzRoy, and their assistants collected this finch (whose beak was 10% larger than the largest *G. magnirostris* found in the archipelago today) on Chatham and Charles islands. *Opuntia megasperma* once grew in abundance on both these islands.

Another bird that Darwin collected and that has since disappeared from Chatham Island is the vermilion flycatcher (*Pyrocephalus rubinus dubius*). It was last seen on Chatham in the 1980s. Fortunately, this "Scarlet-breasted" bird and "its yellow breasted female"[165] can still be found on several other islands in Galápagos. Darwin noted that the vermilion flycatcher, like the Galápagos flycatcher (*Myiarchus magnirostris*), occupies "both the arid and rocky districts near the coast, and the damp woods in the higher parts of several of the islands in the Galapagos Archipelago."[166] He also noted that these "insectivorous birds," though widespread in both "the low dry country & high damp parts," were rare compared to the seed-eating finches and doves.[167] Darwin thought their low density due to "an exceeding Scarcity of insects (so much so that the fact is very remarkable)."[168] Today vermilion flycatchers are found principally in the highlands of the large, inhabited islands, only frequenting the coast on some of the uninhabited isles. This modern distribution is likely determined as much by the presence of introduced predators as by the distribution of insect prey.

When Darwin returned to the *Beagle* late in the afternoon on September 22 his head reeled with new thoughts. His first week in

Galápagos had provided a feast of firsts: swimming lizards, giant tortoises, eroded tuff cones, and an extinguished lava factory. There was also the extraordinary fearlessness of the land birds and their baffling same-but-different resemblance to species on the South American continent. Indeed, there was much to think about. As he boarded the *Beagle* with his bag full of specimens, he thought his Galápagos collection was more or less complete. His field notebook certainly reads as though he assumed uniformity among the mockingbirds of the archipelago. For at the end of a passage of notes on the geology of Chatham is a sentence about the mockingbirds of Galápagos, written as if he had already visited several of the islands. It reads, "The Thenca [mockingbirds] very tame & curious on these Islands."[169]

Darwin had representatives of new birds, lizards, shells, insects, rocks, and plants. He could pick up duplicates and any organism he might have missed, on the next islands en route. He had not been able to examine or collect in the highlands of Chatham, but there was still the hope of doing so on another island. Little did he know that the very next island in line, Charles Island, would present him with a twist, by revealing that the islands do not have identical organisms. He would not realize the implications of this fact until it was too late to collect full series of organisms from the different islands, but he would be shown his first hint.

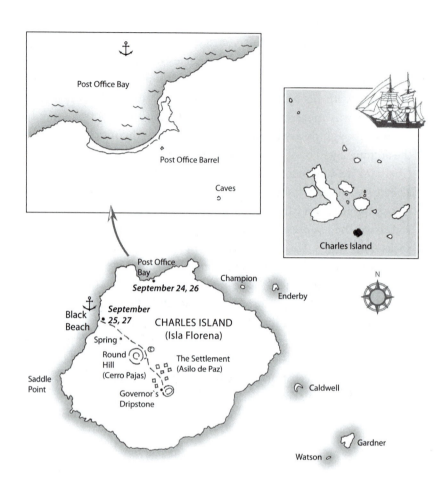

Post Office Bay

Post Office Barrel

Caves

Charles Island

Post Office
Bay

September 24, 26

Champion

Enderby

Black
Beach

*September
25, 27*

CHARLES ISLAND
(Isla Florena)

Spring

Round
Hill
(Cerro Pajas)

The Settlement
(Asilo de Paz)

Saddle
Point

Governor's
Dripstone

Caldwell

Gardner

Watson

N

Chapter V

~~~~~~~~~~~~~~~~~~~~

Charles Island
(Isla Floreana)

*It will not easily be imagined how pleasant the sight of black mud
was to us, after having been so long accustomed to the parched soil
of Peru and Chile.*
—Charles Darwin, *Journal and Remarks*.[1]

It can be said that "Geology carrie[d] the day"[2] on Chatham Island.
But it was zoology that triumphed on Charles. Indeed, Charles Island was the single most important visit of the entire voyage of the
Beagle for the conception of Darwin's theory of evolution. For it was
here that Darwin learned that the mockingbirds and tortoises of the
Galápagos Islands differ among the different islands, and it was this
realization that led him to believe that species can evolve, in isolation
and over time, into new species.

An Oasis in the Desert

The *Beagle* arrived at Charles Island on Thursday, September 24, sailing into Post Office Bay late in the evening. Post Office Bay gets its
name from a wooden box that was nailed to a post, erected behind the
beach and used as a maildrop. No one knows who put it up. It was not
James Colnett, as often assumed, for he never even landed on Charles

Island.[3] It was possibly used by Patrick Watkins (the first resident of Galápagos) to communicate with visiting ships before his departure from the island in 1809 (see Early Human Colonization). Or it may have been started by the crew of a whaling ship. The earliest record of its existence is from Captain David Porter of the *Essex*, who reported finding a box labeled *Hathaway's Postoffice* in which was a letter dated June 14, 1812.[4] Whalers used it regularly to send and receive letters during their long absences from home. Inbound whalers deposited letters for ships already in the islands, and outbound vessels carried letters homeward. It was an effective, if slow, postal system, and one that has been reestablished with a wooden barrel for the enjoyment of modern visitors. It was not in use, however, when Darwin arrived. At that time, while Charles Island was under the governance of José Villamil, all mail came and left from Black Beach, the port of call for his highland settlement of Asilo de Paz (Haven of Peace).[5]

At Post Office Bay (01°14′12.69″S, 90°26′55.75″W) FitzRoy and Darwin were welcomed by the acting governor of the island. It was not Governor Villamil, who was away on business on the Ecuadorian mainland, but rather an Englishman named Nicholas Lawson, who had been left in charge. Lawson had come down from the highland settlement (01°18′36.42″S, 90°27′06.64″W) to greet a whaling ship that had also recently arrived. It was a fortuitous meeting, for Lawson would impart crucial information to Darwin about the Galápagos tortoises and how they differ among the islands. He knew the islands intimately, having explored the archipelago repeatedly over the preceding five years, and was eager to converse with some fellow English speakers.

Lawson boarded the *Beagle* and invited FitzRoy and Darwin to dine with him at the settlement on the following day. So on the morning of Friday September 25, the men took a boat eight kilometers down the coast to Black Beach (01°16′32.67″S, 90°29′18.33″W), where there was a "good path" to the highlands.[6] For the first hour of their ascent inland they sweated through a dusty "thicket of nearly leafless underwood . . . affording a congenial habitation only to the Lizard tribe."[7] As on Chatham Island, the lower woods were dominated by gray incense trees (*Bursera graveolens*). The settlers, Lawson explained, used the sap of this tree as a salve for cuts.[8] *Opuntia megasperma* cacti, yellow cordia shrubs (*Cordia lutea*), and pepper-scented croton bushes (*Croton scouleri*) also grew in abundance.[9] The crotons stained their clothes a permanent mottled brown, as the twigs snapped and daubed them with sap. White blossomed glorybower shrubs (*Clerodendron molle*) and Galápagos lantana (*Lantana peduncularis*) scented the warm air with a "Honeysuckle"[10] like perfume.

Halfway to the settlement the men passed a small spring of fresh water around which stood several abandoned huts. The half-dried-up spring afforded little relief to the parched landscape and the men continued without lingering. Only when they had rounded the base of Round Hill (Cerro Pajas), the gateway to the highlands, did the terrain change. Abruptly. Within minutes they were "cooled by the fine Southerly trade wind & . . . refreshed by a plain green as England in the Spring time."[11] Before them stood little rustic huts "built of poles & thatched with grass"[12] widely scattered amongst plantations of "bananas, sugar canes, Indian corn, and sweet potatoes, all luxuriantly flourishing."[13] The dusty path quickly turned into a runnel of "*black mud* & on the trees . . . mosses, ferns & Lichens & Parasitical plants*" flourished.[14] Lawson explained that during that time of year (the cool season) the highlands were often veiled in clouds and that the trees dripped from a heavy moisture that hung in the air. He was describing what is now locally called "garúa." It was not raining on that particular day,[15] but Darwin was already familiar with the kind of "thick drizzle or Scotch mist"[16] that Lawson was referring to, from his time in Peru. Light though the rain appeared, it could be surprisingly drenching.

Darwin, with his undying love for "Tropical scenery"[17] was elated with "how pleasant the change was."[18] Having spent months in the dry and barren plains of northern Chile and Peru, nothing was more welcome than this scene which resembled, if remotely, the lush Brazilian forests he adored—admittedly without "the lofty, various & all-beautiful trees of that country."[19] To FitzRoy it was a veritable "oasis in the desert."[20]

The governor's house was a damp wooden structure built on the side of a cave-pocked hill (the dominant cone at Asilo de Paz) where a spring dubbed the "Governor's Dripstone"[21] provided water for the settlement. Lawson explained that the 200 settlers, most of whom were political prisoners, led "a sort of Robinson Crusoe life." They hunted tortoises and feral pigs and goats, while their crops thrived with little assistance in the misty climate.[22] Despite an abundance of food the "people [were] far from contented."[23] They missed their families on the mainland, complained "of the deficiency of money,"[24] and lived in a state of perpetual damp, in which "house salt cannot be kept dry, books and paper become mouldy, and iron rusts very quickly."[25]

At the settlement Lawson extended "the welcome of a countryman,"[26] laying out a "Feast"[27] composed of the native and introduced products of the island. Darwin and FitzRoy were offered a "variety of food quite unexpected in the Galápagos Islands, but fully proving their productiveness."[28] There were undoubtedly plates of salted tortoise meat, fish, pork, and goat meat accompanied by chicken and turtle eggs, plantain, pumpkin and sweet potato, loaves of "yuca" (manioc) bread, corn cakes, melons, and bananas.[29] Cold spring water and perhaps even a local tea sweetened with sugarcane were served in pitchers fashioned from the shells of young tortoises.[30] The food was exotic, the setting exquisite and their host as "bright" and "energetic"[31] as the mockingbirds that scampered across the table snatching scraps of meat from the guests.[32] It was a meal to remember, and not just for the food. The mockingbirds, as well as Lawson's enlightening conversation, made September 25, 1835, one of the most significant days of Darwin's visit to Galápagos and, indeed, of the entire voyage of the *Beagle*.

Early Human Colonization

This archipelago has long been frequented, first by the Bucaniers [sic], and latterly by whalers, but it is only within the last six years, that a small colony has been established here.
—Charles Darwin, *Journal of Researches*[1]

When Ecuador separated from Gran Colombia in 1830, General José Villamil (a native of Louisiana, who moved to Ecuador when Louisiana was bought by the United States) suspected that a dispute might arise over ownership of the Galápagos Islands. He prompted Ecuador to claim the archipelago for its own, and at the same time, persuaded President Juan José Flores to grant him exclusive rights to colonize and govern Charles Island. Charles was then renamed Floriana (now spelled Floreana), in honor of the nation's first president.

At the beginning of 1832, Villamil sent a dozen skilled craftsmen, under the command of Ignacio Hernandez, a partner in his colonization scheme, to build the first Galápagos settlement on Floriana. Shortly after arrival, on February 12, 1832, the men conducted an annexation ceremony and Galápagos officially became part of Ecuador. This day is now commemorated every year as Galápagos Day. An extraordinary coincidence makes the day doubly worth celebrating, for February 12 is also Darwin's birthday. Darwin turned 23 the day Galápagos joined Ecuador, and for Darwin it was a miserable day. The *Beagle* was navigating the North Atlantic on her way south to Brazil, and Darwin, "overcome" with seasickness, was forced to spend the whole day "in painful indolence."[2]

During the first few months of colonization, the highland settlement of Asilo de la Paz (Haven of Peace) was stocked with successive boatloads of political prisoners. General Villamil moved to the settlement in September 1832, and governed on site for the next five years. However, he was frequently called away on political business to the mainland, often to defend the continuation of his governing rights, and during these long absences from Charles Island he left an interim administrator in charge. For most of 1835 Nicholas Oliver Lawson, a British sailor from Jamaica who had thrown in his lot with Governor Villamil and the colonization project, filled the post with enthusiasm. Lawson was indispensable to Villamil for he knew the islands well, having explored the archipelago thoroughly in 1830. He was also important to Darwin, for he informed him of the evolutionarily significant fact that the tortoises of Galápagos differ between islands.

Lawson was responsible for introducing goats and pigs to Charles and perhaps some of the other islands. In 1838, when the French frigate *La Vénus* called in at Charles Island, Lawson was still working for Villamil, but this time in the capacity of captain of the colony's schooner. Captain Petit-Thouars was so impressed by Lawson's knowledge of the islands and generous character that he named the southern tip of Charles, Cabo Lawson.[3] The appellation was not retained and the place is now known simply as Punta Sur (South Point). It is not clear how long Lawson remained on Charles Island. By 1844 he had moved to Isla Bolivia (Indefatigable Island/Santa Cruz Island), and become mayor of a small colony of 21 settlers. Villamil had established the colony in 1837, the year he abandoned Charles.[4] Also in 1844 another Englishman, William Gurney, became mayor of a new settlement on Chatham Island.[5]

Today (2009) only about 100 people live on Charles Island. None of them are political prisoners. They live in the dry coastal town of Puerto Velasco Ibarra at Black Beach and some farm the highlands where the settlement used to be. FitzRoy would be pleased by this arrangement. He had seen the shortcomings of a highland settlement and had remarked that "a house on the dry ground, and plantations in the moist valley, would answer better" than living in the damp.[6] Water is now piped to the coast from the old Governor's dripstone and other seeps at Asilo de Paz. The hill itself is a tourist attraction, famed for its evocative "pirate caves." The inhabitants of these caves were never pirates in the true sense of the word, but rather a series of disgruntled whalers who jumped ship to make a hermit's living growing vegetables to sell to passing ships. The earliest known and most notorious inhabitant of this island, was an Irishman named Patrick Watkins. In 1809, after several years living on Charles Island, Patrick decided to abandon his "wretched" lot by stealing a whaleboat and sailing to the mainland. Shortly after arriving he was thrown in jail and his fate was sealed.[7]

Lawson recounted to Darwin and FitzRoy a story of a sailor who had "tired of the world"[8] and lived on the island for several years before it was colonized. This may have been Watkins but was more likely a man called John Johnston. Johnston was a European whom Lawson found living on the island in 1830 and to whom he entrusted the care and propagation of his goats and pigs. Two years later, Johnston officially witnessed the annexation of Galápagos and the foundation of Governor Villamil's settlement. Then he was rudely ordered to leave the island.[9] "So strongly was the old man attached to his cave, that he shed tears when taken away,"[10] recounted Lawson. Johnston could not bear returning to civilization. After a short hiatus on the mainland, vainly seeking justice for his eviction, he returned to Galápagos to eke out an existence on James Island, selling seal skins to whaling ships. He was still there in 1833 when John Coulter, surgeon on board the whaling ship *Stratford*, visited the island.[11] He was probably not there in 1835, for Darwin did not record seeing such a man during his own stint on James.

Olmedo

The settlement on Charles was an agricultural enterprise that failed to thrive beyond subsistence level. As Governor Villamil was seeking a profit-making venture, he sent Lawson to explore the commercial potential of the other islands of Galápagos. Lawson reported back that James Island was promising, for it harbored large numbers of tortoises and plenty of salt. Tortoise oil had "great commercial value"; it was used "in cooking, in preference to lard and beef fat . . ."[12] and could be sold for a handsome profit "on the neighboring continent." Salted tortoise meat (and fish) could also be sent back to the mainland or sold directly to visiting whaling ships. In June 1835, Lawson sent a group of 22 men and women to James Island (Isla Olmedo as Villamil called it), to set up camp and start the tortoise exploitation business. It was a lucrative enterprise while the tortoise population remained high. The record shows that the Olmedo branch of the colony, which comprised about 10 percent of the Charles Island settlers, persisted at least four years, and perhaps as many as eight.[13] While Darwin was camped on James Island he learned much about the tortoises of Galápagos from talking with these men.

Because the freshwater supply on James Island is precarious, the island never maintained a permanent (or large) human population. A few individuals occupied the island intermittently in the 20th century. In 1926, a salt mine was established in James Bay and was worked for two years. It was started up again in 1963 but failed to bring in enough profit to keep it going beyond another couple of years.[14] It was then turned over to the Galápagos National Park, and James Island was declared off limits from further human habitation.

Darwin's "Mocking-Thrush"[33]

If there is one species that symbolizes the importance of Galápagos to Darwin and his theory of evolution, it is *Mimus trifasciatus*, the Charles Island mockingbird. When Darwin was in Galápagos, the mockingbirds (or mocking thrushes as Darwin then called them) were the only group of organisms he appreciated as being distinct on the different islands, and it was the Charles Island mockingbird that jolted him to this awareness. As Darwin dined in the highlands of Charles Island and observed the mockingbirds up close—for they were especially numerous around "the houses & cleared ground,"[34] and, as on all the islands, so "very tame & inquisitive"[35] that "one alighted on a cup of water that [Darwin] held in [his] hand, and drank out of it"[36] —he saw that they were quite different from the mockingbirds he had just collected on Chatham Island. These mockingbirds were larger, more robust, and darker brown than the small, pale gray mockingbirds on Chatham. Their eyes were red brown, they lacked the malar stripe and cheek pattern of the Chatham island mockingbirds, and they had blotchy chest patches that gave them a distinctly scruffy appearance. On Chatham their eyes were yellower, their facial plumage well defined, and they were sleek (see color plate 12). So different were the mockingbirds on these two islands that they prompted Darwin to "pay particular attention to their collection"[37] everywhere he went in Galápagos. He must have wondered which of the two mockingbirds he would find on the next island en route, or whether he would find a completely new mockingbird altogether. Any answer would satisfy him, for the theorist in him hungered for information to explain the distribution of organisms and the ambitious collector in Darwin salivated at the prospect of obtaining new species. The result of his curiosity was that he collected mockingbirds from each of the four islands he visited, and in doing so acquired three new species.

When Darwin visited Albemarle and James (the next islands en route) he found that the mockingbirds on those islands differed yet again (in size, plumage, and eye color) from the mockingbirds he had seen on Charles Island, and even (though to a lesser degree) from each other. They were more similar to the mockingbirds on Chatham Island. In fact, they were so similar that, even though Darwin provided clear locality information for his mockingbird specimens and correctly recognized three distinct mockingbirds among the four islands he visited, he erred in classifying them. Twice. When he wrote up his zoological notes on the *Beagle* Darwin decided that his Chatham and Albemarle specimens were a single variety while his James specimen was on its own.[38] When he arrived home to England, ornithologist John Gould convinced him that his Chatham and James specimens belonged to a single species, which Gould then named *Orpheus melanotis*.[39] Darwin's Albemarle specimen was a distinct species, which Gould named *Orpheus parvulus*.[40]

Current taxonomy says that both Darwin and Gould were mistaken; it is the Albemarle and James mockingbirds that belong to a single species (*Mimus parvulus*), while the Chatham Island mockingbird (*Mimus melanotis*) is a species all on its own.[41] Darwin and Gould were not completely off the mark, however. Recent genetic studies have shown that the Albemarle mockingbird and James mockingbird are distinct at the subspecies level.[42] Furthermore, Darwin and Gould were both correct in declaring the Charles Island mockingbird a separate species. Gould named it *Orpheus trifasciatus*, and today it is known as *Mimus trifasciatus*, the most genetically distinct species of all the Galápagos mockingbirds. Darwin neither saw nor collected the fourth Galápagos mockingbird, *Mimus macdonaldi*, because it is found only on Hood (Española), an island he did not visit.

Thus, it was the Charles Island mockingbird that first alerted Darwin to the fact that the Galápagos mockingbirds exist as "distinct species in the different Is^ds"[43] and that on each island "each kind is *exclusively* found."[44] It was the Charles Island mockingbird that inspired Darwin to collect mockingbirds on all the islands he visited, an opportunity he might otherwise have missed, for he did not consciously collect island series of any other organism. It was also the Galápagos mockingbirds as a group that provoked Darwin's initial conception of what would later be termed "adaptive radiation"—the diversification of a group of organisms from a common ancestor, through geographical isolation.[45]

"How wonderful it is," wrote Darwin in 1844, "that . . . three closely similar but distinct species of a mocking-thrush should have been produced on three neighbouring and absolutely similar islands; and that these three species of mocking-thrush should be closely related to the other species inhabiting wholly different climates and different districts of America, and only in America. No similar case so striking as this of the Galapagos Archipelago has hitherto been observed."[46]

Indeed, so fundamental were the mockingbirds to Darwin's theory of evolution that when he wrote *The Origin of Species* 24 years after his visit to Galápagos, they were the only organisms from Galápagos that he specifically named to illustrate his point. As has been pointed out by Darwin historian Sandra Herbert, the Galápagos mockingbirds deserve to be named "Darwin's mockingbirds" far more than the Galápagos finches merit their name "Darwin's finches."[47]

The Charles Island mockingbird has regrettably become one of the rarest birds in the world. The species population number varies between 200 and 500 individuals (depending on annual fluctuations in food supply) and is distributed over two small islets—Champion and Gardner—located 700 meters and 8 kilometers respectively off Charles Island.[48] It is no longer found on Charles Island at all. Its disappearance from Charles, sometime between 1868 and 1888, has been blamed on human activities, the introduction of rat, cat, and dog predators, and the deforestation of native vegetation by feral goats and donkeys.[49] Mockingbirds coexist with these same nonnative mammals on Chatham, Albemarle, and Indefatigable islands so why did they go extinct only on Charles Island? It has been suggested that the mockingbirds of Charles Island were particularly vulnerable to the introduction of black rats (*Rattus rattus*) because Charles Island, unlike the other three inhabited islands, never supported a population of native rats.[50] Rapid deforestation of the island's *Opuntia* cactus forests (*Opuntia megasperma*) by goats is another likely factor. Although it can be argued that goats have destroyed *Opuntia* forests on Chatham and Hood islands without causing the extinction of their mockingbird populations, *Opuntia* may be more important to *M. trifasciatus* than to the other mockingbird species. *Opuntias* are certainly used heavily for food and nesting sites by *M. trifasciatus* on Champion and Gardner islets, but how dependent the mockingbirds are on the cacti is unknown. Most likely the Charles Island mockingbird's demise on Charles was due to a combination of human-related factors. The preservation of Champion and Gardner islets from invasive species is therefore essential for the continued survival of this historically important mockingbird. Plans are now, in 2009, underway for the removal of all feral mammals and rats from Charles Island, and there are hopes for the eventual repatriation of mockingbirds from Champion and Gardner islets through a captive breeding program.[51]

As for the evolution of the Galápagos mockingbirds, the closest living relative is not *Mimus thenca*, the Chilean species that Darwin first likened to the Galápagos group. Nor is it the long-tailed mockingbird *M. longicaudatus*, found in Ecuador and geographically the closest species. Rather, modern genetics have shown a close affinity with a group of northern hemisphere mockingbirds, and most especially with the Bahama mockingbird *M. gundlachii*, a species that is today found throughout the northern Caribbean Islands. Surprising though this may seem, a Caribbean connection exists among several Galápagos organisms, including the snakes, the flamingo, and the white-cheeked pintail duck. It is thought that climate changes (especially shifts in oceanic currents and prevailing winds) associated with the closure of the Panamanian isthmus 2 to 4 million years ago were responsible for the dispersal of many organisms from the Caribbean region.[52] Darwin's finches comprise another group of Galápagos organisms with a Caribbean link. The ancestral finch arrived in Galápagos between 2 and 3 million years ago.[53] The mockingbirds arrived between 1.6 and 5.5 million years ago.[54]

Tortoise Tales

Acting governor Nicholas Lawson proved to be a talking encyclopedia about Galápagos tortoises, and much of what Darwin learned about these animals came from him, the settlers of Charles Island, and some "Spaniards" (Spanish-speaking Ecuadorians from Charles Island) stationed on James Island, as well as from personal observation. On Charles Island tortoises were rare, and there is no indication that the *Beagle* men saw a single live adult. Lawson informed the men that as a result of decades of hunting, tortoise numbers had dwindled on Charles Island but were still sufficient that "two days hunting will find food for the other five in the week."[55] Still, the rate at which they were being removed was indefensible. Just two years earlier Lawson himself had directed Commodore Downes of the U.S. Frigate *Potomac* to Saddle Point where in the course of three days, 100 crewmen took 600 tortoises off the island.[56] Lawson admitted that the tortoises on Charles Island were unlikely to survive beyond "20 years"[57]—in fact, they disappeared in less than 10.[58] Three months prior to the *Beagle's* arrival, Lawson established a base on James Island (or Olmedo, as the island was locally called) to which he sent a hunting party to exploit the still numerous James Island tortoise.[59] This enterprise kept the settlement well provisioned, produced a commodity (tortoise oil) that could be sold on the mainland, and stocked whaling ships eager to buy food from the colonists. Lawson described the James Island tortoises as "so very large" that a single animal supplied "upwards of 200 £bˢ of meat." He recalled one individual caught in 1830, with the date 1786 carved into its shell. It had "required 6 men to lift it into the boat."[60]

Lawson then told Darwin that the tortoises differ between islands and that from "the form of the body, shape of scales & general size," he and the "Spaniards" could "pronounce, from which Island any Tortoise may have been brought."[61] The significance of this statement escaped Darwin at the time and he, himself, failed to notice any differences in the tortoises he saw on Chatham and James. However, nine months after leaving Galápagos, when considering the distribution of the Galápagos mockingbirds and the asserted differences between the East and West Falkland Island fox, Darwin reflected back on Lawson's words. He wrote, "If there is the slightest foundation for these remarks, the zoology of Archipelagoes —will be well worth examining; for such facts [would *inserted*] undermine the stability of Species."[62] By this time, however, it was too late to verify Lawson's claim. Darwin had not collected any adult tortoises of his own, nor kept a single bone or scute from the 45 (48 according to FitzRoy) Chatham Island tortoises brought on board the *Beagle* for food. Four young tortoises had been collected from three different islands—"Covington's little tortoise from Charles and [Darwin's] from James"[63] plus two acquired for FitzRoy from Hood Island[64]—but they were all too small to distinguish by carapace shape. At a young age all Galápagos tortoises look roughly alike. These young tortoises were kept alive, and during the passage home "one grew three-eighths of an inch, in length, in three months; and another grew two inches in length in one year."[65] Still, it was still not enough to reveal their adult form. When Darwin arrived home and had the little tortoises examined by leading reptile authority John Gray he could only lament, "The specimens that I brought from three islands were young ones; and probably owing to this cause, neither Mr. Gray nor myself could find in them any specific differences."[66] For the time being Lawson's words remained unsubstantiated.

The tortoises of Galápagos (and those of California and Mauritius) were originally lumped by European taxonomists under the same species name as the Indian tortoise, "*Testudo Indicus.*"[67] This loose classification allowed the interpretation that the tortoises of Galápagos had, perhaps, been introduced from elsewhere in the world and then become "naturalized" in the islands.[68] But Cuvier, in his *Animal Kingdom*, acknowledged that "many species [had] been confounded" under this name and it became clear that the giant tortoises of the world warranted closer examination. In the mid 1830s the taxonomy of the Galápagos tortoise was placed under revision. In early 1838, just over a year after Darwin returned home from the voyage, French herpetologist Gabriel Bibron told Darwin he had reason to believe that the Galápagos tortoise was indigenous to the archipelago.[69] To distinguish it from giant tortoises found elsewhere in the world the name of the Galápagos tortoise was changed to *Testudo nigra*. For Darwin this claim of endemism was a giant step in the right direction.

The good news did not end there. Gabriel Bibron also told Darwin that, from having looked at various museum specimens of Galápagos tortoises, he thought there were two distinct species in Galápagos.[70] Darwin pricked up his ears at this; even though Bibron and fellow herpetologist André Marie Constant Duméril could not say whether the species were separated by island, the fact that two types were now recognized from the archipelago gave new credibility to Lawson's words.[71] Some time later Darwin read Captain Porter's 1815 *Journal of a Cruise made to the Galápagos Islands*, in which Porter, like Lawson, asserted in no uncertain terms that the tortoises differ between islands.[72] Porter wrote:

> Those [tortoises] of James' Island appear to be a species entirely distinct from those of Hood's and Charles' islands. The form of the shell of the latter is elongated, turning up forward, in the manner of a Spanish saddle, of a brown colour, and of considerable thickness; they are very disagreeable to the sight, but far superior to those of James' Island in point of fatness, and their livers are considered the greatest delicacy. Those of James' Island are round, plump, and black as ebony, some of them handsome to the eye; but

their liver is black, hard when cooked, and the flesh altogether not so highly esteemed as the others.[73]

By now Darwin was fully convinced of the truth in Lawson's assertion that the tortoises differ between islands. Still, it was not until the 1870s, when Albrecht Günther began a systematic study of the tortoises of Galápagos by examining a growing number of specimens in museums throughout the world, that the extent of Lawson's claim would begin to be investigated.[74] Günther ultimately recognized six species of Galápagos tortoise.[75] Today there is no clear consensus on the taxonomic rank of the various populations of Galápagos tortoise. However, one commonly postulated interpretation, which we adopt here for simplicity's sake, is that there is just one species (*Geochelone nigra*), divided into 14 subspecies, three of which are now extinct.[76] Most of the taxa are confined to separate islands, with one subspecies per island. A notable exception is on Albemarle Island, where several distinct populations occupy the large island's five volcanoes. On this island it is apparent that lava flows between the volcanoes have isolated the populations in effectively the same manner that oceanic channels separate islands. This isolation has allowed at least two tortoise populations to diverge sufficiently to be considered separate subspecies.[77]

As for the origin of *Geochelone nigra*, the closest living relative has been identified as *Geochelone chilensis*, a small tortoise occupying southern South America.[78] The direct ancestor of the Galápagos tortoise was most likely much larger than *G. chilensis*, for gigantism in tortoises is considered a necessary precursor for surviving long distance travel adrift in the ocean.[79] Genetic studies indicate that tortoises arrived in Galápagos between 2 and 3 million years ago, and the fossil record shows that there were indeed giant tortoises extant in South America at this time.

There has been much speculation as to why Darwin did not notice the different forms of tortoises on the different islands while he was in Galápagos, and collect accordingly. In his defense, the tortoises Darwin saw in the wild and examined up close on Chatham and James islands do not look dramatically different from each other. Although the carapace of the Chatham Island tortoise is more flared than that of the James Island tortoise, both are intermediate in form between the classic saddlebacked carapace of the Hood and Charles islands tortoise and the true dome-shaped carapace of the Indefatigable Island tortoise (see color plate 27). But what about the distinct, saddlebacked tortoises of Charles Island? Even if Darwin did not see any live saddlebacked tortoises on Charles Island, surely he saw a carapace or two at the settlement. Or did he? FitzRoy, in his *Narrative of the voyage*, written after the voyage, commented on a "quantity of tortoise shells lying about the ground"[80] that he had seen in "an apology for a garden"[81] by the first spring en route to the highland settlement. Some had been used, in place of flowerpots, as shades to cover seedlings. It is interesting that Darwin never mentioned observing the same. Did he purposely neglect to mention the fact? Was he perhaps embarrassed by his failure to notice the different tortoises while he was in the islands? Or did he simply not see them?[82]

The reason Darwin overlooked the differences between the tortoises while in Galápagos may lie in the simple fact that only Chatham Island tortoises were brought on board the *Beagle*. Darwin never had a chance to look at two different carapace types side by side. A chink in this argument is the fact that the *Beagle* crew collected tortoises from two different locations on Chatham, and it has since been suggested that the tortoises in these areas were distinct from one another, perhaps even at the subspecies level.[83] An examination of specimens collected from the southern end of Chatham in 1906 (shortly before the population's disappearance through overhunting) has revealed that the southern tortoises had a flatter carapace than the tortoises that exist today in the north of Chatham.[84] Nonetheless, it is unlikely that Darwin saw members of the two populations side by side because of the lapse in time between the two *Beagle* collections. The 15 to 18 tortoises that were collected from Pan de Azucar on September 18 were most likely consumed (and the carapaces disposed of) during the 24 days preceding the acquisition of the second lot of 30 tortoises from Fresh Water Bay on October 12. Darwin was camped on James Island in October and would not have seen the second batch until even later, after he was picked up on the 17th.

So, although it can be argued that two different carapace types may have been brought on board, Darwin is unlikely to have seen them together. On the other hand, if the two populations of Chatham Island tortoise were truly distinct, and obviously so, a new and interesting possibility emerges. The northern Chatham Island tortoises look fairly similar to the James Island tortoises, but the extinct southern Chatham Island tortoises may have looked quite different. It may have been seeing the flat carapaces of the *Beagle*'s second batch of Chatham Island tortoises, after just having experienced hundreds of round backed tortoises on James, that first stimulated Darwin's reflection on Lawson's assertion (and that of the inhabitants of Charles and James) that the tortoises differ on the different islands of Galápagos. It was not until about nine months after leaving Galápagos that Darwin put pen to paper regarding the evolutionary significance of Lawson's assertion,[85] but Darwin may have been pondering his words far earlier than this. After all it was the "Spaniards" who could tell the tortoises apart, and Darwin had just spent ten days in their company on James Island, being informed of all kinds of facts regarding the tortoises of Galápagos (see chapter 7).

A different argument as to why Darwin did not comment on the differences in the tortoises during his time in Galápagos, and one that is repeated in the literature but contested by the authors, presumes that Darwin thought the tortoises to be introduced, and thus not worthy of scrutiny.[86] It is rooted in the fact that Captain FitzRoy dismissed them as such, at least when he was back in England. In his *Narrative*, FitzRoy remarked:

> It is rather curious, and a striking instance of the short-sightedness of some men, who think themselves keener in discrimination than most others, that these tortoises should have excited such remarks as — "well, these reptiles never could have migrated far, that is quite clear," when, in simple truth, there is no other animal in the whole creation so easily caught, so portable, requiring so little food for a long period, and at the same time so likely to have been carried, for food, by the aborigines who probably visited the Galápagos Islands on their balsas, or in large double canoes, long before Columbus

saw that twinkling light, which, to his mind, was as the keystone to an arch.[87]

To support his contention that the Galápagos tortoises were introduced rather than native, FitzRoy drew loosely upon the comments of a previous visitor to Galápagos, William Dampier. "Honest Dampier," wrote FitzRoy, "immediately reverted to the tortoises of the West Indies, and of Madagascar, when he saw those of the Galápagos."[88] But FitzRoy was not being quite so "honest" himself. While Dampier did compare the Galápagos tortoise with various tortoises he had seen in the West Indies and those he had heard lived in Madagascar, he did less to equate them than to point out their differences. "These Tortoise in the *Gallapagoes*," wrote Dampier, "are more like the *Hecatee* [of the West Indies], except that, as I said before, they are much bigger; and they have very long small Necks and little Heads."[89]

There is nothing in Darwin's writings to suggest Darwin subscribed to FitzRoy's reasoning. Nothing in the notes Darwin wrote in Galápagos or on board the *Beagle* on the way home, or even later in England, suggests that he was, at any time, persuaded that the tortoises of Galápagos were introduced. If anything, his words imply the contrary.

Throughout the voyage Darwin was interested in the geographical distribution of different species, and often questioned whether certain species—commensal rodents and animals of use to man, such as wild goats and horses—might have been introduced by humans.[90] In Galápagos he wondered about the origin of a rat he found on James Island.[91] Had ships brought it to the islands, or was it native? Even though the Galápagos tortoise was a highly "useful" resource by all accounts, Darwin never queried its origin in his notes. Indeed, the closest Darwin came to remarking on the distribution of the tortoises while on the *Beagle* was a single statement about their range within Galápagos. He wrote that the tortoises are "found in all the Islands of the Archipelago; certainly in the greater number."[92] Later, in writing the second edition of the *Journal of Researches*, Darwin used this same statement to argue for their indigenity and against their introduction. He wrote, "There can be little doubt that this tortoise is an aboriginal inhabitant of the Galápagos; for it is found on all, or nearly all, the islands, even on some of the smaller ones where there is no water; had it been an imported species, this would hardly have been the case."[93]

If Darwin had known that the tortoises of Galápagos were lumped under the same scientific binomial "*Testudo Indicus*" as other giant tortoises in the world, and, more importantly, agreed with this classification, he might well have believed they were introduced. But again, Darwin's writings imply the contrary. During the voyage of the *Beagle*, Darwin never once used the name "*Testudo Indicus*" in referring to the tortoises of Galápagos. He did not even use the genus name *Testudo*, a fact that contrasts with his ready willingness to call the marine iguanas of Galápagos by their generic name, *Amblyrhynchus*. (Darwin knew about the name *Amblyrhynchus* from reading Byron's book on board the *Beagle*.[94]) Darwin simply called the tortoises "tortoise," "turpin," or "terrapin." This suggests that Darwin was either unaware of the scientific binomial or thought it too broad a classification to merit use. The *Beagle* carried several volumes of Cuvier's *Animal Kingdom*, but it is not known whether the relevant, ninth volume on reptiles, which spelled out this classification, was among them.[95] Only when he was back in England and had consulted with herpetologist Thomas Bell did he use the name. In the first edition of his *Journal* he wrote, "This [Galápagos] tortoise, which goes by the name of *Testudo Indicus* , is at present found in many parts of the world." It is "not improbable," he continued, "that they all originally came from this [Galápagos] archipelago. . . . If this tortoise does not originally come from these islands, it is a remarkable anomaly; inasmuch [*sic*] as nearly all the other land inhabitants seem to have had their birthplace here."[96]

Furthermore, it must be argued that while Darwin was in Galápagos he was not uninterested in the tortoises. He was *exceptionally* interested in them. He wrote more on the tortoises and their behavior than on any other organism in the islands. If he had thought the tortoises introduced, this is unlikely to have been the case.

Darwin was clearly interested in the tortoises while he was in the islands, but the question remains: Why did he not collect a single adult tortoise from any of the islands? The answer surely relates to their great size and Lieutenant Wickham's understandable obsession with keeping a tidy ship. Wickham was responsible for the smart appearance of the decks and often objected to Darwin's collections, referring to them as "d—d beastly devilment."[97] He would undoubtedly have balked at the idea of cluttering the cramped decks with one large carapace, let alone several. There were already specimens of tortoises from Galápagos back in England; this was not considered a novel animal. Surely one of the Chatham Island tortoises brought on board for food would suffice. Yet not one of these shells was retained as a specimen; perhaps Wickham also saw to that. After all, space was at a premium on the *Beagle* and they had a long way to go before reaching England, with no more opportunities for sending specimens home along the way.

It must also be remembered that Darwin and the *Beagle* crew *did* collect tortoises; the fact that they were young ones may not, at the time of their collection, have been considered a drawback, especially as they were kept alive and expected to grow. If Darwin had had any thought of collecting tortoises with regards to recording interisland differences, he may have assumed that the taxonomists back home would detect the variation.

FitzRoy's two little tortoises were given to the British Museum,[98] but the fate of the two other young tortoises remains a mystery. Stories abound. None is so tantalizing as that of Harriet, a Galápagos tortoise housed, until her death on June 22, 2006, at Australia Zoo in Queensland. Some people think that Harriet traveled from Galápagos to England on board the *Beagle*, and was subsequently taken to Australia on the third surveying voyage of the *Beagle*.[99] Even though DNA analysis shows that Harriet came from Indefatigable Island, an island not visited by the *Beagle*, this does not completely rule out the possibility that she was a *Beagle* tortoise. One of the *Beagle* four may have originated on Indefatigable and been brought to the Charles Island settlement by whalers or settlers. In this case, Covington's tortoise from Charles Island is the most likely candidate. Just as likely (or unlikely) is the possibility that Harriet was an unrecorded fifth tortoise, picked up by one of the crewmen involved in Sulivan's survey of Indefatigable Island in the *Beagle*'s yawl.

1A. The town of Shrewsbury in Shropshire, England, where Darwin grew up.
1B. The industrial town of Wolverhampton, Staffordshire, showing the smokestacks of its metal foundries.
1C. Darwin likened the Craterized District of Chatham Island, Galápagos, to "the Iron furnaces near Wolverhampton."

2A

2B

2A. Darwin attended the University of Edinburgh from 1825 to 1827.
2B. He then studied at Christ's College at the University of Cambridge, from 1828 to 1831.

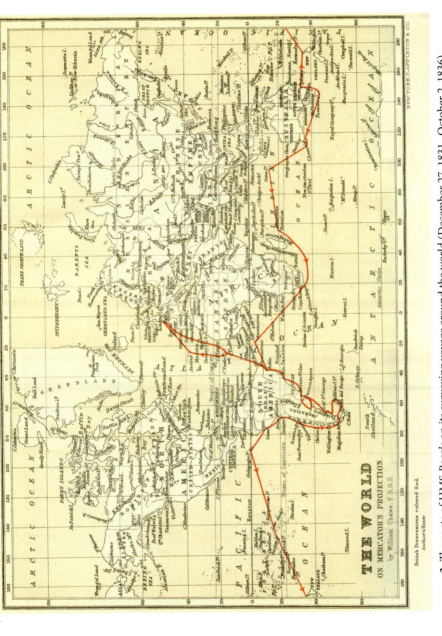

3. The route of HMS *Beagle* on its surveying voyage around the world (December 27, 1831 – October 2, 1836).

4A

4B

4C

4A. The Peak of Tenerife "towers in the sky twice as high as I should have dreamed of looking for it."— Charles Darwin, *Beagle* Diary.

4B. "Delphinus *Fitz-Royi*" (dusky dolphin, *Lagenorhynchus obscurus*) from a sketch by Captain FitzRoy.

4C. HMS *Beagle* was "sorely tried" while rounding Cape Horn on January 13, 1833.

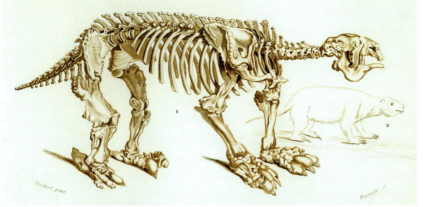

5A

5B

5C

5A. Skeleton of a *Megatherium*, an extinct, giant ground sloth from South America.
5B. South American rheas. Darwin's rhea (*Rhea pennata*) is shown between two subspecies of common rhea (*Rhea americana*).
5C. "*Canis antarcticus*" (Falkland Island fox, *Dusicyon australis*).

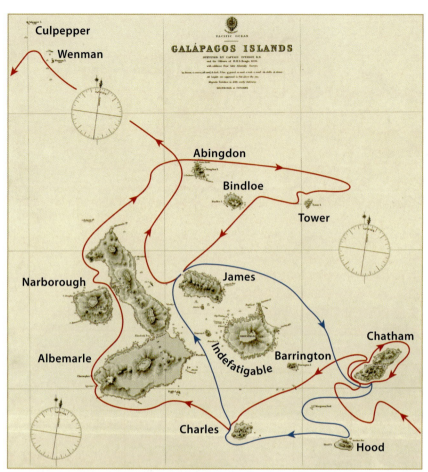

Culpepper

Wenman

PACIFIC OCEAN

GALÁPAGOS ISLANDS

SURVEYED BY CAPTAIN FITZROY, R.N.
and the Officers of H.M.S Beagle, 1836.

Abingdon

Bindloe

Tower

Narborough

James

Chatham

Albemarle

Indefatigable

Barrington

Charles

Hood

6A

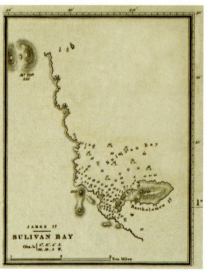

JAMES I.ᵗ
SULIVAN BAY

6B

6A. Route of HMS *Beagle* through Galápagos. Darwin's route is shown in red. The *Beagle's* route while Darwin was camped on James Island is marked in blue.

6B. Detail of Sulivan Bay, James Island, from FitzRoy's map of Galápagos. Although Darwin never landed here, the site is noteworthy for having been surveyed by and named after *Beagle* surveying officer James Bartholomew Sulivan. The fact that the coast line is essentially identical to today's coastline indicates that the apparently recent lava flow that delineates Sulivan Bay predates 1835.

7A. Pan de Azucar, Chatham Island.
7B. Chatham Island mockingbird (*Mimus melanotis*).
7C. Finger Hill (Cerro Brujo), Chatham Island.
7D. Chatham Island rice rat (*Oryzomys galapagoensis*).

8A

8B

8C

8A and **8B.** Marine iguana (*Amblyrhynchus cristatus*).

8C. Pacific green turtle (*Chelonia mydas agassisi*). In Galápagos Darwin found numerous specimens of a barnacle (*Platylepus decorata*) embedded in the soft skin of a green turtle. Darwin described and named the barnacle when he was back in England. It was one of four barnacles that he noted from Galápagos.

9A. Chatham Island tortoise (*Geochelone nigra chatamensis*), La Galapaguera Natural, Chatham Island.

9B. Striped Galápagos snake (*Antillophis steindachneri*), James Island.

9C. Sally Lightfoot crab (*Grapsus grapsus*).

Endemic Galápagos plant species collected by Darwin in 1835.

10A. Darwin's cotton (*Gossypium darwinii*).

10B. Thin-leafed Darwin's shrub (*Darwiniothamnus tenuifolius*).

10C. Macrea (*Macrea laricifolia*).

10D. Wing-fruited lecocarpus (*Lecocarpus pinnatifidus*).

10E. Galápagos carpetweed (*Sesuvium edmonstonei*).

11A

11B

11D

11C

11E

11F

11A. Round Hill (Cerro Pajas), Charles Island.
11B. Bulimulus snail (*B. ochsneri*).
11C. Post office barrel, Post Office Bay, Charles Island.
11D. Saddlebacked Galápagos tortoise (*Geochelone nigra hoodensis*).
11E. Charles Island mockingbird (*Mimus trifasciatus*).
11F. Galápagos rail (*Laterallus spilonotus*).

Arrows point to some of the distinguishing characteristics of the four populations Darwin observed. Note differences in eye color, beak length, and plumage color and pattern.

12A. Chatham Island mockingbird (*Mimus melanotis*).

12B. Charles Island mockingbird (*Mimus trifasciatus*).

12C and 12D. Galápagos mockingbird (*Mimus parvulus*) on Albemarle Island (12C) and James Island (12D).

13A. Small ground finch (*Geospiza fulginosa*).
13B. Medium ground finch (*G. fortis*).
13C. Warbler finch (*Certhidia olivacea*).
13D. Large ground finch (*G. magnirostris*).
13E. Sharp-beaked ground finch (*G. difficilis*).

14A

14B

14C

14A. Eruption of Sierra Negra Volcano, Albemarle Island, October 2005.

14B. Craterized area near Christopher Point, Albemarle Island.

14C. Ropey (pahoehoe) lava, James Island.

15A

15B

15C

15D

15A. Tagus Cove and Darwin Lake.
15B. Female and male cactus finch (*Geospiza scandens*).
15C. Galápagos mockingbird (*Mimus parvulus*).
15D. Lake in Beagle Crater, Albemarle Island.

16A 16B

16C

16D

16A. Female Chatham Island lava lizard (*Microlophus bivittatus*).

16B. Male Charles Island lava lizard (*M. grayi*).

16C. Vermillion flycatcher (*Pyrocephalus rubinus*) on a Galápagos tortoise (*Geochelone nigra porteri*).

16D. Land iguana (*Conolophus subcristatus*).

Shortly after Darwin's visit to Galápagos, around 1837, the Charles Island settlers established a small settlement on Indefatigable (Bolivia), to where it is thought that Nicholas Lawson moved and became mayor.[100] While Indefatigable was visited by whalers prior to this, the camp would have attracted more whalers and facilitated the transport of tortoises to ships. Visiting whaling ships did take tortoises to Australia in the 19th century, and a whaling ship is the most likely transport that carried Harriet away from her birthplace.

While the *Beagle* connection is tenuous at best, the story of Harriet is an interesting account of the survival of a 19th-century Galápagos tortoise. If Harriet really had been on board the *Beagle* this would have made her over 170 years old when she died! It is not impossible. It is known that other species of tortoises kept in captivity can live over 100 years and possibly as long as 200. A Madagascar tortoise (*Geochelone radiata*) picked up from the Indian Ocean by Captain James Cook was given to the Queen of Tonga in 1777. The animal died in May 1966.[101] A Mediterranean Spur-thighed Tortoise (*Testudo graeca ibera*) that was taken from a Portuguese ship in 1854 died in England in 2004.[102] An Aldabra tortoise (*Geochelone gigantea*) living in a zoo in India died in 2006 after living for at least 150 years and possibly 250.[103]

Post Office Collections

Darwin spent Saturday, September 26, at Post Office Bay, where he set to work "industriously collect[ing] all the animals, plants, insects & reptiles from this [area]."[104] He accumulated a number of fish, snakes, lizards, land snails, and land birds on Charles Island but was disappointed by the paucity of insects, both here and on the other islands he visited. Indeed, with the exception of the barren forests of Tierra del Fuego, he had "never collected in so poor a country."[105] If Darwin had visited Galápagos during the first half of the year (the hot, rainy season), instead of the second half (the cool dry season), he would undoubtedly have encountered a far greater diversity and concentration of arthropods. As it was, he attributed the apparent dearth of insect life to the islands having frequent drought conditions and an inconstancy of plant luxuriance.[106] Though he "perseveringly swept under the bushes during all kinds of weather"[107] and in all habitats on each island he visited, he obtained but a few spiders, ants, flies, wasps, moths, and a dragonfly.[108] He fared somewhat better, however, with his favored beetles. By looking "under stones on a hill," in the "branches of [a] dead Mimosa," and "under [a] dead bird"[109] in the "lower sterile land" he amassed a total of 29 species and discovered an endemic genus of flightless tenebrionid beetle—*Stomion*. He collected three of the eleven known species of this genus of "Darwin's darklings," as the beetles are commonly called today, little realizing that they are a prime example of species radiation in Galápagos.[110] Perhaps the most alarming component of Darwin's arthropod collection was a *Scolopendra*, the most feared animal in Galápagos. This giant centipede (*S. galapagensis*) is one of the largest in the world, growing, as the inhabitants warned Darwin, "to 14 inches long"[111] and capable of inflicting an exceedingly painful bite. Two other venomous arthropods in Galápagos, the endemic scorpion (*Centruroides exsul*) and endemic carpenter bee (*Xylocopa darwini*), are surprisingly absent from Darwin's notes and collections.

Darwin came to Galápagos looking, without success, for Tertiary fossils to reveal the history of animal succession in Galápagos. He left never knowing that he had literally walked over hordes of younger, Quaternary fossils on Charles Island. Behind the beach at Post Office Bay lie several gaping lava tubes, which in the 1980s were found to contain a variety of Holocene fossil vertebrate bones, many in the form of ancient barn owl pellets.[112] Darwin never mentioned the caves and it is safe to say that he never entered these hidden treasure troves. Admittedly, being only thousands of years old as opposed to millions, they would not have answered Darwin's questions about what organisms first inhabited Galápagos. Indeed, they are of little significance to modern scientists asking the same sort of questions, for they do not represent the very first organisms to colonize the islands millions of years ago, before their descendants evolved into their present forms. But they are of value as a record of the species that inhabited the island shortly before the arrival of humans. As such they provide a baseline for conservation efforts, and offer an interesting perspective to Darwin's collections. For apart from two organisms that Darwin never collected, nor mentioned seeing on Charles Island—the barn owl and Galápagos red bat—and that may or may not have been present in 1835, the fossils show much the same vertebrate assemblage described by Darwin's collections.

It is a much richer one than today's. Since Darwin's visit at least 10 native vertebrates have disappeared from Charles as a result of humans: the Charles Island mockingbird (*Mimus trifasciatus*), Galápagos racer snake (*Alsophis biseralis biseralis*), warbler finch (*Certhidia olivacea*), Galápagos hawk (*Buteo galapagoensis*), sharp-beaked ground finch (*Geospiza difficilis*), extra large form of the large ground finch (*Geospiza magnirostris magnirostris*), Galápagos rail (*Laterallus spilonotus*), Charles Island tortoise (*Geochelone nigra galapagoensis*), barn owl (*Tyto alba punctatissima*), and Galápagos red bat (*Lasiurus brachyotis*). The Galápagos dove (*Zenaida galapagoensis*) has virtually disappeared from the island. Fortunately, mockingbird, snake, and dove populations still survive on Charles's satellite islets of Champion and Gardner. Lava tubes on some of the other islands reveal similar stories, including the extinction of several species of endemic rats in the archipelago.

Today Post Office Bay is a designated National Park visitor site, the principal attractions being mailing and picking up letters at the post office barrel, in the tradition of 19th-century whalers, and kayaking in the bay. One of the underground lava tunnels is also accessible to tour groups.

Alien Invaders—The Human Factor

We need not marvel at extinction; if we must marvel, let it be at our own presumption in imagining for a moment that we understand the many complex contingencies, on which the existence of each species depends.
—Charles Darwin, *The Origin of Species*[1]

Darwin arrived in Galápagos at a propitious time. He was early enough to see the ecology of the islands for the most part intact, before much permanent damage had been effected through human activity in the islands. But on Charles Island he also got a glimpse of the future of Galápagos, of what devastation humans and introduced organisms can exact on a pristine place.

The four Galápagos Islands visited by Darwin have changed dramatically since 1835. In fact, of all the islands in the archipelago they, and South Seymour (Baltra) and Indefatigable (Santa Cruz) Islands, have undergone the greatest ecological transformations. The story behind the changes is the human history of the archipelago. Since humans first occupied Galápagos—as pirates, as whalers, as settlers—they favored these islands for their sources of fresh water.[2] As a result they impacted most heavily upon them. On these islands they killed native animals for food and sport, imported domestic plants and animals, and converted natural habitat into farmland, roads, and houses.

Today there are towns on three of the islands Darwin visited (James is no longer inhabited) and they all battle an ever-burgeoning number of ecological threats. A huge variety of exotic plants and animals continue to arrive, accidentally or deliberately, with devastating consequences. About 60 percent of the plant taxa in Galápagos and 25 percent of the invertebrate fauna are imported exotics and these percentages continue to rise as more introductions arrive.[3] In the highland agricultural zones invasive plants such as guava (*Psidium guajava*), hill raspberry (*Rubus niveus*), red quinine (*Cinchona pubescens*), and elephant grass (*Pennisetum purpureum*) have taken over large tracts of land and literally choked out many native species. Domestic mammals turned feral have long since spread beyond national park boundaries. Goats (*Capra hircus*), as prolific as they are destructive, are among the most damaging. They devour the native vegetation and deplete the food supply of the native herbivores—the tortoises and land iguanas. Donkeys (*Equus asinus*) graze on native plants and trample nests. Pigs (*Sus scrofa*) are another serious problem; they eat the eggs and young of reptiles and ground nesting birds. Dogs (*Canis familiaris*) and cats (*Felis catus*) are a constant dilemma for they prey on several native bird and reptile species and yet the local residents foster them as pets. Introduced arthropods, such as the red fire ant (*Wasmannia auropunctata*), yellow paper wasp (*Polistes versicolor*), and cottony cushion scale (*Icerya purchasi*), harm the native fauna and flora by attacking or competing with native species. These are just a few of the escalating number of exotic plants and animals that are altering the ecosystems of the inhabited islands and threatening the same on all the islands. It is a different Galápagos from the one Darwin experienced.

This is not to say that Galápagos was pristine in 1835. Humans had known of the archipelago's existence for 300 years and had been regularly occupying its waters in pursuit of whales and seals for the previous four decades. By the time Darwin arrived mariners had killed tens of thousands of giant tortoises and an unknown quantity of land iguanas for food. Countless fur seals and sea lions had also been slaughtered for their pelts and oil. But worst of all, several nonnative mammals had already arrived in the islands. The first of these was the black rat (*Rattus rattus*), also known as the ship's rat. It was introduced to James Island in the 17th century, undoubtedly by pirates careening their ships and putting infested supplies such as "flour, sweet meats & sugar . . . on shore against a time of scarcity."[4] Later, over the 19th and early 20th centuries, black rats made it to some other islands (Charles, Chatham, Albemarle, and Duncan/ Isla Pinzón) by mariners "smoking" their ships on or close to shore, and early settlers bringing supplies to Galápagos from the mainland. Rats were a dreadful nuisance for ships in those days. David Porter, captain of the USS *Essex*, which cruised Galápagos waters in 1813, spelled out just how problematic they were. In doing so he revealed just how easily rats became introduced to oceanic islands.

I intended to clean my ship's bottom, overhaul her rigging, and smoke her to kill the rats, as they had increased so fast as to become a most dreadful annoyance to us, by destroying our provisions, eating through our water-casks, thereby occasioning a great waste of our water, getting into the magazine and destroying our cartridges, eating their way through every part of the ship, and occasioning considerable destruction of our provisions, clothing, flags, sails, &c. &c. It had become dangerous to have them any longer on board; and as it would become necessary to remove every thing from the ship before smoking her . . . I believed that a convenient harbour could be found among one of the groups of islands that would answer our purpose.[5]

The rats are blamed for the extinction of several endemic rats in Galápagos and for crippling already weakened tortoise populations by eating their hatchlings.

Goats, with their voracious appetites for native vegetation and prolific propagation, came next. The first record of goats in Galápagos is from 1813, when Captain Porter put four goats and a Welsh ram ashore "to graze" on James Island.[6] The animals promptly escaped. We know they survived for at least a year, for Captain Pipon of the HMS *Tagus* reported seeing "4 goats which had been left by the U S. Ship of war Essex"[7] in James Bay on July 30, 1814. John Coulter, surgeon on the British whaling ship *Stratford*, reported "shooting goats" on James in 1833.[8] However, Darwin did not mention seeing any goats on James in 1835 and they were not reported again until the mid 1900s.[9]

Although Darwin did not see goats on James, he discovered just how destructive they could be when he visited Saint Helena Island in the South Atlantic Ocean nine months later. There the land was "utterly desert" and at first he could not believe that the interior had once been covered in forest. According to a "well attested" and "official" historical account of the island,[10] goats and pigs had been introduced in 1502 and over the next 200 years had effectively eaten and trampled their way through the island. The result was that much of the native flora and fauna went extinct. The feral animals were exterminated in the 1730s but by that time "the evil was complete and irretrievable."[11] "It is not too strong an expression to say," Darwin later wrote, "that the introduction of a single mammal might change the whole aspect of a district, even to the minutest living details."[12]

Human impact on Galápagos increased when people started living on the islands. When Darwin arrived, two islands were inhabited to some extent. Charles had an agricultural settlement of about 200 and James a working camp of 22. Both, however, were new establishments, just four years[13] and four months old, respectively. Foreign plants and animals had been introduced to Charles Island with the colonists, and indeed even before that, but few had spread beyond the settlement or done excessive harm to the wildlife. Darwin was just in time to see most of the wildlife intact, certainly as intact as it ever would be.

Introductions on Charles Island

In the first decade of the 19th century Patrick Watkins, the first known inhabitant of Charles Island, and indeed of all of Galápagos, "succeeded in raising potatoes and pumpkins in considerable quantities, which he generally exchanged for rum, or sold for cash" to passing ships. Apparently he also kept chickens, for when he deserted the island in 1809 he left inside his abandoned hut a written plea: "Do not kill the old hen; she is now sitting, and will soon have chickens."[14]

John Johnston was Pat's successor, and lived on Charles Island for at least five years before Governor Villamil's men arrived in 1832. During this European's reign (one report has Johnston down as a "citizen of *Altona*,"[15] another as a Swede[16]) domestic livestock arrived. Johnston was given "two asses" by the captain (Captain Locke) of a whaling ship to help with "his farming operations, and the bringing of the terrapin out of the bush to the 'pen,' when he had them convenient to sell, or use himself."[17] Villamil took over possession of these donkeys when Johnston left the island, and Admiral Petit-Thouars rode such a beast to the settlement when he visited the island in 1838.[18]

In 1830 Captain Nicholas Lawson introduced "some domestic animals, such as goats, sheep, and pigs" to Charles Island, and possibly to some of the other "fertile" islands as well, with "the well-founded hopes of deriving, after a little while, great advantages."[19] He entrusted his animals to Johnston with the condition that the stock not be destroyed. Two years later, Lawson joined forces with Governor Villamil's colonization of Charles Island, helping, among other things, with the introduction of even more domestic mammals. Captain FitzRoy recorded one of these early introductions. He noted the arrival of a small schooner at Black Beach on October 16, 1835, in which there "were some emigrants; who brought cattle, and information that the governor, Villamil, might be expected to arrive in a few days, with a vessel laden with animals, and supplies for the settlement."[20]

According to Darwin the settlers were already hunting the "wild pigs & goats with which the woods abound."[21] FitzRoy, however, wrote that these animals were still "scarce and wild, not having yet had time to increase much." "[T]hey are hunted with dogs," he acknowledged, but due to their scarcity thought "it would be wiser to let them alone for a few years."[22] Overhunted they were not, for when Abel du Petit-Thouars visited Charles Island just three years later, he calculated there to be "one hundred and thirty head of cattle, cows, bulls, beef or veal, some horses, a fair number of asses, several hundred goats, and one estimates the number of hogs over two thousand."[23] Unfortunately, the dogs had also propagated; Petit-Thouars noted an "excessive number of dogs . . . a veritable calamity for the tortoises, which they kill."[24]

Farm animals were not the only exotics adapting to life on Charles Island when Darwin arrived in 1835. Several foreign plants were already flourishing in the highlands. Watkins and Johnston had cultivated potatoes, pumpkins, and melons, and Lawson and Villamil had introduced wheat and barley, orange, "Quito orange," pomegranate, fig, papaya, banana, and lemon trees, coconut palms, Indian corn, sugar cane, and cotton from Guayaquil.[25] While few of these plants became invasive, the cotton did supremely well at mid elevation. It hybridized with the native cotton and the hybrid is now the dominant variety in the arid zone around the town of Puerto Velasco Ibarra today.

Introductions on Other Islands

There are many records of exotic species arriving on Charles Island prior to 1835. In contrast there are few records of man-assisted introductions to other islands. It is difficult to determine whether alien species, apart from black rats on James, were on the other islands Darwin visited. It is possible that donkeys and dogs, the two "beasts" useful for tortoise collection, accompanied the Charles Island settlers to their sister base on James Island in 1835. Lawson also probably released goats and pigs on to more islands that just Charles, in 1830. John Coulter, ship's surgeon on board the *Stratford*, wrote of seeing "several groups of reddish coloured goats"[26] on Chatham Island and of shooting goats on James in 1833, yet Darwin never mentioned seeing domestic mammals on any island except Charles.

Chatham and Albemarle Islands were gradually colonized by humans during the 1840s and 1850s. When Commander William Edgar Cookson of HMS *Peterel* visited the archipelago in 1875, 40 years after Darwin, he recorded mammals on all four islands. There were cattle, horses, donkeys, pigs, and goats on Charles, all but the horses on Chatham, cattle and donkeys on Albemarle, and large numbers of donkeys and pigs (but interestingly no mention of goats) on James.[27]

The future of Galápagos is not as dismal as it may seem from this gloomy catalog of introductions. Several of the uninhabited islands (Daphne, Narborough, Tower, Wenman, and Culpepper, to name a few) remain completely free of introduced mammals and virtually free of damaging plants and insects. Conservation efforts for the most part are working, and there have been successes in the eradication of some populations of introduced species, such as fire ants from Barrington and Bindloe Islands. In 2006 James Island became a success story all of its own: after an intense eradication program that lasted several years, all the pigs (removed by 2000), donkeys, and goats (gone by the beginning of 2006), were declared eliminated from the island. The same project removed all goats from northern Albemarle Island, a feat previously thought impossible due to the vast area (about 250,000 hectares) afflicted.[28] It is estimated that a total of 140,000 goats were destroyed on the two islands in a period of five years. In 2007 donkeys and goats were removed from Charles Island, and plans are now (in 2009) underway to try to eliminate the introduced rats and mice from at least some of the islands. The islands will never be the same as when Darwin saw them, but for many native species these conservation efforts have been and continue to be a lifesaver.

Views from the Top

Late in the afternoon of September 26, the *Beagle* moved to Black Beach, where she dropped anchor that evening. The 27th was a Sunday and all hands were allowed on shore. With several crewmen Darwin ascended Round Hill, the highest peak on the island, finding it "covered in its upper part with coarse grass and Shrubs."[113] In 1833, John Downes, commander of the U.S. Frigate *Potomac*, had described the hill in much the same way, noting that the grass, or "paja" (in Spanish), was "used by the inhabitants for covering their houses; and for which reason they have named it the *Serra de la*

Paja."[114] Neither Darwin nor FitzRoy adopted this name in their writings, but the name persisted and today the hill is known by all as Cerro Pajas (01°17′43.17″S, 90°27′20.49″W). The coarse grass (most likely not a grass but Andersson's sedge, *Cyperus anderssonii*) has since gone.[115] The hill is now covered with a highly invasive introduced ornamental lantana shrub (*Lantana camara*).

Cerro Pajas typifies the huge changes in the island's native flora that have resulted from human occupation of the island. Charles Island formerly supported a forest of cat's claw trees (*Zanthoxylum fagara*), but much of it is now gone.[116] Cat's claw trees were vital habitat for the sharp-beaked ground finch (*Geospiza difficilis*), a species collected by Darwin that has long since disappeared from Charles (and Indefatigable), although it is still found on other islands. The extra-large form of the large ground finch (*Geospiza magnirostris magnirostris*), collected by Darwin here and on Chatham, disappeared when its food, the cactus *Opuntia megasperma*, was razed by goats and donkeys. When Byron sailed past Charles Island in 1825 he described the island as being covered with "the prickly pear."[117] In 1833 the crew of the U.S. frigate *Potomac* remarked that the lowlands were scattered heavily with prickly pear trees, and that the trees formed a considerable portion of the diet of the introduced hogs and goats.[118] Today there are very few opuntias left standing. Mature stands of *Opuntia megasperma* survive on nearby Champion and Gardner islets but they have not been sufficient to maintain a population of the large ground finch.

Peering through damp branches on the top of Round Hill Darwin "counted in different parts of the Island . . . from 39–40 hills, in the summit of all of which there is a more or less perfectly circular depression."[119] As all the hills were thickly covered with vegetation and there were no recent lava flows to be seen, Darwin wrote little else about the geology of the island. Nor did he collect a single rock specimen from Charles. Ironically Charles Island has the best cinder cones in Galápagos, two prime examples of which are Cerros Pajas and the dominant cone at Asilo de Paz (01°18′51.65″S, 90°27′13.39″W). Cinder forms explosively like tuff, but instead of erupting in water and fracturing into tiny dust- or sand-sized particles, the magma erupts in air and is torn into larger fragments, or pyroclasts, by the violent release of volcanic gases. The caves above the Governor's dripstone testify to the loosely associated nature of the cinders, allowing them to be easily eroded and carved. Darwin, however, wrote about none of this.[120]

From the top of Round Hill, and assuming fair visibility,[121] Darwin would have been able to see Charles' satellite islands to the east. He did not visit any of the islets himself but *Beagle* surveyor Chaffers climbed several and from one of them brought back "a fossil shell: which he extracted from Volcanic Sandstone [tuff] at the height of 400–500 ft [(122–152 meters)]."[122] Darwin identified the location as Champion Island, but Champion is a scoria crater and only about 45 meters high. Neighboring Enderby Islet is a taller tuff cone, and therefore a closer match. Darwin thought the shell, which he identified as "a Murrex [*sic*],"[123] had been uplifted. Uplifted fossil gastropods (including murex), dating from the Pliocene (5.3–1.8 million years before present) and Pleistocene (1.8 million to 11,500 years before present) eras can be found on Seymour, Albemarle, and Indefatigable islands,[124] but these occur in distinct fossiliferous strata no higher than 45 meters above sea level, and never in volcanic tuff. A more likely explanation for Chaffer's murex is that it was ejected during the eruption itself, for ejected shell and coral fragments can indeed be found on top of Enderby. Alternatively, and if the shell was indeed whole, it may have been carried to the top of the island by a seabird. Frigatebirds have been observed picking up shells, sticks, and plastic debris from beaches, and carrying them "in play" before discarding them. The presence of an occasional shell or piece of plastic in the interior of uninhabited islands is generally attributed to such behavior.[125]

Chaffers later reported finding many "fossil" oyster shell fragments and extracted "3 species of shells" from a steep tuff crater on Bindloe (Marchena) Island. The crater was most likely one of the eastern flank volcanoes, where fragments of ejected oysters and *Acroporal* corals have since been found.[126] This time Darwin allowed that Chaffer's shells might have been ejected during the eruption, though he still favored uplift as an explanation. He wrote, "I am aware that volcanic mud eruptions might well eject shells: but I think there is at least an equal probability that the sandstone is a submarine deposition and has been subsequently raised to its present elevation of about 300 ft [91 meters]."[127]

Chaffer's shells from Bindloe, his murex from Enderby, and Darwin's own shells extracted from Frigatebird Hill on Chatham Island prompted Darwin to wonder how much land elevation had played a part in shaping the Galápagos Islands. Ultimately, Darwin dismissed uplift as having a significant role in the formation of the islands. The shells lost much of their meaning. Eventually they themselves became lost, and are now known only from Darwin's notes. Fortunately, many of Darwin's geological specimens from Galápagos avoided this fate and can now be found at the Sedgwick Museum of Earth Sciences in Cambridge, England.

A Confusion of Finches

It is only a very wise man or a fool who thinks he is able to identify all the finches which he sees.
—Michael Harris, *Birds of Galápagos*[1]

Depending on how they are classified there are 14 or 15 species of Darwin's finches.[2] All are endemic to the Galápagos Islands, except for one that is found only on Cocos Island, 650 kilometers to the northeast. The degree to which Darwin's finches have diversified is unequaled by any other group of birds in Galápagos. Among birds worldwide, they are surpassed only by the Hawaiian honeycreepers, about 50 different species of which evolved in the last few million years. For this reason Darwin's finches are considered a textbook example of adaptive radiation, and "a classic model of speciation."[3] But how important were they to Darwin?

Interestingly, while Darwin was in Galápagos he did not find the drab "little birds [that] quietly hopped about the Bushes"[4] particularly noteworthy, except for their ubiquity and remarkable fearlessness. For they "were not frightened by stones being thrown at them,"[5] would alight on the backs of land iguanas "with the utmost indifference,"[6] and could be easily caught "with a cap or hat."[7] Darwin wrote little else about the finches while he was in the islands, but his apparent disinterest is belied by the 31 specimens he collected—almost half of his entire bird collection from the Galápagos. Why did he (and his shooter, Syms Covington) collect so many? It was not just because they were easy to snag. After all, he took just one specimen of Galápagos dove, a bird even more numerous and fearless than the finches. Nor was it because he was trying to collect a complete set of species from each island he visited. He did not bother to label most of his finch specimens by locality. Nor did he collect finches from every island he visited, for he left Albemarle Island out entirely. A more likely reason for the disproportionate number of finch specimens he collected is due to the "inexplicable confusion"[8] he, and indeed all the collectors aboard the *Beagle*, encountered in trying to tell "the little birds" apart. Because there was "much difficulty in ascertaining the Species"[9] there was no way of limiting the collection to just one of each. Whatever the reason behind it, the result was an excess of finch specimens for the experts back home to sort out.

When looking over the specimens during the voyage home to England Darwin catalogued them into the following groups: several true finches ("Fringilla"), an "Icterus like finch," some "Gross-beaks," and a "Wren."[10] But this improvised classification raised doubts in his mind, for along the range of specimens there appeared to be "a gradation in [the] form of the bill,"[11] which blurred the very boundaries he imposed. He was also puzzled by the variation he saw in the plumage of the birds. Some were brown, others black, some streaked, and still others were pale brown with dark heads. Was the coloration a species indicator? Plumage pattern was the primary identifying factor for classifying bird species in those days[12] and Darwin had quickly keyed in to the color variation between the mockingbirds of Chatham and those of Charles. He suspected, however, that among the finches, color primarily differentiated sex and age, not species, and "that only the old Cocks possessed a jet black plumage."[13] "On the other hand," he noted cautiously, "—Mr. Bynoe & Fuller assert, they have each a small jet black bird of the female sex."[14] This incorrect observation made by his shipmates did nothing to help his confusion. To further confound matters Darwin remembered that while in Galápagos he had seen "no possibility of distinguishing the ["Fringilla"] species by their habits, as they are all similar, & they feed together (also with doves) in large irregular flocks."[15]

Shortly after Darwin returned home, ornithologist John Gould examined and described Darwin's bird specimens. In January 1837 he astounded Darwin with the news that the "Fringilla," "Gross-beaks," Icterus, and wren were really 13 species of one new and closely related group of ground finches.[16] Actually, Darwin collected only 9 species (he missed the woodpecker finch, medium tree finch, large cactus finch, and mangrove finch), but Gould split the species rather too finely. Gould also confirmed what Darwin already suspected: his Galápagos mockingbirds comprised three distinct species and all but one (the bobolink) of his 26 Galápagos land birds (22 in modern terms) were new species, closely allied to species on the South American continent, yet found nowhere else in the world. (Today the two owls, vermilion flycatchers, and yellow warbler are considered endemic only at the subspecies level.) Darwin, who had already been pondering evolution as an explanation for the interisland differences among the mockingbirds and tortoises of Galápagos, was now convinced it explained the existence of all the peculiar species of Galápagos.

Now, for the first time, Darwin had "reasons to suspect that some of the species of the sub-group Geospiza [may be] confined to separate islands."[17] After all, it was clearly physical isolation that had led to the divergence of the mockingbird species in Galápagos. He could say with confidence that the mockingbirds replace each other on the different islands, but he collected only three species. If he could show island representation among the 13 (9) species of finches he had collected, it would greatly bolster his argument for evolution. But how could he do this? He had failed to label many of his own finch specimens by island locality, so had no idea which finch came from which island. He therefore asked to borrow the more carefully labeled finch specimens that Fitz-Roy, Fuller, and Covington had collected in order to match theirs with his, and try to determine which species occurred on which island. Despite his efforts Darwin never was able to say much about the geographical distribution of the finches. They have a far more complex distribution than the mocking-birds, many co-mingling on the same islands. Deciphering the origins of the finches was proving to be no easy matter.

It was an enormous relief when, in 1845, Joseph Hooker described Darwin's plants from Galápagos and pronounced "the aboriginal plants of the different islands wonderfully different."[18] They complemented "the marvellous [*sic*] fact of the species of birds [i.e., mockingbirds] being different in the different islands of the Galapagos"[19] and backed up Darwin's "assertion on the differences in the animals of the different islands, about which," he confessed, "I have always been fearful."[20]

Darwin felt free to drop the subject of the finches after this. But the general question of why the organisms of Galápagos should have diversified so much in the archipelago continued to niggle at him. He posed the problem in *The Origin of Species*, when he queried, "How has it happened in the several islands situated within sight of each other, having the same geological nature, the same height, climate, &c., that many of the immigrants should have been differently modified, though only in a small degree[?]"[21] His difficulty in answering the question lay in his erroneous view of the Galápagos Islands as being all pretty much the same physically. This is not to say that he thought each island identical in size, shape, height, local climate, and position in the archipelago relative to the prevailing wind and ocean currents, but he certainly underappreciated the differences with respect to their having a diversifying effect on the archipelago's flora and fauna. He compensated for his oversight by recognizing that "the nature of the other inhabitants, with which each has to compete, is at least as important, and generally a far more important element of success" than "the physical conditions of a country," in the process of species modification.[22] He could certainly see how interspecific competition might work in modifying plants. A plant genus, he wrote, is often represented by different species on different islands because individual plants "have to compete with a different set of organisms" for the "best-fitted ground."[23]

How this idea applied specifically to the evolution of the finches, however, remained a mystery. For Darwin could not see how competition had driven them to diversify. His gross error was that he believed the finches (except for the cactus finch) all fed in the same way and on the same dietary items,[24] so he never linked "the perfectly graduated series in the size of their beaks"[25] to differences in their diet. If they all ate the same things "indiscriminately," how could competition for food have shaped their different bills? And if not food, what else? So, despite their startling degree of diversity, and their undeniable role in convincing Darwin of evolution, just how the various finches had diversified remained a complex, confusing, and tantalizingly curious subject that Darwin never fully understood.

One hundred years after the launch of the voyage of the *Beagle*, ornithologists began to tackle the mystery of the finches. In 1931 Harry Swarth reclassified the finches, and included five new species collected since Darwin.[26] Then in 1936, Percy Lowe suggested that the finches were not actually separate species but just "hybrid swarms" that had originated from a few ancestral species.[27] Lowe coined the name "Darwin's finches." Two years later David Lack went to Galápagos and conducted the first field study of the group. He concluded that the finches comprise distinct species, and that although they are found in various combinations on the islands today, they originally diverged in isolation. Despite disagreeing with Lowe on the idea of hybrid swarms Lack adopted his term "Darwin's Finches" and popularized it in the title of his resulting book. More importantly Lack opened a new area of research by suggesting that a finch's beak, contrary to Darwin's opinion, is highly adaptive to its diet. He also believed that it is shaped by the finch competitors sharing the island. Unfortunately, Lack was unable to test his ideas because he conducted his study mainly during the rainy season when finches do tend to eat the same foods—leaf buds, seasonal caterpillars, spiders, moths, and other arthropods.

In the 1970s Peter and Rosemary Grant picked up where Lack left off and began to look closely at finch feeding behavior during the dry season, when arthropods are in short supply and seeds dominate finch diets. They were soon able to demonstrate that the different species were adapted to eating different types of food.[28] When Darwin, who also visited Galápagos in a dry season, observed flocks of finches feeding together on the ground, he presumed that all the species were eating the same things indiscriminately. But the Grants were able to show that they were being far pickier than that. Even within species there was discrimination.

The Grants' long-term, ongoing study of the ecology and evolution of Darwin's finches is one of the most famous evolutionary studies in the world. From studying the finches of Daphne Island annually since 1973 the Grants have been able to demonstrate the process of evolution in the island's population of medium ground finch (*Geospiza fortis*). The first 20 years of their study was well documented in Jonathan Weiner's Pulitzer Prize winning book *The Beak of the Finch*. But the study has been growing since then, and is now well into its third decade. Briefly, the story goes like this. By measuring the beaks of *G. fortis* on Daphne the Grants found that they vary considerably in size. In this one species the beaks vary in depth to a greater proportionate degree than adult humans vary in height. If the smallest beak is equivalent to a 4-foot-tall human, then the largest beak would be comparable to an 8-foot-tall human. In the finches the variation is measured in mere millimeters and is often barely detectable to the eye, but even half a millimeter makes all the difference to what a finch can and cannot eat. And this can be a matter of life or death, when living on a small dry island that is subject to large fluctuations in annual rainfall and therefore food supply. It is a game of survival that the Grants have been documenting—by counting finches, and by measuring beaks, seeds, and rainfall—year in, year out for the past 35 years.

The study started in 1973 but the drama really began in 1977 when a severe drought killed off 85 percent of Daphne Island's *G. fortis*. The population plummeted from about 1200 individuals in January to just 180 by the end of the year. The deaths were not random. The individuals that died had, on average, smaller beaks than the ones that survived. They had starved because as the stocks of seeds dwindled, they were unable to feed on the large seeds that remained. In particular, they were unable to crack open the mericarps of the caltrop fruit (*Tribulus cistoides*), the seeds of which sustained the large-beaked birds. Natural selection acting on beak size resulted in a surviving population of large-beaked *G. fortis*. Because beak size is heritable, the evolutionary effects of this selection event were seen in the next generation. That is, the average beak size of the survivors' offspring was significantly larger than that of the population of *G. fortis* living before the drought. This was the first time measurable evolution had ever been shown in an undisturbed (that is, unmanipulated) natural population of organisms.

Five years later the pendulum swung the other way. A year of unusually heavy rainfall in 1982–1983, the result of a massive El Niño event, turned the dry and dusty islet into a jungle of vegetation, which produced a plethora of seeds.

With plenty of food to go around the finches multiplied prolifically. The years that followed (1984 and 1985) were droughts, but many of the birds survived the dry conditions by feeding on the stocks of seeds left over from the El Niño. But once again, survival was not random. As the large seeds were depleted and the remaining seeds became increasingly difficult and costly to access, buried as they were under mats of dying vegetation, the large-beaked birds began to die off. While large-beaked birds can feed on small seeds, they are less efficient at manipulating them than are small-beaked birds. They also need many more seeds to maintain their larger body size. This time selection favored the birds with the smallest beaks.

Since the 1980s the Grants have shown, through continued measurement of beak size and food supply, several more selection events. In 2005, selection on beak size was comparable in magnitude to the event measured in 1977. Droughts in 2003 and 2004 caused another massive die-off of finches. As in 1977, there were about 1200 *G. fortis* individuals at the beginning of the drought. By early 2005, when the rains returned, a whopping 93 percent had perished; there were now only 83 *G. fortis* left on the island. Unlike in 1977, when predominantly small-beaked *G. fortis* died, this time it was the large-beaked individuals that perished. Why? New finch competitors had altered the game. In 1982, a few large ground finches (*G. magnirostris*) flew to Daphne from a neighboring island. The genetics show that they came from Jervis Island (Rábida) or possibly James Island (Santiago). By 1983, they had established themselves as a breeding population on Daphne and over the next few years they increased in numbers until they were effectively competing with the largest of the *G. fortis* for the largest seeds. For many years there was enough food to support both large-beaked *G. fortis* and *G. magnirostris*. Then the drought hit. With their wrench-like beaks the *G. magnirostris* had an easier time cracking open the mericarps of the *Tribulus* seeds and soon depleted the stocks. The largest of the *G. fortis* were the first to suffer. This time it was the small-beaked *G. fortis* that survived, bred, and passed on their small beaks to the next generation. Competition between two finch species had caused the average beak size of *G. fortis* to take a nosedive.[29]

To this day the Grants continue to measure evolutionary responses to fluctuations in food supply caused by a combination of environmental changes and the population dynamics of competing finch species. They are also studying the genetics involved in beak formation, and how information within the finches' DNA determines how new forms evolve. They have calculated that under certain conditions, with unidirectional selection, it could take as little as 200 years for one finch species to evolve into a new species. They have not shown speciation (no one has), but they have shown evolution. Or, as Jonathan Weiner put it in *The Beak of the Finch*, the Grants have shown "the evolutionary process not through fossils but directly, in real time, in the wild: evolution in the flesh."[30]

Darwin would be delighted. Despite never having equated finch beak size with diet, never having identified the selection pressures that determine beak size, he had recognized that the structure of a bird's beak must be "of service" to its life in some way and not just an ornament. He had even hypothesized on the process by which a bird's beak could change. As he brilliantly put it in a letter to rector and history professor Charles Kingsley in 1867:

> When speaking of the formation for instance of a new sp. [*sic*, species] of Bird with long beak Instead of saying, . . . a bird suddenly appeared with a beak [particularly] longer than that of his fellows, I would now say that of all the birds annually born, some will have a beak a shade longer, & some a shade shorter, & that under conditions or habits of life favouring longer beak, all the individuals, with beaks a little longer would be more apt to survive than

> those with beaks shorter than average. The preservation
> of the longer-beaked birds, would in addition add to the
> augmented tendency to vary in this same direction.[31]
>
> The Grants have proven Darwin right.[32]

Darwin slept restlessly that night, or so it is tempting to imagine. The fresh food had nourished his body and the lush landscape had been a veritable sight for sore eyes, but there was something disturbing about Charles Island. It was not only because the island was a political penal colony. The men and women had seemed friendly enough, and Lawson had been downright hospitable. The highlands themselves had been delightfully refreshing, and the weather, for the most part, had been fair. Darwin could not quite put his finger on what was troubling him. Perhaps he was merely homesick. It had been a pleasure meeting such a lively fellow Englishman as Lawson but his voice must have made Darwin long for England. Indeed, much about the highlands of Charles Island would have reminded Darwin of England. The "chacras,"[128] or small farms, were as green as northern pastures and the weather was decidedly British in its misty coolness. Even the minute crickets, moths, and flies had reminded him of temperate England! He had dashed off a nostalgic letter to his sister Caroline and left it at Black Beach, but he had an eerie premonition that it would never be sent. Indeed, there is no record it ever arrived.[129]

Perhaps Lawson's tales of tortoise exploitation had generated a disturbing feeling that the island was changing at the hand of man before Darwin could quite grasp its meaning. Or was it something else that Lawson had said about the tortoises? Syms Covington had managed to obtain a young tortoise from the island; a look at his servant's acquisition the next day might jog his memory. The mockingbirds also haunted him; why should they be so different on this island from those on Chatham Island, just 90 kilometers away, yet both resemble the mockingbirds he had seen in Chile?

Darwin had managed to increase his Galápagos collections admirably for such a short visit but the geology of Charles had been a veritable disappointment. He wrote only a page and a half of geology notes about Charles and its islets, which he would pithily expand to just two more pages on the way home to England. Chaffers had not helped matters by presenting him, that very evening, with a "fossil" murex chipped from the top of a nearby islet. Was he missing the best geologizing? Darwin needn't have worried. The next island would bring him lava in abundance, and such spectacular geological formations that even England would be wiped clean from his thoughts.

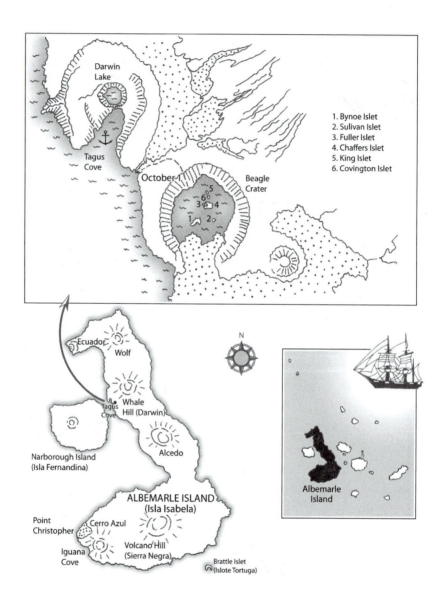

Darwin
Lake

Tagus
Cove

October 1

Beagle
Crater

1. Bynoe Islet
2. Sulivan Islet
3. Fuller Islet
4. Chaffers Islet
5. King Islet
6. Covington Islet

5
6
3   4
1   2

N

Ecuador
Wolf

Whale
Hill (Darwin)

Tagus
Cove

Alcedo

Narborough Island
(Isla Fernandina)

ALBEMARLE ISLAND
(Isla Isabela)

Point
Christopher

Cerro Azul

Iguana
Cove

Volcano Hill
(Sierra Negra)

Brattle Islet
(Islote Tortuga)

Albemarle
Island

Chapter VI

~~~~~~~~~~~~~~~~~~~~~~~~

Albemarle Island
(Isla Isabela)

*The inhospitable appearance of this place was such as I had never
before seen, nor had I ever beheld such wild clusters of hillocks, in
such strange irregular shapes and forms, as the shore presented,
except on the fields of ice near the South Pole . . .*
—James Colnett, *A voyage to the South Atlantic and round Cape Horn*[1]

Vulcan's Playground

The *Beagle* left Charles Island on Monday morning, September 28.
She was well stocked with vegetables and pigs from the settlement,
but was running low on water. A small amount had been "conveyed
down the hill by Bamboo Pipes, but [the spring's] supply was quite
precarious in 1835."[2] FitzRoy gave orders to continue on, without filling the ship. He put all hands on strict water rations—"½ a Gallon for
cooking & all purposes"[3] —and steered the *Beagle* toward Albemarle
Island, where he hoped to find more water at Tagus Cove. The trip
took three days. The change in weather from mostly cloud to "Vertical sun"[4] only intensified the growing discomfort on board. Darwin
had one consolation—the geology to be viewed en route.

On the way to Albemarle the *Beagle* passed close by the crescent-
shaped islet of Brattle (Tortuga), the widest tuff ring in Galápagos.
Brattle was a perfect example of a phenomenon Darwin had first seen

in Pan de Azucar on Chatham Island, and that he and the surveying officers of the *Beagle* observed repeatedly throughout the archipelago: all the tuff craters in Galápagos are broken down on their southern side. Officers Stokes, Sulivan, and Chaffers, "who necessarily paid more attention to the figure than the constitution of the land," helped bring this pattern to Darwin's attention.[5] Darwin listed 28 cases (12 islets and 16 insular craters) and could find no exception to the rule. His explanation was as lyrical as it was true. "Throughout the islands of the archipelago, both the sea, from the trade wind & the long swell of the Great Ocean constantly unite their unwearied forces against Southern shores. Hence, that side, especially in the more exposed Islands, is bold and precipitous, whilst the northern shores approach the sea with a gradual slope."[6]

After seeing the same phenomenon in the terrestrial craters of Ascension Island ten months later, Darwin hypothesized an additional cause for their oblique appearance. On Ascension the prevailing winds had "no doubt" given rise to a higher buildup of "ejected fragments and ashes" on the leeward rim during the eruption itself.[7] Darwin then reasoned that "this same power might . . . also, aid in making the windward and exposed sides of some of the [Galápagos] craters, originally the lowest."[8]

Modern day visitors to Galápagos can see an asymmetrical tuff cone as soon as they fly into Baltra airport, simply by looking out of the window at the offshore islet of Daphne Major. Not only is Daphne Major a prime example of a tuff cone, it is famous as the place where evolution has been demonstrated in a natural population of Darwin's finches (see A Confusion of Finches). But the most recognized lopsided tuff ring in the world is not found in Galápagos at all, but on the Hawaiian island of Oahu. Like the tuff cones of Galápagos, Oahu's Diamond Head looms higher on one side than the other, but there the trade winds blow from the northeast. The highest side is therefore the southwestern rim.

The *Beagle* sailed west along the southern coast of Albemarle and on the afternoon of Tuesday, September 29, stopped briefly at Iguana Cove (Caleta Iguana). Mellersh and King were then sent off in a whaleboat to survey Elizabeth Bay and the western shore of Narborough Island. They were to be picked up four days later at Tagus Cove. There is no evidence to suggest that Darwin, or indeed any of the *Beagle* men, landed at Iguana Cove. They found the little cove with its plunging cliffs an unwelcoming, "wild-looking place — with such quantities of hideous iguanas as were quite startling!"[9] The water was decidedly frigid, the coldest anywhere so close to the equator. It measured 58.5 degrees Fahrenheit (14.7°C), a low not recorded since leaving the Peruvian coast.[10] They tarried but a couple of hours to survey the area before continuing along the cold, "unearthly shore."[11]

Near Point Christopher their "eyes and imagination were engrossed by the strange wildness of the view; for in such a place Vulcan might have worked."[12] Here the land was "craterized by cones just like at Chatham"[13] but with "even a more *work-shop* appearance."[14] As at Chatham the scene impressed Darwin with its apparent youth; the landscape could have been formed just yesterday. He was eager to explore but "a calm prevented [the *Beagle*] anchoring for the night"[15] and they were compelled to continue on lest swells bear them ashore.

As the *Beagle* now sailed northward up the western coast of Albemarle, Darwin marveled at the island's immense shield volcanoes. They were "surmounted by enormous Craters, from which bare & black streams of Lava [could] be traced down their sides."[16] He watched a small jet of "steam issuing from [a] Crater high up" on the "2^d mound from the South"[17] daring to hope it would burst into something more dramatic. Darwin called the mound "Volcano hill"[18] in his geology notes and was clearly referring to the volcano that today is known as Sierra Negra. Alcedo Volcano, the third shield volcano from the South, also has a history of fumarole activity, and steam can still sometimes be seen rising from its top by yachts navigating the Bolivar Channel between Albemarle and Narborough islands. However, Captain FitzRoy's account supports the idea that Darwin was describing Sierra Negra, not Alcedo. In his *Narrative* he wrote, "On the south-eastern height of Albemarle, smoke was seen issuing from several places near the summit, but no flame."[19]

Darwin was hoping to see an eruption. In 1825 the crew of HMS *Blonde* had seen volcanoes "burning around us on either hand"[20] when they were anchored at Tagus Cove. Eleven years earlier, in 1814, Captain Pipon of HMS *Tagus* "saw two in action"[21] on Narborough. The Galápagos archipelago is one of the most active groups of volcanic islands in the world but the eruptions occur erratically and rarely last longer than three weeks. Darwin happened to arrive smack in the middle of a 19-year period of apparent inactivity; at least there are no known written records of an eruption in Galápagos between 1825 and 1844.[22] He never did see "the sublimity, the majesty, the terrific grandeur" of an awakening Galápagos volcano, "shooting its vengeful flames high into the gloomy atmosphere, with a rumbling noise like distant thunder" and "belch[ing] forth its melted entrails in an unceasing cataract,"[23] such as that witnessed and described by Captain Benjamin Morrel of the U.S. schooner *Tartar* in 1825. Sadly, the wretched wisp of steam rising from an active fumarole was the closest Darwin came to witnessing such a spectacle in Galápagos. He would have been happy just to look down into one of the "immense Cauldron like Craters"[24] that he guessed plumbed the top of the volcanoes, but he was given neither the time nor the means for such a climb. It was a disappointment he could not easily forget. Twenty-two years later he lamented to geologist Charles Lyell, "I always regretted that I was not able to examine the great craters on Albemarle Is^d."[25]

On October 22, 2005, Sierra Negra, the same sleeping giant that was puffing steam in 1835, erupted in a dramatic explosion. It had lain dormant for 26 years. The eruption lasted one week and covered 14 square kilometers (over 20%) of the floor of the largest caldera in Galápagos (7 × 10.5 km), with upward of 50 million cubic meters of fresh lava.[26] Rivers of lava also poured down the northern flanks of the volcano, toward Elizabeth Bay. The authors were fortunate to witness the eruption up close from the rim of the caldera, and can testify that Morrel's seemingly exaggerated descriptions have a vibrant ring of truth![27]

A Little World within Itself

> *The place is like a new creation: the birds and beasts do not get out of our way; the pelicans and sea-lions look in our faces as if we had*

> *no right to intrude on their solitude; the small birds are so tame*
> *that they hop upon our feet; and all this amidst volcanoes which*
> *are burning around us on either hand altogether it is as wild and*
> *desolate a scene as imagination can picture.*
> —George Anson Byron, *Voyage of HMS Blonde*[28]

At sunset on Wednesday, September 30, the *Beagle* sailed into Tagus Cove or Banks Cove (Blonde Cove as it was also known to those on board[29]), and moored at 6:18 p.m. Ten years earlier Captain Byron of HMS *Blonde* had navigated the same entrance. He, or one of his companions, had written, "[A]s we shot into the cove we disturbed such a number of aquatic birds and other animals, that we were nearly deafened with their wild and piercing cries."[30] Surely the darkening walls of the tranquil, bowl-shaped harbor echoed with barking sea lions, braying Galápagos penguins, and groaning flightless cormorants as they do today. But Darwin recorded none of this. The animals that are arguably the main tourist attraction of Tagus Cove today are completely absent from his notes. Byron saw penguins here in 1825, and the crew of the *Essex* caught "shags" (cormorants) on the island in 1813.[31] The fact that Darwin overlooked these species is therefore surprising—surprising until one appreciates how preoccupied Darwin was with the geology of the area. Just sailing into the breached entrance of this phenomenally layered extinct tuff crater may well have taken all his attention away from the very animals that bring the cove to life. He was bowled over by the geology of "this Crater Harbor."[32]

The Galápagos Islands are famous for several bird and mammal species that because of their uniqueness or ubiquity in the archipelago have become almost synonymous with the word Galápagos. Among these are the red-footed booby (*Sula sula*), swallow-tailed gull (*Creagrus furcatus*), red-billed tropicbird (*Phaethon aethereus*), waved albatross (*Phoebastria irrorata*), Galápagos fur seal (*Arctocephalus galapagoensis*), Galápagos sea lion (*Zalophus wollebaeki*), flightless cormorant (*Phalacrocorax harrisi*), and Galápagos penguin (*Spheniscus mendiculus*). Some historians have understandably imagined Darwin marveling at these animals, their behaviors, and their adaptations, in much the same way modern visitors do today.[33] But the truth is, Darwin never mentioned seeing a single one of them. He

did acquire a tobacco pipe reportedly made from the tibia of an albatross obtained in Galápagos, but that does not mean he saw a waved
albatross in the islands. In fact, there is some question as to whether
the pipe was made from a Galápagos bird at all. The stem is certainly
too long (320 mm) and slender (7–8 mm) to have been fashioned
from an albatross's leg bone, but it was possibly made from the ulnas
of a waved albatross (*Phoebastria irrorata*) or radial wing bones of a
brown pelican (*Pelecanus occidentalis urinator*). The femur of a greater
flamingo (*Phoenicopterus ruber*) is also a possibility, among the birds
of Galápagos.[34] The pipe was exhibited at the Darwin/Wallace centenary reception in 1958 and then presented to the Linnean Society of
London, where it remains today.

It is probable that Darwin saw at least some of the species listed
above, and their absence from his notes merely indicates his priori-
ties; first and foremost geology, and second, novel terrestrial organisms unique to Galápagos. Nineteenth-century naturalists made their
names by finding organisms new to science, not by remarking on
those that had already been described. By virtue of their ability to
travel long distances, Darwin may not have regarded the seabirds and
sea mammals as being confined to Galápagos. He certainly dismissed
the blue-footed booby (*Sula nebouxii*) and brown pelican (*Pelecanus
occidentalis urinator*) as being the same as species (the Peruvian
booby *Sula variegata* and the Peruvian pelican *Pelecanus thagus*) that
he had seen in Peru and he failed to collect specimens that would
have told him otherwise.[35] Marine mammals were entirely excluded
from his Galápagos notes and he considered the few "Marine birds"
he took to be "the most indifferent part of [his] collection."[36] Consequently, he was caught off guard with the lava gull (*Leucophaeus
fuliginosus*); his specimen was examined by John Gould and found
to be new. Darwin responded, "Considering the wandering habits of
the gulls, I was surprised to find that the species inhabiting these islands is peculiar, but allied to one from the southern parts of South
America."[37] Darwin also failed to take note of the very feature of the
flightless cormorant, celebrated in its name, that makes it unique
to Galápagos, and so evolutionarily interesting. Of course, he may
never have seen the cormorants at all. This is likely the case with the

swallow-tailed gull, as it is the simplest explanation for why Darwin asserted that Galápagos has only one gull, the lava gull.[38] He did not mistake the swallow-tail gull for a tern (as might be supposed), for Darwin also wrote that Galápagos has only one tern,[39] the brown noddy (*Anous stolidus galapagensis*). He apparently did not see the sooty tern, *Onychoprion fuscatus*, which today breeds on the northern islands of Wenman and Culpepper.

Before arriving in Galápagos, Darwin had seen and written about various species of boobies, gulls, albatross, sea lions, cormorants, and penguins encountered during different stages of the voyage. The behaviors that Darwin described for these species are in many cases characteristic of their Galápagos relatives. One has only to read Darwin's earlier notes to see how taken he must have been with the delightful antics of the marine animals of Galápagos. After all, this was a man who recognized that "no one can watch the Flying fish, Dolphin & Porpoises without pleasure."[40] So, while it appears that Darwin ignored some of the most entertaining Galápagos species, we can infer that he was no less immune to their delightful performances than are most people who visit the archipelago today.

In the Falkland Islands Darwin admired the way a "Jackass" penguin (most likely a Magellanic penguin *Spheniscus megellanicus*) "uses its wings very rapidly & looks like a small seal: from its low figure in the water & easy motion [it appears] crafty like a smuggler."[41] He had also enjoyed watching a cormorant "catch a fish, let it go & catch it again 8 times successively as an otter does a fish or Cat a mouse."[42] In the Chronos Archipelago he was "amused by the impetuous manner in which [a] heap of seals [sea lions], old & young, tumbled into the water as the boat passed by. They would not remain long under, but rising, followed [the boat] with outstreched [*sic*] necks, expressing great wonder & curiosity."[43] And near Montevideo, Uruguay, he painted a scene that could well have been taken straight from Galápagos, if one only replaces the word "porpoises" with "dolphins":

A wonderful shoal of Porpoises at least many hundreds in number, crossed the bows of our vessel. —The whole sea in places was furrowed by them; they proceeded by jumps, in which the whole body was exposed; & as hundreds thus cut the water it presented a most extraordinary spectacle. —When the ship was running 9 knots these animals could with the greatest ease cross & recross our bows & then dash away right ahead.[44]

From Darwin's description it is likely that these "porpoises" were actually common dolphins or bottlenose dolphins, species that also occur in Galápagos. Alternatively, they may have been dusky dolphins, a specimen of which was later "harpooned from the *Beagle*" off Patagonia's Valdés Peninsula. Although this species is not found in Galápagos it is noteworthy in being the only organism collected on the voyage to be named after Captain FitzRoy. The Zoological Society of London's museum curator George Robert Waterhouse named it "*Delphinus FitzRoyi*."[45] Today it is known as *Lagenorhynchus obscurus*.[46]

Darwin spent just one day (Thursday, October 1) on shore on Albemarle Island, but wrote the same amount on the geology of the area as he did on James Island, where he spent ten days (October 8–17). He filled 24 manuscript pages (13 of which were most likely written in Galápagos and 11 on the way home to England), with observations from Tagus Cove and Beagle Crater. Beagle Crater, which at the time was unnamed, was a perfect example of a tuff ring and both craters contained a great variety of tuff features. Darwin rationalized his excessive note taking. "I have particularly described these Craters because I do not recollect having read of an exactly parallel case. —nor indeed of a large Crater entirely composed of Volcanic Sandstone under any circumstances."[47]

Darwin concentrated on Beagle Crater. Instead of landing in Tagus Cove, as do visitors to the area today, he was taken down the coast to a small valley between Tagus and Beagle craters where fresh water was "draining continually through the rock"[48] into little pits in the tuff (00°16'23.45"S, 91°21'58.61"W). The amount was despairingly inadequate for the ship's needs but "sufficient to draw together all the little birds in the country. —Doves & Finches swarmed round its margin."[49] The green oasis of the landing site made a pretty but deceptive welcome. The terrain inland is rough and rutted. Deep furrows and a tapering finger of lava from the mountain behind (what Darwin called "whale hill"[50] and what is now known as Volcan Darwin) form a hilly network of ridges and gullies all covered with a thick tangle of brush. Beagle Crater rises steeply at the coast and a favorable ascent can be found only by navigating the maze-like topography of the valley inland to where the slope is more gently inclined.

From the landing site Darwin bushwhacked his way to the edge of an "immense stream of Lava."[51] It was "many miles broard [*sic*], [and] almost entirely destitute of vegetation." "I believe I must except 2 or 3 plants of a Cactus," he cautiously added in his notes.[52] Viewed from a distance the lava field appeared invitingly even and unfissured, and Darwin surmised it had flowed smoothly and quickly like water. Up close, however, it was deeply serrated. It was far more jagged than the roughest lava he had come across in the "Craterized District" of Chatham Island, and impossible to walk over. Looking up the flanks of Whale hill Darwin could see that an even younger "ribbon of blacker Lava [had] flowed from a minute & perfect Crater high up on [the] sides of [the] mountain."[53] It had frozen in a pleasing circular pattern resembling the tracks of a plough. Due to the "excessively rough"[54] nature of the greater lava flow, however, it was as inaccessible as the fields of England it would have called to mind.

Picking his way out of the tangled gullies Darwin walked along the knife-edged wall of the lava field toward the back of Beagle Crater. Then, on the northwest side, where the air is cloistered from all sea breezes but the slope is gentle, he started his climb. It was like ascending an enormous corrugated roof, or in Darwin's words, "the shell of a pectan or scallop."[55] The outer slopes of Beagle Crater form a ribbed skirt of "longitudinal doons [*sic*, dunes] (from 8 to 20 & even 40 ft wide) which are separated from each other by shallow gullies."[56] Superficially these gullies appear to be water channels carved by erosion, but they are an original feature of the eruption. Beagle Crater is a rare example of a base surge eruption, one that goes off something like a nuclear explosion. The eruption produces a mushroom cloud and turbulent clouds of ground-hugging gasses and ash, which avalanche down the sides of the volcano at terrific speeds. The accompanying winds propel the ash particles into elongate ridges, in the same way that sand grains get blown into dunes. But unlike sand dunes that form over several years in 30 kilometer per hour winds, these volcanic dunes form in seconds, in winds raging between 150 and 350 kilometers per hour.[57] The volcanic dunes are also, in places, hollow.

Darwin did not evoke quite such a violent explosion to account for the corrugations but he correctly surmised that the gullies formed during the eruption and only later became "deepened by alluvial action."[58] He imagined a giant mud eruption, "the boiling thick mass . . . pour[ing] over in narrow streams."[59] High up on the flank of the crater the dunes were more like "arched hollow channels . . . [and in] size, structure and appearance . . . *identical* with" the basalt lava tubes he had seen in the Craterized District of Chatham Island.[60] He reasoned that during the eruption the upper surface of these streams must have hardened through "evaporation and cooling" while the innards "flowed onwards."[61] But unlike the lava tubes on Chatham Island, the dunes were "not . . . nearly so continuously hollow."[62]

As Darwin walked up the rippled mantle of Beagle Crater, his footsteps reverberated dully on the vaulted, subterranean passages of the tuff tubes. Ascending the hollow casings was a slippery ordeal, for the ground itself was like a plastered floor, "the plaster being old, cracked and frequently separating in plates."[63] It did not help that little "pisolitic . . . balls [of tuff] . . . from size of shot to small bullet"[64] continually undermined his tread. In appearance as in effect they were much like ball bearings. Called accretionary lapilli, these balls are another indication of a base surge eruption. As ash clouds cool, condensation makes the ash particles wet and sticky so that they roll together (snowball) in the tumultuous winds. Darwin grabbed a chunk of substrate studded with these ash balls for his collection bag, and perhaps as a souvenir of his climb.

At the lip of the crater the dunes turn into "arched hollow channels"[65] and open into little hooded doorways. When Darwin reached the top he peered into their entrances, expecting to find tunnels descending into darkness. But to a one they were sealed with volcanic debris and overgrown with grasses and shrubs, giving the distinct impression of Lilliputian huts locked up and abandoned to the weeds. One of these "weeds" was a slender long-stemmed plant with needle-like leaves and yellow pompons looking much like vibrant versions of the ash balls strewn about the ground. Darwin cut off a spray of flowers and in doing so added an endemic genus, *Macrea*, to his Galápagos collections.

There are seven endemic angiosperm genera in Galápagos and Darwin collected representatives of four of them; *Scalesia, Darwiniothamnus, Lecocarpus,* and *Macrea*. He described a fifth, *Brachycereus,* and almost certainly saw the sixth, *Jasminocereus*. The last, *Sicyocaulis,* is known only from Indefatigable, an island Darwin did not visit. *Macrea* is named for James Macrae, botanist on board HMS *Blonde,* who collected a specimen of *Macrea laricifolia* ten years before Darwin did. At the time Darwin was unaware of this fact, and forgivably so. Macrae's specimens were not described for twenty years; that is, not until ten years after the voyage of the *Beagle,* when Joseph Dalton Hooker examined and named them along with Darwin's much larger collection of Galápagos plants.

Parched and burning from his climb, Darwin sat down on one of the "little hoods"[66] to rest, and was rewarded with a cooling breeze and an even more welcome vista. Before him stretched one of the grandest tuff craters in Galápagos, and in the basin, about 150 meters below, lay a large, glistening lake, almost one and a half kilometers in diameter (00°16′53.13″S, 91°21′07.53″W). It did not take long for Darwin to respond to the sirens call. As he later recorded in his diary, "The day was overpowringly [*sic*] hot; & the lake looked blue & clear. —I hurried down the cindery side, choked with dust, to my disgust on tasting the water found it Salt as brine."[67]

Darwin recognized that Beagle Lake had once resembled Tagus Cove in being open to the sea on its southern side. Unlike Tagus Cove, an "immense stream of Lava from the mountain behind ha[d] crossed the mouth [of Beagle Crater] & ha[d] poured part of its contents inwards and part towards the Sea,"[68] thus plugging the crater's breached entrance. Now closed off, and fed only by ocean seepage, Beagle Lake formed an isolated world, complete with a lonely archipelago of six islets of its own; a model Galápagos in a miniature sea. Further isolated by the ocean on the one side of Beagle Crater and a sea of barren lava on the other, it may have been this very scene, recalled later, that inspired Darwin to call Galápagos "a world within itself,"[69] a world cut off from the neighboring continent of South America, "whence it has derived a few stray colonists, and has received the general character of its indigenous productions, the greater number of its inhabitants, both vegetable and animal, being found nowhere else."[70]

After filling his head with geologic imagery, measuring angles of strata, and pocketing a handful of rocks, Darwin retraced his steps out of Beagle Crater, through the valley, and back to the ship in Tagus Cove. In navigating the labyrinth of gullies between Beagle Crater and the crater containing Darwin Lake (00°15′23.43″S, 91°22′20.29″W), he would have passed the remains of two huts made some time ago by marooned sailors—a not uncommon occurrence in the history of Galápagos. It is easy to get lost in the islands and some sailors simply did not return from a jaunt on shore. Others deliberately escaped a rough life on board or were jettisoned by their captain. The *Beagle* crew had rescued a sailor stranded on Hood Island, and at James Island they would hear that "a man [was] missing, belonging to an American whale ship, and [that] his shipmates were seeking for him."[71] Darwin knew of the Tagus huts from reading Byron's party's account of their visit to the area in 1825: "Our Sandwich Island chiefs landed on our anchoring, and having found two huts left by some former visitors they remained in them to enjoy the pleasures of fishing and bathing according to the customs of their own country, while we staid in harbour."[72]

While Darwin never did mention seeing the huts, he did borrow from Byron's account in other respects. When describing the marine iguanas he used the choice words "imps of darkness"[73] and when depicting certain lava flows he, like Byron's men, likened the surface to that of an ocean. For one of Byron's men wrote, "As far as the eye could reach we saw nothing but rough fields of lava, that seemed to have hardened while the force of the wind had been rippling its liquid surface. In some places we could fancy the fiery sea had been only gently agitated; in others, it seemed as if it had been swept into huge waves."[74]

And Darwin wrote, "The outline of the field of Lava as compared to the Basaltic one of Chatham Isd is much smoother. —there is not that appearance of huge frozen billows, or nearly so many fissures of contraction. On the contrary the surface itself is excessively rough. —I should compare the one to the ocean, the latter to a lake violently agitated by a Storm."[75]

His thirst for geologizing somewhat slaked, Darwin opened his eyes to the wildlife. The yellow land iguanas (*Conolophus subcristatus*) grabbed his immediate attention. They littered the undergrowth, basking in criss-crossed, leafy patches of sunlight, and occasionally nodding a territorial warning before thundering to their burrows "with [a] quick & clumsy gait"[76] as Darwin strode through the area. In general structure as in numbers they resembled Byron's "imps of darkness," which draped the shoreline, but they deserved a brighter description. Darwin provided a colorful one. As if he had been surreptitiously peaking into acting-artist Philip Gidley King's paintbox while he was away surveying Narborough Island with Mellersh, Darwin penned, "Their colors are, whole belly, front legs, head 'Saffron Y [yellow] & Dutch orange' —upper side of head nearly white. —Whole back behind the front legs, upper side of hind legs & whole tail 'Hyacinth R [red].' This in parts is duller, in others brighter passing into 'Tile R [red].'"[77] This color matching wasn't actually a mere whim on Darwin's part. The fancy names came from Patrick Syme's *Werner's nomenclature of colours*,[78] a book Darwin consulted whenever recording the hues of the organisms he observed and collected on the voyage.[79]

That day the crew of the *Beagle* caught 40 land iguanas. Darwin sampled their white flesh and wrote a cryptic opinion of it: "[B]y those whose stomachs rise above all prejudices, it is relished as very good food."[80] Whalers commonly took iguanas, and their eggs, for food during this period and it is surprising that land iguanas can still be found in the area today. While they are no longer hunted, their numbers are kept sadly diminished by feral cats (introduced after Darwin's time), which eat the young.

Darwin noted that the little "Muscivorous Lizards" (lava lizards) that were common on all the islands "perform the same gestures" as the land iguanas. Both reptiles defend their territories by "raising themselves as if in defiance on their front legs, [and] vertically shak[ing] their heads with a quick motion."[81] To Darwin's cost he did not notice that the lava lizards, like the mockingbirds, differ between the islands. From his specimen lists and the first edition of his *Journal* it is apparent that Darwin did suspect there were two species of lava lizard on Chatham and Charles. He did not, however, notice that they were separated by island. The males, females, and young of each species look very different from one another; only the females have bright orange (yellow or red on some islands) facial pigmentation, only the large males have black throats, only the young have dorsal stripes. In addition, young males can exceed adult females in length and Darwin may well have thought the different sexes were separate species. He apparently ceased collecting lava lizards after Charles Island, presumably when his attention was diverted to the land iguanas of Albemarle and James. Thus, he collected only two of the three species of lava lizard that he observed, and of the seven that make up the entire genus *Microlophus* in Galápagos.[82]

The lizards' importance to Darwin took a further step backward when Thomas Bell, the taxonomist who described Darwin's lizards, recognized only one species among Darwin's collections, explaining that while "Mr. Darwin obtained numerous specimens, one only . . . is fully adult."[83] This was an unfortunate conclusion, because from Darwin's specimen list it is apparent that he had at least two adults. There was a female (specimen #1281), clearly *Microlophus bivittatus* from Chatham because of its "belly cream-color, with band on each side of orange: patch of do. [same] beneath throat."[84] The other was a male from Charles (specimen #1296). Only the male was illustrated in the *Zoology of the Beagle*, and it became the type specimen for *Leiocephalus* (now *Microlophus*) *grayii*. Darwin was persuaded that there existed only one species of lava lizard in Galápagos.

A different herpetologist erred in describing the five specimens of snakes Darwin collected from Charles and James islands. While Darwin initially believed he had several species of endemic snakes,[85] Gabriel Bibron convinced him they were all identical to a single species found in Chile.[86] No one challenged Bibron's judgment for years. Then in 1860 Albrecht Günther looked at a specimen and disagreed. The Galápagos snake, he declared at a meeting of the Zoological Society, is peculiar to the islands. With his own lack of expertise, Darwin avoided taking sides and allowed that there might be two species of snakes in the islands: Bibron's Chilean *Psammophis temminckii* and Günther's endemic *Herpetodryas biseralis*.[87] Today four different snake species are recognized in Galápagos, all of which are endemic. The species Darwin collected from Charles Island is called the eastern Galápagos racer, *Alsophis biseralis biseralis*. It has since disappeared from Charles Island but can be found on the nearby islets of Champion and Gardner. The same species also occurs on Chatham.[88]

Two snake species are known from James Island: the striped Galápagos snake *Antillophis steindachneri* (which Darwin collected) and the central Galápagos racer *Alsophis biseralis dorsalis*.[89] Darwin missed collecting the banded Galápagos snake *Antillophis slevini* and the western Galápagos racer *Alsophis biseralis occidentalis*, both of which occur on Albemarle Island.[90] He also missed the Hood Racer, *Philodryas hoodensis*, because it exists only on Hood (Española), an island he did not visit.

If Darwin was unlucky with the lizards and snakes he was decidedly jinxed with the geckos. He completely missed collecting geckos in Galápagos despite looking in the right habitat (under rocks) while searching for beetles. There are six endemic leaf-toed gecko (*Phyllodactylus*) species in Galápagos. There exists only one endemic species per island, except on Chatham Island where there are two species. Like the lava lizards, some of the species are found on more than one island. Darwin could potentially have seen and collected four different species: *Phyllodactylus leei, P. tuberculosus, P. bauri,* and *P. galapagoensis*.

The fact that Darwin overlooked many of the reptile species in Galápagos is significant. Darwin's understanding of evolution arose from a progressive awareness that the Galápagos Islands do not have identical animal and plant species on each island, but rather, possess series of closely allied organisms that are often island specific. His best evidence among the animals of Galápagos were the mockingbirds, for he collected specimens from each of the four islands he visited and was able to distinguish three different species. Like the mockingbirds, the lava lizards, geckos, and snakes also differ between the islands, and had Darwin noticed this while in Galápagos, and collected accordingly, he may have reached an evolutionary standpoint while he was there. If nothing else he would have left the islands with more substantive evidence to support his argument for evolution.

Despite missing many species, Darwin considered the reptiles to be "the most striking feature in the zoology of these islands."[91] For, although the species, by his estimation, were not numerous, "the number of individuals of each kind [was] extraordinarily great." Just how great he had not yet fully appreciated, for on James Island he would encounter numbers of land iguanas and tortoises as never before. He would then write, "when we remember the well-beaten paths made by the many hundred great tortoises—the warrens of the terrestrial Amblyrhyncus [*sic*]—and the groups of the aquatic species basking on the coast-rocks— we must admit that there is no other quarter of the world, where this order replaces the herbivorous mammalia in so extraordinary a manner."[92]

If Darwin had left Charles Island homesick and melancholy, Albemarle had perked him up with just the right medicine—a hefty dose of geology. He had examined a fascinating tuff volcano that differed in many aspects from the tuff crater he had climbed on Chatham. He had gazed upon enormous black lava flows crossed by yet younger flows, one of which had dammed a cove and formed a lake. The landscape alone must have wiped clear his longing for England. It had clearly distracted him from the haunting zoology of the islands. If he saw a tortoise, and it is possible that he did, for whalers commonly caught tortoises in the vicinity of Tagus cove, he made no record of the fact. Superficially the subspecies (*Geochelone nigra microphyes*) resembles the Chatham Island tortoise (*Geochelone nigra chatamensis*) and would not have inspired Darwin to think any more of Lawson's assertion that the tortoises differ between islands. The land iguanas were the only animals to capture Darwin's attention, because they were new, and because it was impossible to ignore such colorful creatures. As for specimens, Darwin hardly bothered collecting. He came away with only about ten plants, four rocks, an insect or two, and a mockingbird.

It is tempting to think that when Darwin returned to the *Beagle* late in the afternoon, hot and dusty and his clothes ripped and stained with *Croton* sap, he dived into the water to cool off. The sun-heated, hypersaline, shallow lake at Beagle Crater would have offered no such relief and Tagus Cove is an irresistible swimming hole. Upon hitting the water Darwin would have let out a yawp. Here the water is often cold and that evening it was a chilling 62°F (16.6°C), several degrees colder than the temperatures that had recently been measured at Chatham and Charles. Darwin would one day send Alexander von Humboldt the Galápagos sea surface temperatures (96 in total) that were being recorded daily by the *Beagle*'s crew.[93] They were all unexpectedly low for the Equator, though few quite this low. The Humboldt Current, studied by and named after the great explorer, is responsible for bathing Galápagos with cold water from the southern latitudes, and for keeping the islands particularly arid. However, local upwelling from the Cromwell current (an undercurrent that flows in from the west) accounts for the especially cold temperatures to be found on the leeward shores of the western islands.

Captain FitzRoy believed the cold sea temperatures in Galápagos explained a phenomenon that Darwin had noticed and puzzled over; namely an apparent absence of coral reefs in the islands. "Capt Fitz Roy has suggested," Darwin acknowledged in his geology notes, "that the cold water, which occasionally is brought up from the under currents & reduces the temperature of the Sea, on these shores, to a degree, perhaps unparalleled in such a Latitude, may account for this absence of a tribe of Animals, which seems only to flourish where the heat is intense.— How far this ingenious idea may be of general application will require extended observation."[94]

Darwin also wondered whether the archipelago might have an unusual deficiency of calcareous matter,[95] something Lyell claimed was necessary for coral reef development and abundant on volcanic isles.[96] Later, in his coral reefs book, Darwin contended there was little foundation for the idea that lime evolving from volcanic islands is necessary for coral development because many coral reefs, like those in Australia, are not found near volcanic land.[97]

Contrary to Darwin and FitzRoy's conclusions, reef-building (hermatypic) corals are present in Galápagos. As they depend on photosynthesizing zooxanthellae, a symbiotic plant, they grow primarily in shallow, protected areas where the water is heated by the sun to several degrees above ambient temperature. Although the corals are found scattered throughout the archipelago, it is not surprising that Darwin and FitzRoy missed seeing them, for the highest densities of the predominant genera (*Pocillopora, Pavona,* and *Porites*) occur in places not visited by the *Beagle*: the east side of Albemarle, the east side of Bindloe (Marchena), the north side of some of the central islands, and the far northern islands of Wenman (Wolf) and Culpepper (Darwin).[98]

A swim is always an invigorating way of ending an exhausting day hiking in the hot sun, and if Darwin took a plunge he would have emerged rejuvenated. Baths of frigid water would do more than just refresh Darwin in later years—they would revitalize him. A few years after his return to England Darwin developed such a debilitating stomach disorder that one of the few effective palliatives for his violent bouts of vomiting, was Dr. James Gully's water cure, a circulation-stimulating treatment involving heating the patient and then dousing him with ice water.

Knowing how much the physical exertion of exploring terra firma agreed with Darwin, FitzRoy struck a bargain with his landlubber friend.[99] If he should fail to find sufficient fresh water to fill the ship at James Island (the next stop en route) he would allow Darwin to remain there while he took the *Beagle* back to Chatham Island and the gushing waterfalls of Fresh Water Bay. It was "an admirable plan for the collections"[100] and a likely scenario, for Darwin had read Colnett's account of the poor quantities of water to be found on James. He also knew that the two other freshwater sources marked on Colnett's map (on Chatham Island) had both proven to be dry. On the other hand, he was comforted by the knowledge that there would surely be enough water on the island for his own daily needs, as acting Governor Nicholas Lawson had a small party of tortoise hunters living there. Thus buoyed by the prospect of a week of uninterrupted exploration, to counteract "the jading feelings of constant hurry,"[101] Darwin must have looked forward to the upcoming passage to James. After all, the past week of sailing had been surprisingly calm, the bulk of Albemarle having shielded the ship from the strong winds that whip the southern seas of the archipelago during the dry season.

Nor would he mind a prolonged look at the raw landscape along the rest of this 135-kilometer-long island. Nowhere, he marveled, had he seen such a vast expanse of land "so entirely useless to man or the larger animals."[102] So desolate and lonely... His mind wandered back to the little archipelago in the lake of Beagle Crater that he had gazed upon earlier that day. How removed from the rest of the world it had seemed. How geographically isolated. That fact seemed somehow the essence of life in Galápagos . . .

The *Beagle* sailed from Tagus Cove at 10 a.m. on October 2. It took a full week to reach James Island, a trip that is nowadays motored overnight. After being becalmed for a whole day between Narborough and Albemarle islands, much of the way "was most unpleasant passed in struggling to get about 50 miles [80 kilometers] to Windward against a strong current."[103] The passage also included a long, intentional detour past the northern islands of Bindloe (Marchena) and Tower (Genovesa). Despite the discomforts presented by this long-drawn-out route, Darwin was at least able to get a close look at the northern islands. By night they were especially beautiful, bathed as they were in the silver light of a full moon.

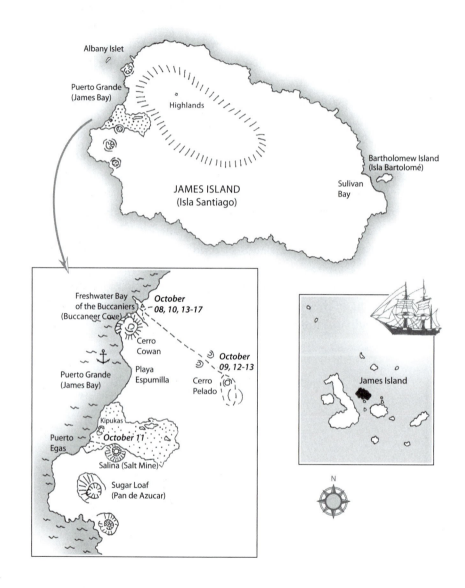

Albany Islet

Puerto Grande
(James Bay)

Highlands

Bartholomew Island
(Isla Bartolomé)

Sulivan
Bay

JAMES ISLAND
(Isla Santiago)

Freshwater Bay
of the Buccaniers
(Buccaneer Cove)

*October
08, 10, 13-17*

Cerro
Cowan

*October
09, 12-13*

Puerto Grande
(James Bay)

Playa
Espumilla

Cerro
Pelado

Kipukas

Puerto
Egas

October 11

Salina (Salt Mine)

Sugar Loaf
(Pan de Azucar)

James Island

N

Chapter VII

James Island
(Isla Santiago)

At every place where we landed on the Western side, we might have
walked for miles through long grass and beneath groves of trees. It
only wanted a stream to compose a very charming landscape.
—James Colnett, *A voyage to the South Atlantic and round Cape Horn*[1]

The Sky for a Roof, the Ground for a Table

When, on Thursday October 8, the *Beagle* arrived at the northern end
of James Island, it was immediately hailed by some men in a small
fishing skiff. The fishermen belonged to a group of 22 Charles Island
settlers sent by Nicholas Lawson, on Governor Villamil's orders, to
tap the natural resources of James Island. Their main "employment
was salting fish [and tortoise meat] and extracting oil from terrapin."[2]
They sent these products back to Charles Island and the mainland,
or sold them to passing ships. The men and women that comprised
the colony answered to a "Mayór-Domo,"[3] presumably the same person who had led the colonists to the island four months earlier—José
Maria Troncoso.[4] So starved were the people for anything other than
the fish, tortoise meat, and iguana eggs they had been living off since
June, "that half a bag of biscuit (50 lbs.) which [the *Beagle* men] gave
them, appeared to be an inestimable treasure, for which they could
not sufficiently thank [them]."[5]

It was a busy day. Darwin and Covington, the surgeon/naturalist Benjamin Bynoe, and FitzRoy's steward Harry Fuller were landed at "Freshwater Cove of the Buccaniers [sic]"[6] (Buccaneer Cove) with a few essential supplies. They would spend the next nine days bivouacked there, "living in the open air, with the sky for a roof, and the ground for a table."[7] Lieutenant Sulivan and his party returned on board from having spent the previous three weeks surveying the central islands, including most of James Island. Chaffers was then dispatched in the yawl with seven sailors to survey the northern islands of Abingdon (Pinta), Bindloe (Marchena), and Tower (Genovesa). At last the Beagle set sail for the southern islands, for by now they were in urgent need of water and FitzRoy had come to realize that "water for shipping can only be procured at the south side of Chatham Island."[8]

For their own needs Darwin and his party employed the local fishermen to collect water from a "miserable little spring"[9] in the next valley. This water source has been known since the 17th century, when pirates William Dampier and William Ambrose Cowley used it for their needs. It lies on the western side of Cerro Cowan, where a deep rain-sculptured ravine opens into the sea (00°10′53.6″S, 90°50′10.6″W). Groundwater seeps from the walls of the ravine in several places, and in one spot, just above sea level, collects in a "little well,"[10] a narrow, meter-long, oblong catchment at the base of a cliff. Roughly 15 meters up the ravine from this "well" the authors discovered the names of a ship (SHIP HALARD), a man (CAMERON HATHAWSON), and a date (APRIL 4 1804) etched in the tuff wall of a shallow cave. The engravings were most likely the work of a whaler or sealer left on shore to collect water for his ship, or perhaps for his own survival. The writing represents some of the oldest surviving (if not the oldest) legible engraving in Galápagos.[11] It is also remarkable in being unmarred by the modern painted graffiti that blemishes much of the mid to late 19th-century engravings found at Tagus Cove on Albemarle Island.

Upon arrival at Buccaneer Cove the first goal was to set up camp (00°10′02.1″S, 90°49′28.1″W). It was not an easy task. So numerous were the land iguanas that Darwin and his party "could not for some time find a spot free from their burrows, on which to pitch [their] tent."[12] Darwin found that the "lizard *warrens*" were an "annoyance [for] the tired walker"[13] because "the soil perpetually [gave] way"[14] under foot. The iguanas, however, were highly entertaining. Seeing that they were "very fond of cactus,"[15] Darwin cut up bits of *Opuntia* pads, tossed the pieces to a pair of iguanas and watched the ensuing fight. Both iguanas tried "to seize [each piece] & carry it away as dogs do with a bone."[16] The winner then ate "very deliberately, without chewing"[17] while a "small finch picking from [the] same piece"[18] hopped "with complete indifference on its back."[19] It was remarkable how the lizards could bite right through the dense spines that covered the pads. Although they attempted to remove the worst of the spines by scraping the pads with their feet, most of the bristles remained. Yet barely a quill caught in their tough, flaking skin, let alone pierced the tongue. Darwin, on the other hand, surely paid for hand feeding the iguanas. As anyone who has done field research in Galápagos knows, it is not the avoidable needle-like projections of the *Opuntias* that are so troublesome, but the hair-thin yellow glochids that stick quickly, invisibly, and irritably. The frustration of locating and then tweezing out tiny painful cactus spines from skin and clothing can linger for days.

As with the marine iguanas of Chatham Island, Darwin could not resist catching the land iguanas by their tails. On one occasion Darwin watched a female digging its burrow, "till half its body was buried." "I then walked up and pulled it by the tail," he recounted in his *Journal*. "[A]t this it was greatly astonished, and soon shuffled up to see what was the matter; and then stared me in the face, as much as to say, 'What made you pull my tail?'"[20]

A behavior that Darwin found even more comical was their territorial posturing. When confronting an intruder the iguanas would "curl their tails, & raising themselves as if in defiance on their front legs, vertically shake their heads with a quick motion." Darwin acknowledged, "This gives them rather a fierce aspect, but in truth they are far the contrary."[21] Darwin had merely to stamp the ground and, "down go their tails, and off they shuffle as quickly as they can."[22]

Land iguanas have long since disappeared from James Island. Darwin collected the only scientific specimen known, and because he did not label it by locality there is some question as to whether it came from James Island at all. Its specimen number (#1315) is second to last in a list of Galápagos organisms that were bottled in "Spirits of Wine" aboard the *Beagle*, suggesting it was taken right at the end of Darwin's stay in Galápagos, on James Island. However, there remains the possibility that it was collected on Albemarle Island, where Darwin first encountered land iguanas, and then was labeled later.[23] Darwin was not the first Galápagos visitor to record seeing land iguanas on James Island—James Colnett (1793–94) and David Porter (1813) preceded him—but as far as it is known he was the last. When scientist Louis Agassiz visited the island in 1872 he could not find a single individual in James Bay.[24]

It is believed that land iguanas disappeared from James Island due to a combination of human-related factors. Population numbers probably declined rapidly after Darwin's visit, as Charles Island settlers continued to use James Bay as a base, and whalers increasingly hunted the island for tortoises. The iguanas were prized as food—one early visitor extolled them as "superior in excellence to the squirrel or rabbit"[25]—and their "numerous large elongated eggs"[26] were also sought by "the inhabitants . . . to eat."[27] Coincidentally, Darwin visited James at a choice time for the settlers to harvest iguana eggs— the onset of the animal's nesting period. The females were excavating burrows and several that Darwin dissected on the spot were found to be gravid and ready to lay. Human consumption of adult iguanas and their eggs must have quickly diminished the population, but pigs and goats, introduced after Darwin, undoubtedly caused the iguanas' permanent demise. Pigs are known to root up iguana nests and eat the eggs, and goats compete with adult iguanas for food. Fortunately, land iguanas still exist on several other islands in Galápagos. *Conolophus subcristatus* is found on Indefatigable (Santa Cruz), South Plaza, South Seymour (Baltra), Albemarle (Isabela), and Narborough (Fernandina) islands. It was also introduced to North Seymour and Venecia islands. A separate species of land iguana (*Conolophus pallidus*) occurs only on Barrington Island (Santa Fé). A third, genetically distinct (but as of yet unnamed) species has recently been discovered on the summit of Wolf Volcano, Albemarle Island. A striking difference between it and the other two species is its pink coloration.[28]

17B

17A

17C

17A–17C. During the 19th century, large numbers of Galápagos tortoises (*Geochelone nigra*) were killed for food by the crews of whaling ships and other vessels. The *Beagle* took close to 50 individuals for consumption from Chatham Island in 1835.

18A

18B

18A and 18B. Kicker Rock, off Chatham Island.

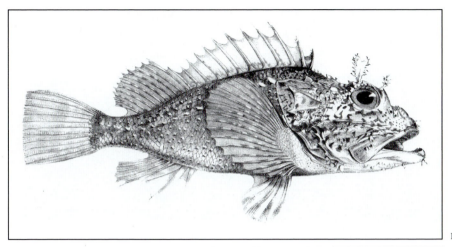

19A

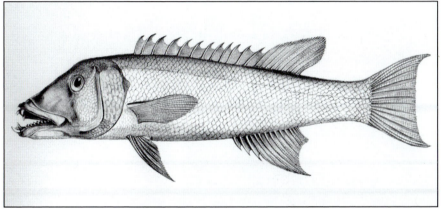

19B

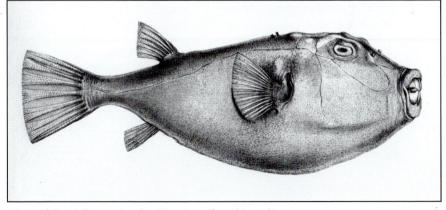

19C

Three of the 15 fish species that Darwin collected in Galápagos.
19A. *"Scorpæna Histrio"* (bandfin scorpionfish, *Scorpaena histrio*).
19B. *"Cossyphus Darwini"* (Pacific red sheephead, *Semicossyphus darwini*).
19C. *"Tetrodon Angusticeps"* (concave puffer, *Sphoeroides angusticeps*).

20A

20B

20C

20D

20A–20C. Darwin likened the Craterized District of Chatham Island (20A) to these illustrations of the Chaine des Puys of Central France (20B), and the Phlegræan fields of Italy
(20C).

20D. Finger Hill (Cerro Brujo), with its basalt core exposed to the Southeast, viewed from the
Craterized District of Chatham Island.

Chatham Basalt. which on its margins. thins out &
covers the Sandstone. — The Basalt in central
3235 part of stream is very compact. blackish grey: contains
cysts of red Olivine(?) . — The upper & inferior
surfaces are cellular to some depth. —
This Basalt must have existed as a pool of liquid
matter within the Basin of the Crater. —
The Kicker rock lies a few miles out at
sea. from this point. — It is a most
singular form. — a flat topped mass. is
surrounded by absolutely perpendicular cliffs. & which
from the depth of water. must be continued
beneath the sea. — On one side is an
equally abrupt spire. — Whole height is 400 ft
the whole consists of a Volcanic Sandstone
____ of the last described. — I can

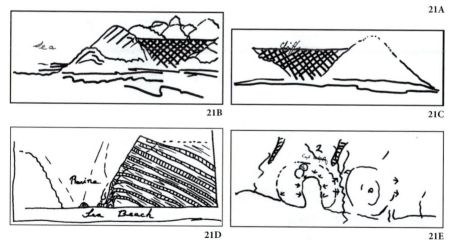

21A. Excerpts from Darwin's notes on the geology of Galápagos.
21B and 21C. Finger Hill (Cerro Brujo), Chatham Island.
21D. "Darwin's layer cake" in Freshwater Cove of the Buccaniers (Buccaneer Cove), James Island. A rendition of the mirror image of this sketch appears in Darwin's 1844 "Geological observations on volcanic islands."
21E. Map of Tagus Cove, Albemarle Island.
Darwin's sketches redrawn by K. Thalia Grant.

22A

22B

22C

22A–22C. The highlands of Charles Island, looking toward the Governor's dripstone. The original illustration (22A) was made by *Beagle* midshipman Philip Gidley King. The more commonly reproduced rendition (22B) by E. de Bérard misrepresents the scene with savages and spears. Leaving Charles Island (22C).

23A

23B

23C

23D   23E   23F

The three species of mockingbird that Darwin observed in Galápagos, in order of their collection.
23A. Chatham Island mockingbird (*Mimus melanotis*).
23B. Charles Island mockingbird (*Mimus trifasciatus*).
23C. Galápagos mockingbird (*Mimus parvulus*).
Radiographs of **Bulimulus** *snail shells from Galápagos.*
23D. *B. darwini* from James Island.
23E. *B. nux* from Chatham Island.
23F. *B. planospira* from Champion Islet off Charles Island.

24A

24B

24A. Whaling ships off Point Christopher on the Southwestern coast of Albemarle Island. The *Beagle* passed by this area on September 29, 1835.

24B. Whaling was rampant in Galápagos during the first half of the 19th century. Countless sperm whales and other cetaceans were slaughtered for their oil.

25A

25B

25C

25A. Darwin observed large numbers of land iguanas (*Conolophus subcristatus*) on James
Island, where they are no longer found, and on Albemarle Island.
25B and 25C. Marine iguanas (*Amblyrhynchus cristatus*) were numerous on all the islands.

26A

26B

26A and 26B. Darwin observed Galápagos tortoises (*Geochelone nigra*) on Chatham and James Islands.

27A

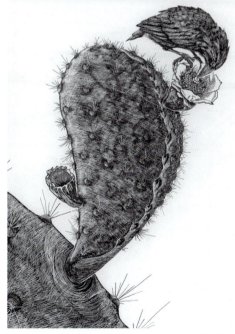

27B

27A. A male lava lizard (*Microlophus grayi*), yellow warbler (*Dendroica petechia aureola*), and cactus finch (*Geospiza scandens*). Darwin collected specimens of each of these species.

27B. Cactus finch (*Geospiza scandens*) feeding on the flower of a prickly pear cactus (*Opuntia echios*).

28A

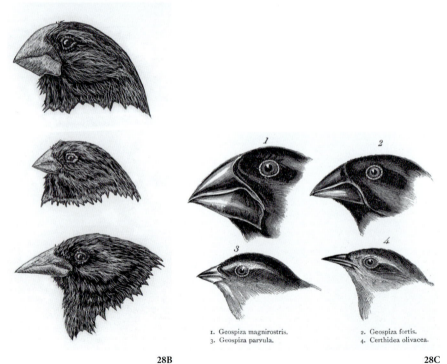

1. Geospiza magnirostris.
2. Geospiza fortis.
3. Geospiza parvula.
4. Certhidea olivacea.

28B   28C

28A. On October 17, 1835, Darwin returned to HMS *Beagle* after ten days on shore at James Island. The *Beagle* sailed from the archipelago three days later.

28B and 28C. When the *Beagle* returned home to England, ornithologist John Gould examined Darwin's bird specimens from Galápagos and described 13 species of a new family of ground finches.

29A. After the voyage, Darwin frequently consulted the ornithological collections of the British Museum, London.

29B. Darwin came up with natural selection as an explanation for evolution, in part by studying artificial selection in domesticated animals. Later in life he took up pigeon breeding, to further study variation and selection.

30A

30C

30B

30D

30A–30D. Down House (30A), where Darwin lived and worked from 1842 until his death, at the age of 73, in April 1882. Experimenting in his greenhouse (30B). His study (30C). His favorite walk (30D).

31A
31B

31A and 31B. Alfred Russel Wallace, pictured here in 1869, hit upon the concept of natural selection independently from Darwin, while on a collecting expedition through Indonesia and Malaysia between 1854 and 1862.

32. Charles Robert Darwin, MA, FRS (1809–1882)

A Highland Marathon

Remembering the "luxuriance"[29] of the highlands of Charles Island, Darwin was keen to explore the even more green and lofty interior of James Island. On Friday, October 9, the fishermen guided Darwin to "a couple of hovels"[30] in the highlands where two[31] of their companions were "employed in hunting the Tortoise."[32] The hovels were located at about 610 meters (2000 feet) elevation where "the country begins to show a green color."[33] This was just over two-thirds the distance from the coast to the summit.

The walk "was a long one,"[34] most of it gradually ascending the shimmering heat of the lowlands and winding through woods of massive Palo Santo trees (*Bursera graveolens*), some over two meters in circumference and far taller than the stunted variety that grew on Chatham. "Mimosa" trees (*Acacia* and *Prosopis*) were also common, mercifully so. Darwin quickly found that "the shade from [their] foliage was very refreshing, after being exposed in the open wood to the burning Sun."[35] The land iguanas were equally attracted to these trees, but for a different reason. Darwin found many "clinging to the branches of the Mimosa"[36] and "quietly browsing"[37] on the leaves. They were seemingly as oblivious to the thorns protecting these trees as they were to *Opuntia* spines.

As the elevation increased, the giant Palo Santos gave way to smooth-barked, curvy-limbed, fruit-pungent "Guayavita" (*Psidium galapageium*) trees, which soon dominated the landscape.[38] When Darwin saw (and smelled) this graceful and strongly aromatic tree, he stopped in his tracks for he could not recall having seen it on Charles Island. It seemed odd that a tree pervading the mid-elevation of this island should be completely absent from another island. After voicing this observation to the tortoise hunters they replied, "Many of the islands possess trees and plants which do not occur on the others."[39] For Darwin's collecting purposes this information came too late, for James was the last island he visited in Galápagos. "Unfortunately, I was not aware of these facts till my collection was nearly completed," he wrote. "[I]t never occurred to me, that the productions of islands only a few miles apart, and placed under the same physical conditions, would be dissimilar."[40] Later Darwin downplayed the importance of "the simple fact of one isl^d having a species & another isl^d not having it." It was the fact that the islands had "close *representative* species of the same genus" that was so significant.[41] Nonetheless, recognition that there were differences at all in the habitat composition of the different islands was an important step in Darwin's thinking, and one that was certainly brought home on James Island.

The crooked branches of the guayavita trees (today called guayabillo) were heavily laden with plump yellow berries and thickly draped in "a pale green filamentous Lichen, which hangs like presses from the boughs of the trees."[42] Darwin watched land iguanas and tortoises alike pulling great mouthfuls of berries from the drooping trees. Seeing how they apparently relished the fruit Darwin popped a fat berry into his mouth—and promptly spat it out. Such an "acid & Austere,"[43] "astringent and turpentinic"[44] taste was clearly fit only for reptiles, and the hardy settlers.[45] Perhaps the men and women made a jam from the berries, for as such they are very palatable. The tortoises' diet of berries cloaked in lichen certainly helped impart a favorable flavor to their meat. When Darwin returned for an overnight trip to the highlands on October 12, he dined exclusively on tortoise meat, finding that "the breastplate roasted (as the Gauchos do *carne con cuero*), with the flesh on it, is very good; and the young tortoises make excellent soup."[46] This was far better praise than when he had first tasted tortoise meat at Chatham Island.

Darwin collected three different species of lichens from the highlands of James.[47] Botanist Joseph Hooker identified the beard-like species as "Usnea plicata, the most common lichen decidedly in the world."[48] However, the lichen draping the guayabillo trees at mid-elevation and eaten by the tortoises there was more likely the common, look-alike species *Ramalina usnea*.[49] Darwin was impressed by the fact that lichens seemed to form a "considerable portion" of the diet of the tortoises in the upper damp regions of the island.[50] Hooker agreed it was curious, and responded that as far as he knew the tortoise was "the only animal" in the world to eat the "Usnea."[51]

There was no mention of land iguanas feeding on lichen in this discussion, or for that matter in any of Darwin's notes. But 25 years later (in 1860) Darwin remembered them doing so. He used this recollection to support his supposition that the marine iguana and land iguana of Galápagos had evolved from a common terrestrial ancestor. In a letter to Charles Lyell, Darwin wrote:

> With respect to the Amblyrhynchus of Galapagos; one may infer as probable, from marine habit being so rare with Saurians & from the terrestrial species being confined to a few central islets, that its progenitor first arrived at the Galapagos; from what country it is impossible to say as its affinity (I believe) is not very close to any known species. The offspring of the terrestrial species was probably rendered marine. Now in this case I do not pretend I can show variation in habits; but we have in the terrestrial species a vegetable feeder (in itself a rather unusual circumstance) [feeding] largely on lichens [(composite organisms consisting of algae and fungi)], & it would not be a great change for its offspring to feed first on littoral algae & then on submarine algae.[52]

Lichens are not recognized as a food source for land iguanas (or tortoises) today. Darwin most likely observed individual land iguanas ingesting *Ramalina usnea* while they were feeding on guayabillo berries at mid-elevation. He may have observed iguanas feeding on a different type of lichen closer to the coast, or perhaps muddled his memories of tortoise and land iguana feeding behavior. The three species of lichens Darwin described in Galápagos[53] (*Usnea dasypoga* var. *plicata, Pseudocyphellaria aurata,* and *Anaptychia leucomelaena*[54]) are all highland forest species, growing in damp areas frequented by tortoises, but perhaps not by the land iguanas. At least, Darwin did not report seeing land iguanas at the top of the island. Unfortunately, there is no way of knowing if lichens comprised a substantial portion of the land iguanas' diet because the James Island population is now extinct.

Whether or not Darwin observed land iguanas intentionally eating lichen in Galápagos, the idea of a lichen link in the evolution of the two iguanas is certainly food for thought. Modern genetics tells us that the marine iguana and the land iguana arrived in Galápagos between 8 and 15 million years ago, on now-submerged islands, and diverged over 10 million years ago.[55] Whether the split occurred on the mainland or after the colonizing ancestor arrived in Galápagos is not clear from these dates. Darwin's idea that they diverged in the islands is the favored hypothesis for its simplicity; only one colonization event is required. It is also supported by the fact that the gut bacteria of the land iguana is more closely related to that of the marine iguana (which feeds on marine algae) than it is to the intestinal flora of the Galápagos tortoise (which has a similar diet to the land iguana) or even to that of the spiny-tailed iguanas (*Ctenosaura*) of Central America, which are the iguanas closest living relatives.[56]

Upon reaching the hovels, Darwin and his companions hiked "an additional 1000 ft [305 meters] elevation"[57] to the top of the island. They followed wide compressed paths—the tracks of giant tortoises— up an ever-steepening slope. The guayabillos gave way to a lush forest of epiphyte-laden *Scalesia* (*Scalesia pedunculata*) and cat's claw (*Zanthoxylum fagara*) trees, which dripped moisture onto a thick undergrowth of sedges and ferns. The damp air became heavily spiced with a new "aromatic smell"—the peppery fragrance of a "succulent, clinging" epiphyte today known as *Peperomia galioides*.[58] Darwin wrote in his diary:

> During the greater part of each day clouds hang over the highest land: the vapor condensed by the trees drips down like rain. Hence we have a brightly green & damp Vegetation & muddy soil. –The contrast to the sight & sensation of the body is very doubtful after the glaring dry country beneath. So great a change with so small a one of elevation cannot fail to be striking.[59]

As they neared the summit, misty clouds swept the ground. The moisture, trapped by the surrounding forest, maintained "large beds of a coarse cyperus [a sedge], in which great numbers of a very small water-rail lived and bred."[60] Though shy and elusive, these endemic scarlet-eyed Galápagos rails (*Laterallus spilonotus*) announced their presence by "uttering loud & peculiar Crys [*sic*]"[61] from the undergrowth. Darwin had first spied "this small Water Hen"[62] in the highlands of Charles Island, and noted that it appeared to be the only land bird "confined to the damp region."[63] Sadly the Galápagos rail was last seen on Charles Island in 1983, but it still survives on James and on Albemarle, Narborough, and Indefatigable Islands.

The top of the island, a region today called "Jaboncillos," was exceedingly damp. Darwin and his companions found "Springs" trickling into ravines and, on the far side of the central ridge, an expanse of muddy pools (00°12′41.9″S, 90°46′55.3″W). The pools covered the bottom of a wide bowl-shaped depression that has been loosely termed the "caldera" of the island.[64] Swarming the vicinity of these watering holes were the giant tortoises whose tracks the men had followed. On October 12 Darwin would return to the highlands with a "blanket bag to sleep in"[65] and spend a full day observing the tortoises, sweeping for insects, and collecting the products of this lush wonderland. On October 9, however, he was pressed for time, for he had to return to the coast by nightfall. In addition, the geologist in him was distracted by a "large & perfect Crater" whose "channels, by which the Lava has flowed over the rim [were] yet visible."[66] The floor of the crater was "wooded," the "very precipitous" interior slopes were "chiefly composed of bright red & very glassy scoria," and on the outside "nothing, but Trachytic Lava [was] found."[67]

In 1996 the authors, in their efforts to determine where Darwin had explored on James Island, succeeded in identifying this crater, a key destination in Darwin's route through the highlands. The highlands are literally peppered with "old broken down," well-vegetated cones and pit craters, but there is only one crater near the summit of James Island that matches Darwin's descriptions. Indeed, it is a noticeably distinct crater and, for a geologist, simply impossible to ignore. This is Cerro Pelado, literally translated as Peeled Hill, for its streams of raw, slag-like trachytic basalt lying waste to the immediate surroundings (00°12′22.17″S, 90°47′03.10″W). Among the craters in the highlands it is unique in having the combination of precipitous walls of red scoria, a well-wooded interior, and "channels" of gray trachytic lava that breach the rim on the northern and southern sides. However, due to inconsistencies in Darwin's notes, the identity of this crater, and even the location of the tortoise pools and springs, have been difficult to establish. Darwin placed both the crater and the springs near the top of the island, "in the central highest part"[68] of the island, close to 3000 feet (just over 900 meters) elevation.[69] But he also put them "about 8 miles [13 kilometers] inland"[70] from Buccaneer Cove,[71] a distance that goes well beyond the summit (which is about 4 miles or 6.5 kilometers from the coast), to the considerably lower region of "La Tragica," lying at about 2000 feet (610 meters).

As tortoises and tortoise pools are found in this wet lower region today it is tempting to believe that Darwin walked this far. However, Darwin made his hike in one day, with a stop at the "hovels" and time for collecting and observing along the way. Eight miles inland plus another eight miles back is simply not realistic for such a one-day hike. An eight-mile round-trip hike from Buccaneer Cove to Jaboncillos, on the other hand, is still a very "long" hike but a far more plausible one. The discrepancy can be reconciled by the fact that Darwin tended to exaggerate distances inland. For instance, on Chatham Island he doubled the stretch of land between the coast and Pan de Azucar, and on James Island he added more than a third to the length of the James Bay lava flow.

Furthermore, Darwin made a point of noting that there were no tree ferns ("nor palms") in Galápagos,[72] yet 6-meter-tall Galápagos tree ferns (*Cyathea weatherbyana*) can be found growing on the sides of craters and sinkholes midway between Jaboncillos and La Tragica. The fact that he did not see any tree ferns in Galápagos makes it unlikely that Darwin explored this far.

Less easy to explain is Darwin's approximation of the diameter of the crater. Cerro Pelado is smaller (0.2 kilometers across) than Darwin's estimates of first one-quarter of a mile (0.4 kilometers) and then "⅓ of [a] mile [0.5 kilometers] diameter."[73] However, the discrepancy can be accounted for by the apparent fact that Darwin did not measure the crater while he was there, but recorded approximations of its diameter, from memory, at a later date. He wrote a qualitative description of the crater in his geological notes at the end of a section dated October 10 (that is, the day after he viewed it), but only recorded measurements in his field notebook, in a section dated October 16 (when he was at Buccaneer Cove, awaiting the return of the *Beagle*). Between these dates he would have gained a different perspective of Cerro Pelado. From the northern rim of the lowland salt mine crater, that Darwin visited on October 11, Cerro Pelado appears like a distant jagged silver crown in the green hills that form the island's summit. Significantly, from this angle the crater appears to be of equal width to the salt mine crater, which Darwin correctly recorded as being one-third of a mile (half a kilometer) across.[74] This may help explain Darwin's memory of Cerro Pelado being the same size as the salt mine crater.

Perhaps the most concrete evidence in support of Cerro Pelado being the crater in question is the following clue. Just 100 meters east of Cerro Pelado, and equidistant to the "caldera" of the island, (which the authors believe was the site of the tortoise pools that Darwin described) is a rare outcrop of soda trachyte. Darwin collected a sample of this type of rock in the highlands of James; it was the last of four rocks he collected in a southeast line from Buccaneer Cove. However, the sample and sample location have since been the subject of a debate, best described by historian Sandra Herbert in her book *Charles Darwin: Geologist.*[75] Briefly, the facts are as follows. During a survey of James Island in 1969, geologists Alexander McBirney and Howel Williams failed to find any soda trachyte on the island. They contended that the rock must have been a mislabeled specimen taken from elsewhere in the world and erroneously added to Darwin's Galápagos collection. In July 2007 the authors teamed up with Sandra Herbert and a small team of geologists from the University of Cambridge[76] and the University of Idaho[77] to pinpoint where Darwin collected his igneous rock specimens from James Island.[78] In the process it was discovered that another geologist, Hartmut Wolfgang Baitis, had collected rocks throughout the highlands of James Island in the 1970s and had found soda trachyte in one specific area. The group's rediscovery of this protrusion of soda trachyte located next to Cerro Pelado, reinforces the authors' construal of Darwin's movements in the highlands of James, and the identity of his "perfect" crater.[79]

The base of Cerro Pelado on its western side was clearly used as a bivouac site for tortoise hunters, perhaps not when Darwin was there, but certainly later in the 19th century. Until recently a stand of half a dozen jaboncillo trees (soapberry trees, *Sapindus saponaria*) grew in this place (00°12'17.1"S, 90°47'05.0"W). The trees gave the summit region its modern name, but they were presumed to have been introduced and so were cut down by Park wardens at the beginning of the 21st century. In 1996 the authors found the area under and surrounding the jaboncillo trees littered with historical debris. Goat grazing had denuded the area of its ground cover and uncovered pieces of broken cast iron cooking pots, earthenware gin flasks, porcelain crockery, clay pipes, and black glass bottles. A research trip to the Museum of London with photographs of the artifacts revealed that they heralded from Britain, Europe, and North America and dated from the mid 1800s to the early 1900s. It is easy to see why this strategically located area would have been chosen as a tortoise hunting camp; in the steep terrain it is relatively flat, it commands a clear view of the northern and southeastern coastlines, there is a freshwater spring just 500 meters to the south, and opposite this spring (00°12'36.6"S, 90°47'06.0"W), on the far side of the ridge that forms the crest of the island, is the shallow "caldera," which most likely contained the tortoise ponds Darwin described.

Today Jaboncillos is relatively dry compared to the rest of the highlands, and the hard-floored "caldera" pools with water only after heavy rainfall. Goats have removed the scalesia trees (*Scalesia pedunctulata*) and cat's claw trees (*Zanthoxylum fagara*), which once kept the area damp with cloud condensation, and the area is now mostly pastureland, reminiscent of the sheep-grazed fells of Britain. How do we even know that the area was once wooded? Aside from Darwin's descriptions, botanist Alban Stewart from the California Academy of Sciences visited James Island in 1905 and 1906 and described the interior of the island as covered by a dense forest, dominated by scalesias.[80] In addition, patches of native forest remain in the area today, having been fenced off in the 1970s by scientists wanting to study the effect goats were having on the vegetation, and to measure rates of reforestation.[81] Outside these quadrats a handful of thorny cat's claw trees stand sentry in the pastures. They look like garden fountain statues, catching wispy clouds in what is left of their foliage and the epiphytic liverworts they support, and weeping onto lonely mats of moss.

The good news is that an eradication campaign, initiated at the turn of the 21st century, has successfully removed all the goats, pigs, and donkeys that plagued the island since their introduction in the 1800s. There is now great hope that with control of the introduced plants on the island, the native vegetation can recover. With a little luck the native "fog drip" trees will once again rule the highlands of James and the "caldera" will revert to a muddy pool frequented by large numbers of tortoises. In 2007 the authors found that passive reforestation was already taking place; the cat's claws were regenerating readily and a few scalesia trees had started to grow outside one of the quadrats. Like a prophecy of the hoped-for changes to come, a tortoise was even spotted in the "caldera" on two separate occasions.

Igneous Ingenuity

Undoubtedly tired from his long hike to the highlands and back, Darwin chose to stick around camp on October 10. Besides, he was eager to examine Buccaneer Cove's enticing geological formations, and to write up his geological observations from the day before. Over the next few days Darwin filled 11 manuscript pages with geology notes focused primarily on Buccaneer Cove, and the adjacent bay of Puerto Grande (James Bay). He also jotted geological observations into his small field notebook, and dated them October 12 to 17. When back on board the *Beagle* and "totally rewriting"[82] his geology notes from Galápagos on the way home to England, Darwin expanded his James Island observations to fill 14 additional manuscript pages.

Darwin wrote his notes on the geology of Chatham, Charles, and Albemarle islands on board the *Beagle*, most likely in the evenings or while traveling between sites and islands. The sentences are generally complete and well phrased, unlike the brief jottings that comprise much of his field notebook. This suggests he had dedicated time for writing, something he would not have had on land, where every second was taken up observing, collecting, and hiking. Only at James Island, where Darwin was camped for ten days, could he write at length on land. It is clear that Darwin wrote the first section of his James Island geology notes on site, rather than waiting until he was back on board the *Beagle*, from the way he began it. With an active voice in the present tense he penned, "Freshwater Bay, where we are bivoaced [*sic*] is formed on its Western side by [a] Large circular hill with [a] Crater on top from which the Strata dip in folds on all sides."[83]

The hill Darwin described in this sentence is today called Cerro Cowan. It is a massive (270 m) eroded tuff cone, the outer skirt of which resembles a giant, moth-eaten tapestry of undulating ash layers studded with numerous lumps of a darker, harder palagonite. Darwin was intrigued by the fact that some layers were composed entirely of these angular fragments. He was also perplexed by the fact that Cerro Cowan appeared to consist of separate hills joined together, "as if the Sandstone was one envelope."[84] He queried whether the tuff substrate had been produced through a series of "mud eruptions," as at Beagle Crater on Albemarle Island, or by some other means. Perhaps quantities of ash had fallen and then become "consolidated by subsequent torrents of rain."[85] Both explanations would account for the blanket of tuff that covers the lavas along the northwestern coast of James. It was the presence of "longitudinal arches" and a pisolitic structure in some of the tuff on Cerro Cowan that persuaded him to accept the first argument.

Remarkable though Cerro Cowan was as yet another variety of tuff cone, Darwin was more inspired by the "trachytic" rocks he encountered inland, "beyond the influence of the [coastal] sandstone craters."[86] During his walk from Buccaneer Cove to the highlands the day before, Darwin had noticed a general change in the rocks as he rose in elevation. The lavas of the lowlands were "very cellular" and abounding with "Crystals of glassy Feldspar very large & abundant." In the highlands the lava was "more compact, the base Blackish grey with scarcely any Crystals, or they [were] abundant & small, the basis itself being Crystalline."[87] These observations, the four rock specimens he collected along his route (a southeast line from Buccaneer Cove to the summit of the island[88]), and additional rocks he examined at Buccaneer Cove inspired some of his most ingenious thoughts on igneous rock formation.[89]

Lava crystals were, at that time, believed to be preexisting fragments of granite, partially melted by and caught up in the lava flow.

Darwin, however, was "unwilling to take up this opinion."[90] Some of the basalts he examined were so packed with crystals that they formed "the more feasible part [of the matter]!" To explain the high crystal content of some types of lava he surmised that certain mineral ingredients in molten lava aggregate and form globules of crystals while the rest of the melt remains fluid.[91] He knew from experience that crystals vary in specific density according to their mineral content. Feldspar crystals, for example, are lighter than olivine crystals. He theorized that as magma (in the magma chamber) rises and begins to cool, large and heavy crystals sink, while small, light crystals rise to the top of the relatively denser melt. Such gravitational settling of crystals within molten lava would explain how different types of lava can be created from the same melt and extruded from the same vent. Darwin's ideas on this subject, inspired principally by the crystals he examined in the promontory of Buccaneer Cove, form the basis of the modern theory of fractional crystallization. The theory is used by geologists to explain how different types of igneous rocks are created around the world.

Geologist Paul Pearson has pointed out that Darwin's ideas on crystal formation show that he used the same line of reasoning to explain igneous rock diversity as he did in pondering the multiplicity of life. For Darwin sought to understand the processes by which both rocks and species are formed by studying relationships, looking for similarities and seeking intermediates between disparate forms. There is a clear (though limited) analogy between igneous rock formation and organic evolution, says Pearson, for they both start with homogeneity and end up with diversity. In the one case, a homogeneous melt produces a variety of different rocks via the driving forces of cooling, crystallization, and gravity. In the other, a single species evolves into a variety of new species under the guiding hand of natural selection and reproductive isolation. Darwin developed his theories more or less concurrently, and the one may well have influenced the other. As Pearson explains, "[Darwin's] success at finding the common origins of diverse geological objects, including among the category of volcanic rocks, may have predisposed him toward pondering the inter-relation of species."[92]

On October 10, Darwin put his extraordinary talents as a geological detective to the test. He attempted to decipher the history of the crater that encompasses both Buccaneer Cove and the adjacent cove to the east, before it was breached by the ocean and torn apart by waves. Darwin quickly surmised that the "promontory" that forms the northeastern arm of Buccaneer Cove and separates it from the eastern cove, is the residual wedge of the crater—something like a remaining slice of pie in an otherwise empty pie tin. The ingredients of a pie can be easily identified from its leftovers, but the constitution of the promontory was far more difficult to decipher. Darwin found "much confusion in its composition."[93] A crumbling mixture of assorted pyroclastic material, produced by seemingly incongruous eruptions, surrounded a large chunk of basalt that represented the core, or "bosom of [the] Crater."[94] When Darwin examined this remnant lava lake he found the lower portion, where it lay on "Volcanic debris" was densely packed with large "glassy feldspar" crystals, while the upper portion contained but few, minute crystals. Here, in miniature, was the same pattern of crystal deposition he was observing on the much grander scale of the island as a whole.

Darwin had better luck examining the base of the promontory, where it slopes inland and curves behind the beach. Here, a part of the cliff juts out at right angles from a small debris-filled ravine. It further stands out for its vibrant colors and bold layering. Composed of many thin streams of dark brown, crystal-studded basalt, sandwiched between thicker strata of red scoria, it looks somewhat like a multi-tiered chocolate cake (see color plate 17c). For this reason the authors dubbed the feature "Darwin's layer cake" (00°10′02.6″S, 90°49′34.1″W).[95] Darwin probed its origins by carefully measuring the length and thickness of the various streams and studying their angle of flow. The streams, between 20 and 35 centimeters thick, traveled upward for "10–15 yards [9–14 meters]" and then sharply downward, "showing their course over [the] rim of [the] Crater."[96] Darwin concluded that this spot was once "a minor point of eruption"[97] on the flank of the extinct crater, and the streams were the product of a series of events separated by "small intervals of time."[98] He drew a sketch in his geology notes (black and white plate 21D), the mirror image of which was later published in his book on volcanic islands.[99]

In various places in the promontory were "*thin* layers of pumiceous dust"; the fallout from the tuff eruptions of Cerro Cowan. Darwin originally believed these represented eruptions within Buccaneer Cove that had originated in the sea. Yet after finding one layer composed of "small rounded particles of cinders abounding with Bulimi [*Bulimulus* land snails] in an extreme state of decomposition,"[100] he deduced that the eruption responsible for this layer must have had a "terrestrial origin."[101]

Darwin's Snails

The whole subject of the distribution of pulmoniferous Mollusca [lung-bearing snails] seems to me very interesting.
—Charles Darwin. Letter to William Henry Benson, 1855[1]

Bulimulus land snails are one of the best examples of adaptive radiation and within-island speciation of all the organisms in Galápagos. They are by far the most speciose. A whopping 71 species of endemic *Bulimulus* snails have been recorded from various habitats throughout the Galápagos archipelago, and 9 potentially new species are currently under taxonomic revision.[2]

Darwin collected 15 species of the genus by looking "on bushes"[3] and "beneath stones"[4] on Chatham, Charles, and James islands. The identity of only three of Darwin's specimens is certain from print: *B. rugulosus* (Sowerby) from Chatham Island, and *B. sculpturatus* (Pfeiffer) and *B. darwini* (Pfeiffer) from James Island.[5] The rest of Darwin's land snails were mixed in with the land snails that Hugh Cuming collected from Galápagos in 1829, and so it is not clear which species were collected by Darwin and which by Cuming.[6] Darwin's land snail collection also included an unspecified "*Helix*" and a "*Paludina*"[7] freshwater snail from Charles

Island. The paludine was also recorded by Abel du Petit-Thouars in 1838, but has not been seen since.[8]

The *Bulimulid* snail that bears Darwin's name is still common throughout the highlands of James Island. *B. sculpturatus* is also present on the island. Many other species, however, have been less fortunate. On the inhabited islands of Galápagos numerous species are now extinct or extremely rare, having succumbed to sweeping habitat changes wrought by human activities. In the 1970s, *Bulimulus* snails were once so common on Indefatigable Island (Santa Cruz) that local farmers described them falling off the *Scalesia* trees into their rubber boots. Today it is rare to find a single *Bulimulus* snail in the highlands, and many of the 25 species recorded from this island have simply disappeared. No one knows exactly what caused their demise, but it is certain that the destruction of native habitat, the accidental introduction of harmful invertebrates (like the red fire ant *Wasmania auropunctata*), and competition from introduced snail species were contributing factors.[9] The sharp-beaked ground finch *Geospiza difficilis*, which once fed on these snails, has also disappeared from this island. In Galápagos as a whole, the continued existence of only 26 species of *Bulimulus* snails, just over a third of the total number of described species, has been confirmed in recent years.[10]

The *Bulimulus* clade, like the Darwin's finches, is an excellent illustration of how species diversify as they adapt to different ecological niches. The snails also demonstrate, at the species level, a natural law that Darwin hit upon many years after his visit to Galápagos while contemplating "the classification or arrangement of all organic beings" and pondering biodiversity.[11] Darwin called his law *the principle of divergence*, and used it to explain the tree of life—why organisms split into different species, branch into different genera, diverge into separate families, and so on.

Darwin's initial conception of evolution arose when he recognized that geographical isolation can drive organisms into new varieties and new species. Geographical isolation explains the evolution of species between islands but does not always account for the divergence of species within islands, within continental areas, or of biodiversity in general. In the early 1850s Darwin struggled to reconcile this problem. As he later recounted his dilemma to Professor Karl Semper, "when I thought of the Fauna and Flora of the Galapagos Islands I was all for isolation, when I thought of S. America I doubted much."[12] He could understand how a "species splits into two or three or more new species . . . [when there was] nearly perfect separation." But he could not see how "a species becomes slowly modified in the same country (of which . . . there are innumerable instances)."[13] How does natural selection generate "very diverse forms"?[14] he asked himself. Why should "organic beings descended from the same stock diverge"[15] away from each other? What causes organisms to radiate and branch? Darwin hit upon an answer by placing natural selection in an ecological framework, by considering environmental factors such as "food, climate, &c."[16] He argued that within species, especially successful, prolific ones, competition for resources tends to push variant offspring into unoccupied ecological niches, or, in Darwin's vocabulary, "to become adapted to many and highly diversified places in the economy of nature."[17] Natural selection thus favors the evolution of specialized varieties that diverge away from the parent form.

Darwin did not apply his principle of divergence specifically to Galápagos organisms. He used it to explain evolutionary divergence in general and to understand the branching points of taxonomy. But he recognized that it applied to small areas as well as large, and even within geographically isolated areas.

Darwin may, at this point, have suspected that his 15 *Bulimulus* land snails filled distinct and narrow ecological niches within the islands. However, there was little he could say about their distribution for he had not paid careful attention to, or recorded, where he had collected them on each island.

Only in recent years has the phylogeny and ecology of the *Bulimulus* snails of Galápagos been examined in detail and the premise behind Darwin's ideas put to the test. According to Darwin's principle of divergence, areas with a high number of ecological niches should contain a larger number of species than areas with a poor variety of ecological niches. As expected, species richness in the *Bulimulus* clade is indeed positively correlated to habitat diversity. The largest, oldest, and highest islands have the greatest number of plant species and vegetation zones, and therefore the greatest diversity of niches. They also have the largest numbers of *Bulimulus* snail species to fill them. These results give support to the idea that from an original stock of a few colonists, the snails multiplied rapidly under favorable conditions and then diverged to fill a wide variety of ecological niches.[18]

Darwin's land shell specimens, and the fact that they were closely allied to *Bulimulus* species on the South American continent helped support his premise that much of the Galápagos fauna and flora consists of modified descendants of South American ancestors. However, he found it difficult to explain just how the land snails had traveled from the South American continent to Galápagos in the first place, and then become dispersed among the islands. It was one of his greatest puzzlements.[19] Birds could fly and plants had seeds that in many cases could float or stick to feathers and thus be transported over water. But how were land snails, notoriously sensitive to salt as they were, expected to travel long distances across the sea? He tried floating snail eggs in seawater but, as expected, they quickly sank and died. So did the adult snails that he dropped into his experimental tank. He guessed that snail eggs could "fly" if embedded in mud stuck to birds' feet, but was unable to test this hypothesis. Eventually he discovered that some adult snails could float in seawater unharmed (and for weeks on end) as long as their operculum was shut watertight. He decided that hibernating snails could travel long distances this way, perhaps wedged in the cracks of floating timber.[20] Today, transport by birds and rafting on vegetation are still considered the most likely mechanisms for land snail dispersal across water.[21]

Most visitors to Galápagos are unlikely to encounter live *Bulimulus* snails, though they will surely see the bleached shells of extinct species cluttering the ground at some sites. Not only are the extant snails greatly diminished in numbers, they are difficult to find because they are small, brown, and often hide on the undersides of logs, rocks, leaves, and plant stems. But one place where extant *Bulimulus* snails can regularly be found is around the pirate caves of Asilo de Paz in the highlands of Charles Island. Two species (*B. nux* (Broderip) and *B.unifasciatus* (Sowerby)) also occur in abundance at another, less frequented, visiting site on Charles Island called Cerro Alieri. Due to Cerro Alieri's close proximity to Round Hill (Cerro Pajas), the cinder cone that Darwin climbed on September 27, 1835, it is quite possible that Darwin saw and collected both these species.

Salt and Sugar

On Sunday, October 11, the "Mayór-Domo"[102] invited Darwin and his companions to visit a "Salina" where the settlers gathered salt for preserving fish and tortoise meat. They were taken by fishing boat from Buccaneer Cove to James Bay, or Puerto Grande as the harbor was then called. Much of the bay is bordered by "a field of Lava about 4 miles long [actually closer to 3 miles, or 4.6 kilometers] & 1–2 [1.6–3.2 kilometers] broard [*sic*]"[103] between the modern visitor sites of Puerto Egas and Playa Espumilla. Looking toward the interior of the island Darwin could see that this lava flow had "burst from several small Craters, placed at the base of the high central Trachytic hills"[104] and had smothered the land all the way to the coast in a thick black mantle.

On the way, the men passed a crude beach of wave-rolled "grey-stone" boulders "abounding with Olivine,"[105] behind which stood two sunset-colored kipukas (00°13'52.85"S, 90°50'45.48"W). These kipukas are the remains of an ancient, well-weathered "glassy red Scoria and Greystone" crater that had been "overwhelmed & almost concealed beneath the flood" of the subsequent pahoehoe flow.[106] They stand out like elaborate giant gateposts on a vast and empty asphalt parking lot. Darwin provided two different estimates for the distance traveled from Buccaneer Cove to the landing spot in James Bay—five miles in his geology notes, and six in his diary. It was therefore either here, on the stone beach fronting the kipukas, or further to the south, close to where the James Bay lava flow borders a sandy beach, that Darwin landed.

Unlike the brittle, jagged aa lava flows that Darwin experienced on Chatham and Albemarle islands, the pahoehoe flow of James Bay was firm, solid, and in many places as smooth as pavement. Darwin and his companions walked straight across this bare flow to reach the inland salina, but, despite the road-like appearance of the flow, found the way slow going. The flow, especially on the southern side where Darwin walked, is cut through with wedge-shaped fissures. These cracks formed when the cooling magma became backed up, creating inflated pools of lava that then split apart as they solidified. The lava around the fissures resembles ripped-up asphalt more than smooth pavement, and for this reason Darwin described his hike as "very rough."[107]

Darwin's pace was also slowed by the mesmerizing patterns on the surface of the flow. Anyone familiar with the smooth and ropey lava fields at Sulivan Bay on James Island, at Punta Espinosa on Narborough Island, or the flows of Kilauea Volcano in Hawaii, can easily visualize the "singular ringed & twisted forms" that Darwin described in the flow at Puerto Grande. In some places the lava was rolled into braids and "cables," or coiled up like intestines. In other areas the melt had doubled back on itself like "folds in thick drapery." Further on, the surface had crinkled into a pattern reminiscent of "rugged bark."[108] There were even spots that resembled "Cow dung."[109] Upon close examination Darwin discovered that the lava, like the greystone boulders at the beach, glistened with crystals of green olivine. It was curious, he noted, that on "this island we have thus olivine lava as the latest [i.e., youngest], whilst in Albermale [*sic*], that of Trachyte."[110]

Darwin portrayed the flow as "bare & utterly destitute of vegetation,"[111] but today it is sparsely vegetated with several pioneer plant species, the most common being the lava cactus *Brachycereus nesioticus*. Darwin's assertion of an absence of vegetation implies that the flow was relatively young in 1835. The inhabitants seem to have thought so, for they told Darwin, "A Terapin [tortoise] was caught some years since with its Shell appearing to have been burnt years before . . . [by] Volcanic fire."[112] Darwin, however, was sceptical. As he later explained, "my work in Geology gave me some idea of the lapse of past time."[113] He knew that lava flows could appear much younger than they really were. "Recent as this stream appears" he reasoned, "it is crossed by another one of a slightly darker color but same constitution. —Close to a wall within the marginal fragments of this stream some low but old trees are growing. I conceive these must have been killed if they had been there when the Lava flowed."[114]

He concluded that the tortoise must have been burnt by "accidental fire in wood"[115] and that the "fresh & glassy" appearance of the flow merely showed "how long the surface has resisted decomposition."[116] It should be noted that Darwin apparently did not examine the "darker," browner, and jagged aa flow up close. He would certainly have collected a sample, for it has a different constitution than the smooth pahoehoe lava, which he did sample. He would also have realized that it underlies the pahoehoe flow in some places, and thus is older, not younger as he surmised.

The date of the latest pahoehoe eruption at James Bay is unknown, but from fragments of earthenware jars used by 17th-century buccaneers and found embedded in the lava by Norwegian anthropologist Thor Heyerdahl in 1953, it has been placed sometime between William Dampier's visit in 1684 and Darwin's in 1835.[117] Given that tortoises can live to over 100 years of age and that fires are rare in Galápagos except when vegetation ignites during an eruption, the burnt tortoise could well have been the survivor of an eruption. When Albemarle's Cerro Azul Volcano erupted in 1998, at least one tortoise was burnt by brush fires sparked by the molten lava. Like the tortoise in Darwin's story, this particular individual's scutes were scorched but the animal escaped otherwise unharmed.

After hiking about "a mile & ½ [2.4 kilometers]" (actually closer to 1 mile or 1.6 kilometers) inland over barren lava, Darwin and his companions arrived at the foot of the tuff crater that held the hidden salina (00°14'15.7"S, 90°50'10.3"W).[118] Here, where the pahoehoe lava flow abuts the northern flank of the crater, the slopes were steep and uninviting. But the crater sits on the edge of a much older, well-wooded flow, and on the southeast side of this crater "streams of ancient Olivine Lava [had] nearly brought the country up to a level with the brink."[119] For Darwin and his companions, accessing the rim of the crater was therefore a simple matter of crossing from the pahoehoe flow into the woods and then winding their way up a natural ramp.

In the woods grew "great numbers"[120] of exceedingly tall prickly pear trees (*Opuntia galapageia*). Many years later botanist Joseph Hooker asked Darwin just how tall the cacti grew on James Island. David Douglas, who had visited the island in January 1825, claimed they grew to 40 feet (12 meters), but Hooker was skeptical. "[Douglas] was apt to pull the long bow," he said. Darwin confirmed that he had never seen a cactus that tall on James; 20 feet (six meters) was closer to the mark.[121] While the *Opuntias* on Indefatigable Island (*Opuntia echios*) do grow as tall as 12 meters, Douglas visited only James Island, and there the species (*Opuntia galapageia*) is just half as tall.[122]

It was not only the height of the *Opuntia* trees that struck Darwin as unusual. He also found the shape of this "cactus whose large oval leaves connected together formed branches rising from a cylindrical trunk" most "peculiar,"[123] in other words, unique. He had seen prickly pears on the mainland of South America—and collected a new, diminutive, species in Patagonia that botanist John Stevens Henslow would name after him (*Opuntia darwinii*—now called *Maihueniopsis darwinii*)—but never before Galápagos had he seen one growing in the form of a tree. The three varieties of *Opuntia megasperma* growing on Chatham and Charles islands have thick reddish trunks and, like the *Opuntia galapageia* of James Island, grow to just over 5 meters in height. The species (*Opuntia insularis*) that grows on Albemarle Island is only comparatively shrubby—it grows to 2.5 meters. Surprisingly, Darwin did not record the arborescent characteristic of the Galápagos opuntias until he got to James Island. But once there he drew a simple sketch of the tree in his diary and somehow, despite the obvious impracticalities of handling a plant covered with sharp "hogs' bristles,"[124] managed to collect specimens of the pad, fruit, and flowers.

When Darwin arrived back in England, Henslow described the specimens and declared the cactus to be a new species, which he named *Opuntia galapageia*.[125] Not only was it the first endemic plant species from Galápagos ever to be given a scientific name, it was the only Galápagos plant that Professor Henslow described and named. Joseph Hooker examined the rest of Darwin's plants several years later. Darwin's specimen was probably one of the very first flowers of the season, for opuntias do not generally bloom in Galápagos until later in the year. This may explain why Darwin did not collect opuntias on the other islands, a great shame in light of the fact that among the plant genera of Galápagos, opuntias are second only to scalesias in the degree to which they have diversified.[126] There are fourteen taxa of opuntias recognised in Galápagos, and all are endemic; Darwin could have collected five of them.

The *Opuntia* was also the only genus of cactus that Darwin collected in Galápagos. He mentioned seeing a second cactus in the islands but it was not in flower and thus of insufficient value as a scientific specimen. In describing it to Henslow, Darwin said that it reached "two or three feet in height [and] first . . . takes possession of the newly formed beds of lava."[127] From this depiction it is likely that Darwin was referring to the endemic lava cactus *Brachycereus nesioticus*, a pioneer of pahoehoe lava flows. However, there are indications that Darwin may have been visualizing the immature plants of a third genus of cactus found in Galápagos: the candelabra cactus, *Jasminocereus thouarsii*. In answer to a query by Joseph Hooker in 1845, regarding the two cacti Darwin described from Galápagos, Darwin wrote, "I cannot find out anything more about the other species of cactus: it grew on Albemarle Isd & I *think* on the other islands."[128] Because Darwin only examined an aa flow on Albemarle Island, it is more likely that he saw *Jasminocereus*, not *Brachycereus*, growing there.

Today *Jasminocereus* can be found at most of the sites Darwin visited. *Brachycereus*, on the other hand, occurs at only one of Darwin's sites. This is the James Bay lava flow on James Island, a place Darwin described as destitute of vegetation. The *Jasminocereus* plants growing on the aa lava flow behind Beagle Crater on Albemarle Island, and indeed in similar lava flows elsewhere in the islands, are stunted, and look quite different from the tall, branched, mature *Jasminocereus* trees that grow to seven meters in height elsewhere in Galápagos. Darwin may have thought they were a different species. Why Darwin did not comment specifically on the mature *Jasminocereus* trees is anyone's guess. He may have thought they were the same as the candelabra cacti of Peru and Chile. *Jasminocereus thouarsii*, however, is endemic to Galápagos. It was named after French captain Abel du Petit-Thouars who collected a specimen during his visit to Charles Island on the French frigate *La Vénus* in 1838.

When Darwin arrived at the outskirts of the tuff cone that contained the salina (Salt Mine or Mina de Sal, as it is now called) he was reminded of the very first tuff crater he had examined on Chatham Island. "The whole upper part of the walls is composed of a compact glassy yellowish-brown Volcanic Sandstone; some of [which] has a semi-Resinous fracture & contains small glassy patches & resembles that [Pan de Azucar] of Chatham Isd."[129] However, when he reached the rim and then descended into the bowl of the crater, which was "much deeper than the surrounding country,"[130] he discovered that the crater was more like Finger Hill than Pan de Azucar. It had a basalt interior. Whereas in Finger Hill the basalt existed as a frozen lake, here the "whole bottom of the Crater, in which was a bed of solid Lava had been blown up & . . . subsequently a great emission of ashes beneath the water had formed the upper ⅔ of unbroken rim."[131]

At the bottom of the crater lay a circular lake dotted with white-cheeked pintail ducks (*Anas bahamensis galapagensis*) and strikingly pink greater flamingos (*Phoenicopterus ruber*). It was also "fringed with bright green succulent plants . . . so that the whole ha[d] rather a pretty appearance."[132] Darwin descended to the shore, and waded in to the shallow water. The margins were "soft & muddy"[133] but the center of the lake had solid "layers of pure & beautifully Crystallized Salt,"[134] some of which formed "splendid cubes, sides nearly 2 inches long."[135] It was from here that the salt was "quarried" by the settlers.[136]

The salt accumulates from seawater seeping into the sea-level crater and then evaporating, but Darwin was at first doubtful of this explanation. "Is the Salt a Volcanic exhalation?" he queried.[137] He collected a sample of water hoping that "an analysis of its Brine may throw some light on this."[138]

As Darwin wound his way back out of the salina and down the other side, a white orb lying amongst some bushes caught his eye. It was a human skull. Darwin was told by the settlers that it belonged to a sealing captain, murdered by his crew a few years since. Darwin seems to have accepted the story without question. On such a deserted island there was little sense in investigating the claim. After all it was such "a quiet spot."[139]

The salt mine that Darwin examined is one of several salinas in the Galápagos, and one of the first to be exploited for commercial purposes. It was worked through the 1840s and then again in the 1920s and 1960s. In 1969 it was then incorporated into the body of the Galápagos National Park, at which time all property and mining rights on James were extinguished and the island was abandoned. Today the salina is a National Park visitor site reached by walking, not across the lava field as Darwin did, but on a trail through the woods behind Puerto Egas. A skull or two might be found lying in the bushes, but these of course are not human. They belong to goats, pigs, and donkeys, and testify to a not-too-distant period when these introduced mammals roamed the island.

Salt is now acquired only from salinas on inhabited islands, principally those near the towns of Puerto Villamil on Albemarle Island and Puerto Ayora on Indefatigable Island. Local residents gather it primarily for construction purposes and for salting Bacalao (a type of grouper), the main ingredient of fanesca, a traditional Ecuadorian Easter dish.

From the rim of the salt crater Darwin could see two larger tuff craters rising from the woods "at the distance of 2–3 miles [(3.2–4.8 kilometers)]"[140] to the south. The inhabitants told Darwin that the smaller and more distant of the two contained another small salina, but due to its steep slopes inside and out, it was not worked for salt. The larger tuff cone, then called Sugar Loaf and now known by its Spanish translation Pan de Azucar, caught Darwin's eye not only for its great height of "1200 ft [(366 meters)]"[141] but for the *"perfect smoothness"* of its slopes. "It precisely resembles an immense plastered floor," he wrote.[142] Darwin did not climb Sugar Loaf, but Sulivan scaled the cone earlier in the month while surveying the island and then waiting for the *Beagle's* arrival. He later informed Darwin that "the Crater is dry, but beautifully circular; its depth is about 600 ft [183 meters], the bottom appeared on a level with the surrounding country [and] the S side is a good deal lower than the north."[143]

A Gathering of Giants

On October 12, Darwin returned to his favored highlands and "thus enjoyed [a total of] two days collecting in the fertile region."[144] From the hovels he once again followed the "broad & well-beaten [tortoise] paths"[145] that led to the highland "Springs."[146]

Upon arriving at the pools Darwin sat down on the driest object around—a giant tortoise—and watched an ethereal scene play out before him. The foggy district literally swarmed with an "extraordinary number of Turpin,"[147] some emerging from, others disappearing into the pervasive mist. How "very curious" it was to watch "such numbers of these huge animals, meeting each other in the highways, the one set thirsty & the other having drunk their fill."[148] They traveled purposefully and "far faster than at first would be imagined."[149] Darwin timed and paced their movements, later jotting a simple calculation in his field notebook: "30 yard in 5 minutes, in 1 hour 360 × 24 = 8640."[150] At this rate, he concluded, they "would go four miles [6.4 kilometers] in the day & have a short time to rest."[151] The inhabitants concurred; from having watched marked individuals, they believed the tortoises "would pass over 8 miles [13 kilometers] of ground in two or three days."[152]

Several tortoises lay submerged in a pool "within 2 yards [1.8 meters]"[153] of where Darwin sat. There they buried "their heads *above* the eyes in the muddy water & greedily suck[ed] great mouthfulls [sic], *quite regardless* of lookers on."[154] Darwin noted that they drank "about

10 gulps in [a] minute."[155] Indeed, they consumed such quantities that Darwin, upon tasting the "water in the Bladder" of a freshly killed individual, declared it "only slightly bitter" while the inhabitants insisted, "The water in the Pericardium is ... [even] more limpid & pure."[156]

Most of the tortoises were too large for one man to budge. In trying to get them to move Darwin discovered that if he "got on their backs, and ... [gave] a few raps on the hinder part of the shell, they would rise up and walk away." "[B]ut," he added, " I found it very difficult to keep my balance."[157] The females, being smaller, could be shifted with less difficulty and Darwin overturned a series of the lightest in order to measure them. He was in for a surprise, for when he returned to the first, he found she had already wandered off. Darwin was being taught what the inhabitants already knew: that "[i]n order to secure the Tortoises, it is not sufficient to turn them like a [sea] Turtle, for they will frequently gain their proper position."[158]

The inhabitants did not kill the tortoises indiscriminately; only the fattest individuals were slaughtered. But the manner in which the men determined the gastronomic value of the tortoise and then killed it was brutal. "When an animal is caught," explained Darwin, "a slit is made in the skin near the tail to see if the fat on the dorsal plate is thick; if it is not the animal is liberated & recovers from the wound— if it is thick it is killed by cutting open the breast plate on each side with an axe & removing from the living animal the serviceable parts of the Meat & liver &c &c."[159] The meat was eaten both fresh and salted, and when fried down it gave "a beautifully clear & good oil."

The breeding season of the tortoises had ended but Bynoe found seven white, "sphærical" eggs "laid along in a kind of crack" in the lower elevations.[160] The settlers, doubtless amid raucous guffaws, imparted more information about tortoise reproduction. When tortoises mate, they explained, the pair "remain joined for some hours. During this time the Male utters a hoarse roar or bellowing, which can be heard at more than 100 yards [91 meters] distance." "When this is heard in the woods," they continued, "[you] know certainly that the animals are copulating ... [for] ... [t]he male at no other time, & the female never, uses its voice."[161] Darwin may have thought this a pointless trait, for he noted that " they are perfectly deaf; certainly when passing a tortoise, no notice is taken till it actually sees you— then drawing in its head & legs & uttering a deep hiss, he falls with a heavy sound on the ground, as if struck dead."[162] This reminded Darwin of another question to ask his "informants."[163] How long do they live? Quién sabe [who knows], the "Spaniards" answered.[164] They had "never found one dead from Natural causes," though the hatchling tortoises "frequently falls a prey to the Caracara [Galápagos hawk]."[165]

From a brief note in one of Darwin's specimen catalogues it is apparent that Darwin brought a young tortoise off the island. Perhaps the settlers presented it to him as a gift. Not for eating, they would have admonished, or you never will find out how long it lives! The fate of Darwin's little tortoise remains a mystery, but it could well have outlived Darwin (see chapter 5). It is now known that tortoises can live at least 100 years, perhaps twice that long, and any individual with scutes that are so worn as to obscure all growth rings, is generally considered to be over 70 years old.[166]

Settlers and whalers killed many of the James Island tortoises during the 19th century, but enough individuals survived to maintain the race. Today the population consists of approximately 700 individuals distributed in a horseshoe-shaped region within the center of the island. There are four distinct nesting areas in this zone, three on the eastern flanks of the island and one behind Pan de Azucar and the salt mine.[167] Except for the occasional wanderer, tortoises are no longer found in the areas Darwin explored. They evidently used to nest in the arid ground behind Cerro Cowan, and in the valley behind Buccaneer Cove where Darwin camped, for in 2007 the authors found tortoise egg shells embedded in the tuff sides of several shallow gullies in these areas.[168] The shells were infused with dust, cracked, and clearly very old. As tortoises have not been recorded from this area in recent history, they may even date from the 19th century.

Wandering about Bird Collecting

Darwin left the highlands of James Island on October 13. He spent the entire morning "descend[ing] [from] the highest crater"[169] to Buccaneer Cove, only to find that he and his companions were out of drinking water. While they had been gone the "surf [had broken] over & spoiled the fresh water"[170] in the "well" at Cerro Cowan. Fortunately, the crew of an American whaler came to the men's rescue, providing them with "three casks of water (& made [them] a present of a bucket of Onions)."[171] Darwin was so impressed by their generosity that he felt compelled to write in his diary, "Several times during the Voyage Americans have showed themselves at least as obliging, if not more so, than any of our Countrymen would have been. Their liberality moreover has always been offered in the most hearty manner. If their prejudices against the English are as strong as our's [*sic*] against the Americans, they forget & smother them in an admirable manner."[172]

After the cool highlands it was difficult to adjust to the heat on the coast. On James, the weather was "cloudless & the sun very powerful" and Darwin complained that "if by chance the trade wind fails for an hour the heat is very oppressive."[173] Darwin was not the only one to notice how hot it was on James. "How remarkably different is the climate of the windward and leeward islands of this group!" exclaimed FitzRoy in his *Narrative*. "[At Charles and Chatham islands] we were enveloped by clouds and drizzling fog, and wore cloth clothes. At Tagus Cove and James Island, a hot sun, nearly vertical, over-powered us; —while the south side of Albemarle, Charles, and Chatham Islands, were almost always overshadowed by clouds, and had frequent showers of rain."[174]

The heat was especially intense at Buccaneer Cove, for it lies in the rain shadow of the highlands and even during the cool season, tends to bake in the sun while Puerto Egas, just 6 miles to the south, stays relatively cool and overcast. From October 14 until the 17th when the *Beagle* returned, Darwin made a point of measuring just how hot it was at Buccaneer Cove. Inside the tent the thermometer rose to "93°F [33.9°C]" and outside "in the wind and sun" the temperature reached "85°F [29.5°C]."[175] Placed on brown sand, the thermometer "immediately rose to 137° [58.3°C], & how much higher it would have risen, I do not know," wrote Darwin, "for it was not graduated above that number."[176] On the black sand it was still "hotter, so that even in *thick* boots it was very disagreeable, on this account, to walk over it."[177]

Such statements make one question whether Darwin visited Galápagos during an El Niño year when air and sea temperatures are unusually high. However, the temperatures measured by Darwin can easily be duplicated at Buccaneer Cove during a "normal" year. Furthermore, the sea surface temperatures taken three times a day by the *Beagle* crew ranged from 58.5°F (14.7°C) to 75°F (23.9°C) with a mean of 68°F (20°C).[178] Had it been an El Niño year, the temperatures would have been higher—up to 5°F (2.8°C) higher.

El Niño events generally occur every three to seven years.[179] Their intensity and effect on Galápagos depend largely on the oscillation of two pressure systems, one in the eastern Pacific and one in the Indo-Pacific. Scientists have attempted to map the history of these ENSO events by looking at historical anecdotes as well as at the skeletal growth rate and stable oxygen isotope ratios of corals heads. The results of both methods suggest that 1835, the year of Darwin's visit to Galápagos, was not an El Niño year. Historical log books indicate there was a moderate El Niño event in 1832, and another in 1837, but nothing remarkable in between.[180] Measurements of Galápagos coral heads suggest that the sea surface temperatures in 1831 were unusually warm and indicative of an El Niño, but that those in 1835 were, if anything, slightly cooler than average for the decade.[181]

Darwin spent the last few days at Buccaneer Cove exploring east and west of the harbor, observing land iguanas, and generally "wander[ing] about Bird collecting."[182] On short water rations on a "frying hot Island"[183] it is hard to imagine the men having much energy for collecting. Fortunately, many specimens could be acquired near the camp and with little effort. Galápagos hawks (*Buteo galapagoensis*) were as abundant as they were bold, and on one day there appeared "thirty, which [Darwin] counted standing within a hundred yards of the tents."[184] The hawks were accustomed to following the tortoise hunters, and would "sit watching on the trees when a Tortoise [was] killed."[185] Hawks had shadowed Darwin on all his excursions, on all the islands. On Chatham Island he had pushed one off the branch of a tree with his gun and on Charles Island he was told that they frequently killed the chickens that pecked about the houses. In their habits the Galápagos hawks were similar to the striated caracara (*Phalcoboenus australis*) he had seen in the Falkland Islands, and Darwin was at first convinced they belonged to "that peculiar South American genus."[196] They were one of the first birds to alert Darwin to the "South American character"[187] of the organisms of Galápagos, and thus on to the realization that the species of Galápagos had evolved from South American ancestry. The Galápagos hawk's closest mainland relative is now recognized as the Swainson's hawk (*Buteo swainsoni*). The ancestors of the Galápagos hawks are thought to have colonized the islands as little as 300,000 years ago.[188]

The shorebirds were also easy to catch. Sedentary lava gulls (*Leucophaeus fuliginosus*) and great blue herons (*Ardea herodias*) adorned the landing beach. Ruddy turnstones (*Arenaria interpres*), least sandpipers (*Calidris minutilla*), semi-palmated plovers (*Charadrius semipalmatus*), and wandering tattlers (*Tringa incana*) raced the surf-swept sand as they do today, and made easy pickings for a sure-shot. A greater challenge was shooting the brown noddies (*Anous stolidus galapagensis*). They fluttered like moths over the waves, alighting at times on the brown pelicans (*Pelecanus occidentalis urinator*) that bobbed cork-like on the surface. Galápagos martins (*Progne modesta*) zipped high about the cliffs, and Syms Covington, "Darwin's shooter,"[189] somehow managed to knock one out of the sky. Not to be outdone FitzRoy's servant Fuller shot a migrant bobolink (*Dolichonyx oryzivorus*), which he reverently presented to Darwin, being the only one of its kind. This specimen is noteworthy in that it retains one of the few surviving labels written in Darwin's hand. It was important to Darwin, because it was a concrete link to the mainland. The bobolink was proof that animals other than marine organisms could make it out to Galápagos from the Americas now and then, both as potential colonizers and as vectors for the transport of other organisms (like seeds that might stick to feet and feathers). Bobolinks are considered regular but uncommon visitors to Galápagos today. They have one of the longest migrations of all New World songbirds, traveling more than 8000 kilometers (over 5000 miles) from northern North America to the pampas of Argentina.[190]

About two miles south of Buccaneer Cove, behind a beach that is now called Playa Espumilla, lies a saltwater lagoon that once harbored large flocks of flamingos (*Phoenicopterus ruber*).[191] Darwin collected the stomach contents of a flamingo, found in a shallow "salt-water Lagoon,"[192] and although he did not specify the location, it is most likely that the bird was shot either here or in the salt mine. Today flamingos are occasionally seen in the salt mine but rarely at Playa Espumilla. During the massive El Niño event of 1982–83, large quantities of rain-washed sediment filled up most of the lagoon at Playa Espumilla, and now only small, shallow pools remain. They still, however, provide habitat for white-cheeked pintail ducks (*Anas bahamensis galapagensis*), a specimen of which was also collected by the *Beagle* men. Indeed, Darwin and his companions may well have dined on ducks and flamingos while camped on the island, for along with tortoises, land iguanas, and doves, these waterbirds were considered good eating by other 19th-century visitors to Galápagos.[193]

Perhaps the best collecting was done in the evenings when yellow-crowned night herons (*Nyctanassa violacea*) emerged from the shadows to stalk the camp in search of centipedes, squawking in alarm when they tripped a guy rope or flushed a rat (*Rattus rattus*). The rats were "very common"[194] and easy prey for short-eared owls (*Asio flammeus galapagoensis*), which grabbed them from the bushes and barn owls (*Tyto alba punctatissima*) which swooped silently upon them from overhead. The men collected a specimen of each of these crepuscular and nocturnal animals, saving the barn owl for FitzRoy's personal collection.

The rat, with its long tail and gray hair, was plainly different from the rice rat Darwin had seen on Chatham Island. It resembled the "common black rat" found on the continent, but was "less carnivorous and [did] not appear to be so strongly attached to the habitations of man."[195] Darwin wondered whether or not this rat, given its different diet, was "the same as [the] domestic rat of S. America? Have they been brought by Ships?" he queried.[196] After his return to England, Darwin expanded on these thoughts. He wrote, "This island was frequented, about one hundred and fifty years since, by the vessels belonging to the Bucaniers [*sic*]; so that the common rat might easily have been transported here. And if a very peculiar climate, a volcanic soil, and strange food, can together produce a race, or strongly marked variety, there is every probability of such change having taken place in this case."[197] By this time Darwin was already convinced of the mutability of species, was pondering the subject of natural selection as an explanation for evolution, and was seeing varieties as incipient species. Here, he thought, was an example of an organism that had apparently changed sufficiently within the past 150 years, in response to environmental changes, to be considered a new variety, if not a new species.[198] George Waterhouse, the man who described Darwin's mammal specimens from the voyage, believed Darwin's rat to be a distinct species, and named it *Mus jacobiae*. However, modern analysis has shown that the introduced rat on James Island is *Rattus rattus,* and that it has not diverged significantly to warrant naming it anything else.[199]

Halley's Comet

Alarm with comet-blaze the sapphire plain,
The wan stars glimmering through its silver train . . .
—Erasmus Darwin, *The Botanic Garden*[1]

During his last few days at Buccaneer Cove, Darwin looked forward to the evenings, not only for the relief they brought from the day's intense heat but for some exceptional stargazing. Darwin described the weather as mostly "cloudless"[2] on James Island, the waning moon rose later each night, and Darwin was not one to ignore the wonders of a starry night. Early in the voyage he had written, "To enjoy the soft & delicious evenings of the Tropic; to gaze at the bright band of Stars which stretches from Orion to the Southern Cross, & to enjoy such pleasures in quiet solitude, leaves an impression which a few years will not destroy."[3] It is a sentiment that could well have been written in Galápagos. But in Galápagos Darwin had a chance to view some truly extraordinary astronomy. On October 13, Halley's comet made its closest approach to Earth for the century, and for the next three days it was visible from the northwestern side of James Island.[4]

Shortly after sunset on October 14, after the cooking fire was damped down and the black night was allowed to saturate the sky, a bright smudge, 6 to 10 degrees in length, appeared low (13 degrees) over the ocean horizon just off the northern tip of Albemarle Island. By 8 p.m. it had disappeared. Over the next two nights the comet faded and moved progressively west over Albemarle, but it also appeared higher in the sky and set later during the night, so in effect became more visible. Darwin's last entry in his field notebook for "Friday" October 16 is a single word—"Comet."[5] He wrote nothing else about the comet, either here or in any of his later writings, and it begs the question, why not? Did Darwin really see the comet, or was he perhaps only told about it by one of his more fortunate companions?

Much of Buccaneer Cove affords a clear view of the northwestern horizon and the northern tip of Albemarle. Had Darwin stood anywhere on the beach, there is little doubt he would have seen the comet, as long as the atmospheric conditions were favorable. However, Darwin was camped in a small valley behind a steep promontory that forms the northern arm of Buccaneer Cove, and this promontory blocks the sky from the horizon to an elevation of roughly 40 degrees. The exact location of Darwin's camping spot is impossible to know as it was determined by the distribution of now-extinct land iguana nests, but nowhere behind the promontory would Darwin have had a view of the comet from his tent. He would have had to walk at least 100 meters from camp to obtain the right angle to see it. Apparently he (or one of his party) did this. But why did he not write more than a single word about the comet?

Darwin was certainly interested in comets. He had studied astronomy while awaiting the start of the voyage, for he had supposed "it would astound a sailor if one did not know how to find Lat & Long."[6] He had also been anticipating the arrival of Halley's comet for several years. In April 1833, he had written to his friend Professor Henslow, "We are all very curious to hear *something* about *some* great Comet, which is coming at *some* time: Do pump the learned & send us a report."[7] Four months later Henslow replied, "The comet you speak of is *expected* in 1835, according to calculation—but it seems very doubtful whether the calculation is correct— The papers of course talk nonsense about it, but it is really something out of the ordinary cometical occurrences."[8] Indeed, many perceived the comet as a portent of doom; it was blamed for the great fire that broke out in New York City on December 16, 1835. (But it was also later touted as the "stork" that brought Mark Twain to life on November 30 of the same year.) It is therefore surprising that Darwin never mentioned the comet again, not even in connection with one of his mentors, the great astronomer Sir John Herschel. Herschel is famous, among his other scientific accomplishments, for his enlightening observations of the return of Halley's comet in 1835. Darwin "enjoyed a memorable piece of good fortune in meeting" Herschel at the Cape of Good Hope in June 1836.[9] The two scientists corresponded in later years, but as far as it is known Halley's comet was never mentioned between them.

If Darwin did see the comet, as is suggested by his notebook entry, possibly he was disappointed by its faintness. Even if the skies were cloudless the atmospheric conditions may have been unfavorable. This was certainly the case in England. As in Galápagos the best time for viewing the comet was the middle of October, but as Darwin's sister Catherine later told him, "it was so hazy all the month, that it could not be seen well at any time."[10]

And what about Captain FitzRoy? Did he see the comet? Before embarking on the voyage, FitzRoy was ordered by the Admiralty to monitor and measure the position of any comet that should "be discovered while the *Beagle* is in port [or] . . . at sea."[11] FitzRoy, however, never reported this one.

Here the explanation is simple. For the duration of the comet's visibility in Galápagos, FitzRoy was with the *Beagle* in the southern part of the archipelago where the skies remained constantly cloudy.[12] By the time FitzRoy had picked Darwin up from James Island and the *Beagle* had cleared the northern reaches of Albemarle the comet had most likely faded out of sight. It is safe to say FitzRoy never saw it.

The last time Halley's comet was seen in Galápagos was in 1986. In April of that year, one of the authors (GBE) observed the comet from the top of Sierra Negra volcano, Albemarle Island. The next viewing is scheduled for 2061.

On October 17, the *Beagle* was sighted, its appearance signaling the end of Darwin's time in Galápagos. By four o'clock in the afternoon a sunburnt but invigorated Darwin and his sweating companions were back on board. There was much sharing of news and swapping of tales. Darwin was handed a collecting bottle of "Spirits of Wine" in which a crustacean floated inside. It was a *Macrobrachium* shrimp that one of the officers had collected from a *"fresh water pool near [the] Sea Beach"* at Fresh Water Bay on Chatham Island.[200] Both *M. hancocki*, a beautiful blue shrimp that grows to about 8 centimeters long, and *M. americanum*, which obtains a length of over 25 centimeters, have since been recorded from this stream, at La Honda.[201]

FitzRoy also produced a letter for Darwin that had, "by a string of extraordinary chances"[202] recently arrived on a schooner from Guayaquil. The letter was from his sister Caroline, written seven months previously, in which she did "teaze [sic]" him to come home; the family was greatly concerned for his "health & spirits."[203] Needless to say, the letter was out-of-date and Darwin was only thankful he had not heeded his family's similar pleas from even earlier in the year.[204]

While Darwin was exploring James Island, the *Beagle* crew had been busy taking on fresh water, wood, and a stock of 30 tortoises at Fresh Water Bay on Chatham Island. They had then sailed south to Hood Island, surveying McGowen reef, the hazardous hidden shallows between Chatham and Charles, along the way. After lingering just two hours in Hood Island's Gardner Bay on the morning of October 14 they had headed for Charles Island. October 15 and 16 were spent first at Post Office Bay, then at Black Beach where the crew took on more supplies. In total, including the purchases made in September, the crew of the *Beagle* bought from Charles Island 6 pigs, 13 barrels of potatoes and 8 pumpkins for just over $47; quite a bargain compared to the prices they would find in Tahiti.[205] The *Beagle* then navigated up the eastern coast of Albemarle, veered off to James Island and arrived at Buccaneer Cove by midafternoon on October 17.[206]

Once Darwin and his companions were safely back on board, the *Beagle* continued on its northwestern course. It then struck off toward Abingdon for a planned rendezvous with Officer Chaffers near that island. Chaffers, however, could not be found and by the evening of October 18, when there was still no sign of the yawl, the *Beagle* crew "fired 2 guns & hoisted a signal light"[207] to call him in. Chaffers finally appeared the following morning and by 10 a.m. he had boarded the *Beagle* and presented Darwin with a pocketful of rock specimens from the northern islands of Abingdon, Bindloe, and Tower. Without lingering any further, the *Beagle* advanced yet farther north to Wenman (Wolf) and Culpepper (Darwin) islands for a quick survey of these outer islands. Then, on October 20 the *Beagle* changed course and made all sail southwest to Tahiti, and Galápagos gradually disappeared from view.

From an aesthetic point of view Darwin was not too sorry to leave. The lush highlands of Charles and James had brought a smile to his face, but he had spent the vast majority of his time in the dry lowlands. The Galápagos Islands, he stated in his diary, are generally gloomy and sun-scorched. They have the "arid Volcanic soil, [and] flowering leafless Vegetation [of] an Intertropical region, but without the beauty which generally accompanies such a position."[208]

From a scientific standpoint, however, Darwin was heart-broken. The geology had been awe-inspiring. The biology had been fascinating. How much so, he was only just beginning to realize. For even as the islands faded from Darwin's wistful gaze, they brightened and intensified in his ripening thoughts. The enchanted islands had cast their magic. Darwin was spellbound.

Part 3
After Galápagos

It is the fate of every voyager, when he has just discovered what object in any place is more particularly worthy of his attention, to be hurried from it.

—Charles Darwin, *Journal and Remarks*[1]

Chapter VIII

~~~~~~~~~~~~~~~~~~~~

Homeward Bound

Looking Back, Leaping Forward

During the 5 weeks the *Beagle* had cruised Galápagos, Darwin had spent 19 days (over 162 daylight hours), and 10 nights on four different islands. He had hiked roughly 80 kilometers (50 miles), climbed to over 900 meters (about 3000 feet) in elevation, reached the summit of two of the islands, and examined dozens of craters. With help from his servant and the officers of the *Beagle* he had also amassed a collection of about 500 specimens. Over half of these were plants and their duplicates. The rest consisted of 65 birds (most as lightly stuffed, unsewn skins), 40 rocks, 15 reptiles (excluding the live tortoises), 15 fish, 17 land snails, and 13 "insect" collections totaling approximately 85 arthropods. He had also collected seashells and several marine invertebrates, including a sea slug, a sea fan, a sea pen, and a "most beautiful" "Lake red" sea anemone.[1] Darwin had "worked hard"[2] during his time in Galápagos and as a result of his labors had much to sort through and even more to ponder. The 26-day passage to Tahiti, "pleasantly cross[ing] the blue ocean,"[3] provided him the first opportunity to do so.

Galápagos had surprised Darwin. He had arrived in the islands hoping for tertiary fossils to reveal the history of animal succession in the archipelago just as they had helped elucidate the organic history of the South American continent. He did not find any such fossils. But their absence, in conjunction with his recognition that Galápagos was geologically young made him think to the future when there might be such a fossil record. He suspected that strata were already

starting to accumulate within the Galápagos archipelago. He imagined that one day the shells and the bones of the tortoises, turtles, and
iguanas he was seeing would be found in the Galápagos rock column
while a new, equivalent fauna, existed on the surface.[4]

Darwin had also hoped to see an eruption in Galápagos. While he
was disappointed in this regard, the geology of "that land of Craters"[5]
had been exceptionally rewarding. He had made observations and
collected the raw materials to explain the formation of a variety of tuff
craters and make sense of their asymmetry, to reinterpret the origin
of volcanic crystals, and to differentiate igneous rocks by their crystal
density and patterns of deposition. He had shed light on the decomposition processes of a lava flow, reasoning that where it is exposed
to the air and away from the eroding forces of the ocean, "the surface . . . appears to resist decomposition for a singular length of time."[6]
He had also come to appreciate how significantly the islands differ in
composition and origin from the South American continent. In South
America the predominant rocks were ancient granites and andesite.
Those in Galápagos were mostly young basalts. The Andes had been
shoved up to their present height by tectonic activity spanning eons.
The Galápagos Islands were almost entirely formed from the buildup
of lava emitted from successive eruptions within a geologically recent
period of time. Even though eruptions were common to both places,
overall the Galápagos Islands were clearly much younger.

As a geologist, Darwin left the islands highly satisfied. As a biologist, however, he was puzzled. When Darwin arrived in the islands
and immediately saw that the birds resembled species on the South
American continent his interest was piqued. Despite their similarities
they were not the same; that would have been disappointing. Rather,
they were "closely allied"[7] in a most tantalizing way, and, as his collections show, certainly novel enough to be worth collecting. Colnett
had likened several of the Galápagos birds to species in New Zealand
and the New Hebrides[8] and one of the plants to the cloth tree of Tahiti and the Sandwich Islands. Dampier had evoked Madagascar, the
Mascarene Islands, and even the West Indies (Lesser Antilles) when
writing about the tortoises,[9] and Byron had defined the archipelago
as a whole "new creation."[10]

Although Darwin immediately recognized South America in the birds, and suspected the same for the plants,[11] he was clearly cautious about jumping to conclusions. After all, the physical environments of the South American continent and Galápagos were quite different. And he had not visited any of the other places mentioned by his predecessors. He would not make any definite statement just yet. Instead, he wrote a guarded entry in his diary, "It will be very interesting to find from future comparison to what district or "centre of creation" the organized beings of this archipelago must be attached."[12] After the voyage, when he no longer believed in divine centers of creation, when his specimens were described and he could honestly compare and contrast them with organisms from all over the world, he would loudly declare the South American character of Galápagos. He would then write:

> It is probable that the islands of the Cape de Verd [Cape Verde] group resemble, in all their physical conditions, far more closely the Galapagos Islands than these latter physically resemble the coast of America; yet the aboriginal inhabitants of the two groups are totally unalike; those of the Cape de Verd Islands bearing the impress of Africa, as the inhabitants of the Galapagos Archipelago are stamped with that of America.[13]

Darwin had also arrived in Galápagos thinking that the fauna and flora of one island would be typical of the group. As he proceeded through the archipelago and beyond, it began to dawn upon him that far from being carbon copies, "the different islands to a considerable extent are inhabited by a different set of beings."[14] For his collection purposes the realization came too late. He had collected on all the islands, but excepting the mockingbirds, he had not collected a complete set of organisms from each, nor had he been consistent in keeping his island collections separate. By the time he reached James Island he had already "partially mingled together the collections from two of the islands [Chatham and Charles]."[15] At the end of the five weeks, Darwin finally appreciated that the different islands possess different plant species, but by then it was too late to collect accordingly. And by the time he understood the evolutionary significance of

Lawson's words (and those of the Spanish settlers) that the tortoises were different on the different isles, he had left Galápagos and was on his way home. The realization that many of the organisms of Galápagos are island specific, this single, "most remarkable feature in the natural history of the archipelago,"[16] had come too slowly. One can imagine his sense of frustration at the time. Less of a gentleman and he might well have beaten down the captain's door, begging him to turn the ship around and make all haste back to Galápagos. Instead, Darwin reacted philosophically, refusing to dwell on his mistake. "No hay remedio"[17] (it cannot be helped), as the "Spaniards" would say. Darwin still had his collections from the various islands, many of which (a notable exception being the finches) were well labeled for locality. As he would write later:

> It is useless to repeat here my regrets at not having procured a perfect series in every order of nature from the several islands: my excuse must be, the entire novelty of the fact, that islands in sight of each other should be characterized by peculiar faunas: I ought, perhaps, rather to think it fortunate, that sufficient materials were obtained to establish so remarkable a circumstance in the geographical distribution of organic beings, although they are insufficient to determine to what extent the fact holds good.[18]

Darwin visited a highly diverse subset of Galápagos islands and it is thanks to these choice destinations that he was able to recognize any interisland differences in the wildlife at all. There is little point in speculating what Darwin could have seen, and might have noticed, had he visited more islands in Galápagos—with one exception. The *Beagle* approached Hood (Española) Island on two occasions; once (on September 16) to dispatch surveying officers Chaffers and Mellersh in a whaleboat to chart the island's coastline, and once (on October 14) to sound Gardner Bay. On neither occasion did Darwin have the opportunity to disembark; the *Beagle* took him onward to Chatham the first time, and he was camped on James Island the second time. However, had Darwin been able to explore Hood Island on foot it is quite probable he would have recognized that the islands are tenanted by different but related organisms, and collected accordingly. His evidence of speciation would certainly have multiplied, for the three southern islands of Hood, Chatham, and Charles exhibit the greatest degree of endemism in the archipelago. That much of the fauna and flora on these three islands is island specific can be explained by the unusually isolated position of these islands in relation to the prevailing currents and winds. They are also among the oldest islands in the archipelago.

By visiting Hood, Darwin would have seen the fourth Galápagos mockingbird species (*Mimus macdonaldi*). The mockingbirds were the only birds that Darwin recognized as differing between islands while he was in Galápagos, and he would not have missed noticing the peculiarity of the Hood Island mockingbird, with its large body size and long, decurved bill.

The markedly large and colorful Hood Island lava lizard (*Microlophus delanonis*) would surely have awakened him to the fact that the lava lizards also differ between islands. As it was he missed noticing the more subtle differences between the lava lizards on the islands he did visit.

On Hood Island Darwin could have collected the large cactus finch (*Geospiza conirostris*), a species that exists only on Hood and Tower (Genovesa) islands. The large cactus finch has a substantially larger and thicker bill than that of the cactus finch (*Geospiza scandens*), a species Darwin saw and collected on the other islands. Had he noticed this early on he might well have paid better attention to labeling his finch specimens by locality.

Had he examined the pronounced saddlebacked tortoise of Hood Island (*Geochelone nigra hoodensis*) on site, in the wild, he may have listened more attentively when Lawson first told him that the tortoises are different on the different islands.

Darwin left the Galápagos with one overwhelming and haunting impression—isolation. The sense of remoteness was everywhere—in the sea, in the lava, in the animals, and in the people. He had felt it in the tedious passages between the islands, hindered by contrary winds and currents in some places, dead calms in others. He had seen it in the lava flows that had separated tuff craters from their oceanic birthplace and salt lakes from their oceanic source. The officers of the

Beagle had measured it in the fathomless depths between islands and, in some places, right off shore. Darwin had seen it in the faces of the wanting settlers of Charles Island and the forgotten men on James Island. He had touched it in the fearlessness of the birds, smelled it in the alien scents of the plants, and heard it in the hiss and bellow of the seemingly deaf tortoises. How distant these islands were from the continent, and how removed they were from their neighbors, despite being in sight of each other.

It was this very sense of isolation that would help Darwin recognize the significance of the variation he saw in the Galápagos organisms, and allow him to make sense of it all. He would come to understand that isolation is a passive but powerful force that allows varieties to diverge so much that by degrees they evolve into new species. Many years later, in a letter to Moritz Wagner in 1876 he reflected "it would have been a strange fact if I had overlooked the importance of isola-tion, in seeing that it was such cases as that of the Galápagos Ar-chipelago, which chiefly led me to study the origin of species."[19] But during the voyage, and on an emotional level, isolation was a dis-turbing concept. The Galápagos scenery, with its raw lava fields and "poor vegetation," had not spoken to him of physical beauty, but had impressed him with its "desolate and frightful aspect."[20] "On a forlorn & weather-beaten coast" he wrote at the end of his diary, as he was re-flecting back on his adventures, "the feelings partake more of horror than of wild delight."[21]

Dis-Belief

Over the next 11 months the *Beagle* circumnavigated the globe, sail-ing to New Zealand, Australia, Africa, back to South America, and then on to England. His greatest accomplishment on the way home took place at the Cocos (Keeling) Islands. There he examined a coral atoll and in doing so verified a theory that he had formulated on the west coast of South America when contemplating the vertical move-ments of the earth's crust, and had been itching to prove ever since. Charles Lyell believed that coral atolls form and sit on the rims of shallowly submerged volcanic craters, but Darwin thought that they merely showed where volcanic islands had once been. He evoked his favored topic of uplift and subsidence to explain their formation. The upward movement of the Andes must be counterbalanced by seafloor subsidence, he argued. Oceanic islands should, over time, sink beneath the surface of the waves. A coral reef growing on the skirts of such an island must also sink, and necessarily die as it descends away from warmth and light. But, at the same rate new polyps should keep building on top of the dead skeletons, and keep a living layer of coral in sun-warmed shallow water. The volcano thus keeps sinking and eventually disappears, leaving only a ring of living coral atop an ever-deepening cylinder of dead coral. He had not been able to prove his theory in Galápagos for he had found no corals there. At Cocos (Keeling) Island, however, he was able to examine an atoll up close, and convince himself that he was right. Islands subside and disappear from view, while coral atolls mark their place. When Darwin returned home, he wrote his coral atoll theory into a paper, which he presented to the Geological Society of London on May 31, 1837,[22] and later expanded into an entire book on coral reef formation.[23]

On the homeward journey there were many prolonged sea passages, giving Darwin ample time for dedicated study. He transcribed and elaborated the notes he had taken in Galápagos, and spent many hours sitting at the chart table of the poop cabin sorting through his specimens and thinking. He would need help from the taxonomists back home to describe and classify his collections, but the one thing he could do was brainstorm. And brainstorm he did, forming the first clouds of a tempest that would shake the world.

A few months before the *Beagle* reached England Darwin rewrote some of his zoology notes into a new compilation of ornithology notes. In a section devoted to the Galápagos mockingbirds and the intriguing fact that the different varieties he had seen were confined to separate islands, he penned a somewhat ambiguous paragraph:

When I recollect, the fact that the form of the body, shape of scales & general size, the Spaniards can at once pronounce, from which

Island any Tortoise may have been brought. When I see these Islands in sight of each other, & possessed of but a scanty stock of
animals, tenanted by these [mocking] birds, but slightly differing
in structure & filling the same place in Nature, I must suspect they
are only varieties. The only fact of a similar kind of which I am
aware, is the constant | asserted difference— between the wolf-like
Fox of East & West Falkland Islds.— If there is the slightest foundation for these remarks the zoology of Archipelagoes — will be well
worth examining; for such facts would undermine the stability of
Species.[24]

His words show that he was teetering on the brink of dis-belief. He
still approached the subject from the point of view that species were
divinely created, fixed entities, made up of varieties. He, like Lyell,
attributed variation within species to the ability of organisms to "accommodate themselves, to a certain extent, to a change of external
circumstances."[25] Yet it appears that Darwin was now beginning to
suspect that varieties were actually incipient species; that, under isolation, varieties could change or "accommodate themselves" to such a
degree as to become new species.[26] And if this were the case, a plausible explanation for the origin of the various, and often island-specific,
organisms in Galápagos was not that each had been created by God
on site, but that they had transmuted in isolation from ancestral species. And, as he later wrote, "the subject haunted me."[27] What Darwin
needed to tip the scales of his indecision was for his specimens to be
described by the experts; for the structural relationship of apparently
similar organisms, both within the islands and between Galápagos
and the South American continent, to be worked out. He also needed
to understand how organisms change. Until then, he realized, "it
seemed . . . almost useless to endeavor to prove by indirect evidence
that species have been modified."[28]

Darwin greeted the end of the voyage with mixed feelings. By now
he was weary of the sea and longed for home, yet he was aware that an
exciting chapter in his life was drawing to a close. A dangerous new
one, in which he would be challenging faith itself, was about to begin.
He wrote in his diary that on October 2, 1836, as the *Beagle* "came

to an anchor at Falmouth. —To my surprise and shame I confess the first sight of the shores of England inspired me with no warmer feelings, than if it had been a miserable Portugeese [*sic*] settlement."[29] He started for home by coach that very night, arriving at The Mount two days later. Only in the welcome arms of his family did he once again feel that "the wide world does not contain so happy a prospect as the rich cultivated land of England."[30]

Chapter IX

~~~~~~~~~~~~~~~~~~~~~~~~~~~~~~

A New Voyage

I have every motive to work hard . . .
—Charles Darwin, *Life and Letters*[1]

Darwin never physically returned to Galápagos but for the rest of his life he visited it again and again in his constant deliberation of the diversity of life. When Darwin stepped off the *Beagle* and onto English soil he was embarking on a whole new voyage—a voyage to explore his budding conviction in evolution. He did not know where it would take him or how long it would last, for he was not completely convinced that evolution was real, but he boarded this new ship with enthusiasm. He would never disembark.

Darwin began his second adventure in a frenzy of activity, so that by November 1836 he was panting, "The busiest time of the whole [*Beagle*] voyage has been tranquillity itself to this last month."[2] Darwin was most anxious to see if his specimens had arrived intact. First he dashed off to Cambridge to see his friend Professor Henslow and the specimens he had sent back from South America. He then rushed to London in time to remove the rest of his collections, including all the Galápagos specimens, from the *Beagle* when she docked at Woolwich. Next he found specialists to examine them all. His fossil mammals went to Richard Owen, an outstanding vertebrate zoologist and palaeontologist at the Royal College of Surgeons. These fossils had been shipped to England early in the voyage and Owen had already examined and described many of them. In doing so he bestowed

17A

17B

17C

17D

17A. Freshwater cove of the Buccaniers (Buccaneer Cove) viewed from the general location of Darwin's campsite on James Island.

17B. Galápagos hawk (*Buteo galapagoensis*).

17C and 17D. "Darwin's layer cake" (17C) and pinnacle (17D) in Freshwater cove of the Buccaniers (Buccaneer Cove), James Island.

18A. Prickly pear cactus (*Opuntia galapageia*) on James Island.

18B. Drawing of *Opuntia galapageia*, from a sketch published by John Stevens Henslow.

18C. Flower of *Opuntia echios*.

18D and 18E. Tree scalesia (*Scalesia pedunculata*) (18D). One of Darwin's specimens of
   S. pedunculata from James Island (18E).

19A–19E. Bulimulus snail (*B. reibischi*). Darwin described tortoises (*Geochelone nigra darwini*) feeding on lichen-draped guayabillo (*Psidium galapageium*) berries in the highlands of James Island.

20A. Cerro Pelado, highlands of James Island. Note the barren, goat-grazed hill behind the crater.

20B. A stand of mature scalesia trees (*Scalesia pedunculata*) survives inside a goat-proof fenced enclosure in the highlands of James Island.

20C. After an intense eradication program, James Island was declared goat-free in 2006.

20D. Feral goats (*Capra hircus*).

21A. James Bay lava flow and Pan de Azucar, James Island.

21B. James Bay lava flow and highlands of James Island.

21C. Looking west from the highlands of James Island. Darwin volcano on Albemarle Island can be seen in the distance.

22A

22B

22A and 22C. Tuff craters of James Island: Cerro Cowan (22A) and the Salina or Salt Mine Crater with Pan de Azucar in the distance (22C).

22B. One of Darwin's tuff specimens, from Chatham Island.

22C

23A   23B

23C   23D   23F

23E

Galápagos bird species observed by Darwin.
23A. Galápagos hawk (*Buteo galapagoensis*).
23B. Barn owl (*Tyto alba punctatissima*).
23C. Galápagos dove (*Zenaida galapagoensis*).
23D. Vermillion flycatcher (*Pyrocephalus rubinus*).
23E. Short-eared owl (*Asio flammeus galapagoensis*).
23F. Yellow warbler (*Dendroica petechia aureola*).

24A

24B

24C

24D

24E

24F

24A. Magnificent frigatebird (*Fregata magnificens*).

24B. Yellow-crowned night heron (*Nyctanassa violacea*).

24C. Brown pelican (*Pelecanus occidentalis urinator*).

24D. Striated (lava) heron (*Butorides striata*).

24E. Great blue heron (*Ardea herodias*).

24F. Lava gull (*Leucophaeus fuliginosus*).

25A

25B

25C

25E

25D

A sample of Darwin's Beagle *collections.*

25A. Large ground finch (*Geospiza magnirostris*).

25B. Endemic Galápagos fern (*Ctenitis pleiosoros*).

25C. Fish specimen.

25D. Rock sample from Albemarle Island.

25E. Geological specimens notebook, which includes Galápagos numbers 3220 to 3292.

26A

26B

26C

26E

26D

More Galápagos plant species collected by Darwin.
26A. Castela (*Castela galapageia*).
26B. Matazarno (*Piscidia carthagenensis*).
26C. Hairy Galápagos tomato (*Lycopersicon chees-
          manii var. minor*).
26D. Yellow cordia (*Cordia lutea*).
26E. Radiate-headed scalesia (*Scalesia affinis*).

Galápagos tortoises (Geochelone nigra) showing a gradation in carapace shape, from the saddlebacked form (top) to the dome-shaped form (bottom).

27A. Hood Island tortoise (*G. nigra hoodensis*).

27B. Chatham Island tortoise (*G. nigra chathamensis*).

27C. James Island tortoise (*G. nigra darwini*).

27D. Indefatigable Island tortoise (*G. nigra porteri*).

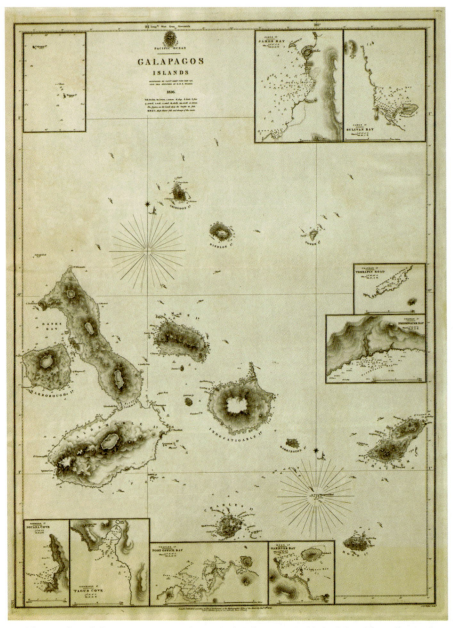

28. The results of the survey. FitzRoy's map of Galápagos. Published by the Hydrographic Office of the Admiralty, 1841.

29B

29A

29C

29D

29E

29F

Several Galápagos species that Darwin either did not see or did not consider unique or otherwise notable. He described the frigatebird, but only its feet and flight, and not the male's remarkable, inflatable red gular pouch.

29A. Waved albatross (*Phoebastria irrorata*).

29B. Sea lion (*Zalophus wollebaeki*).

29C. Galápagos penguin (*Spheniscus mendiculus*).

29D. Flightless cormorant (*Phalacrocorax harrisi*).

29E. Blue-footed booby (*Sula nebouxii*).

29F. Great frigatebird (*Fregata minor*).

30A

30B

30C

30D

30A. Wedgwood jasperware with cameo of Darwin.

30B. Darwin, age 31, four years after returning to England.

30C. Down House where Darwin resided 1842–1882.

30D. The Royal Society's Darwin Medal, awarded to Peter and Rosemary Grant in 2002 for their work on Darwin's Finches in Galápagos. The first recipient was Wallace in 1890.

31A

31B

31C

31D

31A–31D. Tourist yachts in Academy Bay, Indefatigable Island. The resident human population of Galápagos has increased 100-fold since 1835. The Galápagos National Park Service battles an increasing number of introduced, invasive pests such as the Polistes paper wasp which arrived in 1988, most likely with a shipment of bananas for the town of Puerto Ayora.

32A

32B

32C

CHARLES DARWIN
landed on the Galapagos Islands in
1835 and his studies of the distribution of animals and plants thereon
led him for the first time to consider the
problem of organic evolution. Thus was
started that revolution in thought on
this subject which has since taken place. _ Erected September 17th 1935 by
the members of the Darwin Memorial Ex
pedition. _ Victor Wolfgang von Hagen
Alexander R. Brown III Christina Inez
Brooks Christine Inez von Hagen
David Hunter

32D

32A. Street names in Puerto Ayora, Indefatigable Island. Ecuador annexed the Galápagos Islands on "12 de Febrero." February 12 is coincidentally Darwin's birthday.

32B and 32D. Statue of Charles Darwin erected at Puerto Baquerizo Moreno, Chatham Island, on the centenary of Darwin's visit to Galápagos.

32C. Entrance to the Charles Darwin Research Station, Puerto Ayora, Indefatigable Island.

Darwin with recognition as a worthy collector before Darwin even reached home. Darwin's birds went to John Gould, his mammals and most of his insects to George Waterhouse, his reptiles to Thomas Bell, his fish to Leonard Jenyns, and his plants to John Stevens Henslow. Darwin kept the rocks for himself but solicited help from mineralogist William Miller to describe them.

It took almost ten years for the taxonomists to complete the specimen descriptions, with Darwin occasionally assisting and often cheering on from the sidelines. But the results that determined Darwin's persuasion to evolution came back almost immediately. First, Owen's descriptions of Darwin's fossil mammals confirmed Darwin's suspicion that they were closely allied to living mammals. Modern analysis has reconfirmed that the *Hoplophorus* and the *Megatherium* fossils that Darwin collected are extinct relatives of today's South American armadillos and sloths. However, his *Toxodon* and *Macrauchenia* fossils, which Darwin and Owen believed to be related to capybaras and llamas, have since been shown to belong to two distinct South American groups of mammals with no living relatives.[3]

By March 1837, John Gould had confirmed that Darwin's Galápagos mockingbirds were indeed three distinct species, that Darwin's rare rhea from Patagonia was a new species, quite distinct from the common rhea found further north, and that Darwin's "little birds" from Galápagos comprised one new and highly diverse family of finches. To Darwin's new way of thinking this last piece of news sounded like an evolutionary windfall; a diverse group of closely allied species, entirely confined to the Galápagos archipelago, surely pointed to the mutability of species. Clearly, the finches had evolved from a common ancestral species and through isolation on the different islands had diverged over time into an array of new species. Or had they? As Darwin had failed to record the island locality for each finch specimen he was unable to show any kind of geographical distribution for the species, let alone whether any of them were confined to specific islands. Without island specificity Darwin could not support his supposition that the finches had diversified in isolation on the different islands. Ultimately, the finches were of limited use as evidence for transmutation. The most he could say about the evolution of the finches was more of a hint than a definitive statement. In 1845, in the second edition of the *Journal*, he wrote: "Seeing this gradation and diversity of structure in one small, intimately related group of birds, one might really fancy that from an original paucity of birds in this archipelago, one species had been taken and modified for different ends."[4]

While now, in March 1837, Darwin was convinced that species were mutable, his understanding of evolution was very much "under the form of an abstract."[5] He needed a better grasp of the subject, and certainly more evidence to bolster his arguments before going public with his ideas. He had already discovered the vehemence with which others balked at the concept of evolution. When he dared mention to Richard Owen the possibility that old species could give rise to new species, he was severely rebuked. Owen could be excessively arrogant, and Darwin saw him as being "vehemently opposed to any mutability in species."[6] Owen recognized that Darwin's extinct fossil mammals were related to modern mammals, but he did not mean, as Darwin did, that the living mammals were modified descendants of these extinct forms. He simply meant that they were "related" on a Cuvierian scale of classification, by having similar anatomical structures. Darwin's mentors Henslow and Sedgwick were also firm believers in God the "Creator" and fixed species, and Darwin was reluctant to broach the subject with them. Robert Grant, the professor who had introduced Darwin to evolutionary thought back when Darwin was a student at Edinburgh, was now being ridiculed, even ostracized, for his radical views on Lamarckian evolution. Darwin was anxious to avoid the same fate. Speaking openly about his thoughts on evolution at an early stage, before they were fully developed, would put him at risk of being silenced forever. Far better to strengthen his ideas and build his reputation before handing round the blueprints to his thinking. He therefore kept a low profile about his convictions, while ploughing full steam ahead with the construction of his theory. He confided scantily to his family and a few colleagues about his transmutationist thoughts, and consulted widely and voraciously for information to feed his understanding of evolution. It was in this quiet but industrious manner that he spent the next 20 years building his revolutionary theory.[7]

Instead of talking publicly about transmutation, Darwin wrote privately. He jotted down his ideas in alphabetically labeled notebooks, starting in July 1837 with his famous "B" notebook, entitled "Transmutation." In a personal journal, started in 1838, he clarified the date of his persuasion to evolution by writing, "Had been greatly struck from about Month of previous March [1837] on character of S. American fossils —& species on Galápagos Archipelago. These facts origin (especially latter) of all my views."[8]

Publicly, Darwin wrote scientific papers and books about the voyage, busily laying down the groundwork for his acceptance into the scientific community and the eventual reception of his theory. He started with a paper entitled *Proofs of recent elevation on the coast of Chili [sic]*, which was presented to the Royal Geological Society on January 4, 1837.[9] Charles Lyell, president of the society, gave his approval, thus bestowing Darwin with automatic respect among the scientific elite. Next, Darwin transformed his *Beagle* diary into an informative and entertaining summary of what he had done and seen during the voyage. He finished writing this *Journal and Remarks* by June 1837 but it was not until May 1839 that the book was published. It appeared as the third volume of Captain FitzRoy's *Narrative of the Surveying Voyages of His Majesty's Ships Adventure and Beagle between the years 1826 and 1836*, and was published by Henry Colburn of London. The set was considered long-winded and repetitive, and sold poorly. But a few months later Darwin's volume was reprinted on its own as a separate book entitled *Journal of Researches into the Geology and Natural History of the Various Countries Visited by H.M.S. Beagle*. It received much better reviews, the most praiseworthy being a glowing letter of approval from Darwin's hero, Alexander von Humboldt.[10] Not only was Darwin getting noticed, he was forging "a fair place among scientific men."[11]

Darwin then wrote the book he had always wanted to, ever since inspiration struck on the lava fields of St Jago in the Cape de Verd (Santiago, Cape Verde Islands) at the beginning of the *Beagle* voyage. The *Geology of the Voyage of the Beagle* quickly outgrew the limits of one binding, and Darwin was forced to edit it down into publishable sections. The first, *The Structure and Distribution of Coral Reefs*, was published in May 1842, the second, *Volcanic Islands*, featuring Galápagos, was published in November 1844, and the third, *South America*, was published in late 1846.[12] These works brought Darwin much acclaim among the leading geologists of the day. He was elected to the Geological Society in 1838 and promoted to vice-president soon after the publication of *Coral Reefs*.

Darwin also contributed heavily to *The Zoology of the Beagle*, a lavishly illustrated five-part encyclopedia of the fauna encountered during the voyage, financed by a generous government grant. Each part consisted of several volumes. The leading experts provided species descriptions of the *Beagle* specimens and Darwin commented on the behavior of the animals and the habitats in which they were found. *The Zoology of the Beagle* took five years to complete and the order in which the volumes materialized reflects the speed at which the taxonomists worked on the project. The four volumes of *Part I, Fossil Mammalia* (Owen), were completed by 1840; *Part II, Mammalia* (Waterhouse) in four volumes, was published in February 1839; *Part III, Birds* (Gould) in five volumes, was completed in 1841; *Part IV, Fish* (Jenyns) in four volumes, was done by early 1842; and *Part V, Reptiles* (Bell) in two volumes, was out in late 1842.

Theory by Which to Work

Early in his persuasion to evolution Darwin saw that he needed an explanation for evolution. It was one thing to say that species change, another to say how. To better understand the concept of variation in organisms he turned to pigeon fanciers, dog breeders, animal husbandry experts, and horticulturalists, and bombarded them with questions about variation in domestic breeds, in the hopes that "some light might perhaps be thrown on the whole subject."[13] From them he learned that artificial selection of desired traits "was the keystone of man's success in making useful races of animals and plants."[14] He learned, for instance, that by exploiting the natural variation that exists in wild populations of rock pigeons, humans had been able to create over 150 races of domestic pigeons. So different were their external characteristics that they appeared to belong to different genera.

Pigeons with elaborately long tails had been created by selecting individuals with slightly longer than average tail-feathers, breeding them together, and crossing their offspring with more long-tailed progeny. The same process had given rise to races of pigeons with stubby beaks, only this time the chosen trait was short beaks.

How the principle of artificial selection might apply to the natural world, however, remained a mystery to Darwin. Then, in October 1838, Darwin "happened to read for amusement"[15] Thomas Malthus's *Essay on the Principle of Population.* Malthus wrote that humans tend to reproduce at an unsustainable rate and that this superfecundity leads to an excess of offspring. Population growth is kept in check by famine, disease, poverty, and war. Individuals able to resist these agents of "misery and vice" survive to reproduce[16] but others die, and in doing so prevent the population from expanding out of control. Darwin immediately applied Malthus's principle of population, with its elements of struggle and competition, to the natural world, and, as he later wrote, "the idea of Natural Selection flashed on me."[17] What went on in the wild was analogous to artificial selection. Whereas breeders specifically select for a few traits such as fancy feathers or short beaks, Nature selects for traits that assist in "the struggle for life."[18] As Darwin later clarified in his autobiography: "[B]eing well prepared to appreciate the struggle for existence which everywhere goes on from long-continued observation of the habits of animals and plants, it at once struck me that under these circumstances favourable variations would tend to be preserved, and unfavourable ones to be destroyed. The result of this would be the formation of new species."[19] Darwin now had "a theory by which to work,"[20] an explanation for evolution. Four years later, he gave his "principle of preservation" a name—*natural selection.*[21]

"A Wonderful Spot"

Galápagos may not have featured in Darwin's conception of natural selection, but it became a fountain of support for his theory of evolution by means of natural selection. As the *Beagle* collections were examined Darwin learned that almost all his Galápagos specimens of land birds, fish, and insects were new species, allied to species on the South American continent, but peculiar (that is, unique) to Galápagos. Those three little words—"new," "allied," "peculiar" —were infinitely significant to Darwin. Grouped together they formed a veritable road sign that pointed down the path of evolution. Specifically in the case of Galápagos, they told Darwin that organisms had colonized the archipelago from the South American continent and their descendants had evolved into new species as the selective pressures of the new environment had modified them.

With Darwin's growing appreciation of the process of natural selection he did not need distinction at the species level to validate his theory of evolution. He was convinced that varieties were incipient species, and the fact that the boundary between them was often blurred was to be expected. The decision to name an organism as a distinct species, or a variety, he recognized, was sometimes entirely arbitrary. Birds and insects were often being "ranked by one eminent naturalist as undoubted species, and by another as varieties."[22] And this looseness of classification was entirely acceptable to Darwin. If some supposed species "hereafter prove to be only well-marked races" he wrote, "this would be of equally great interest to the philosophical naturalist."[23] For "races," he saw, are the steps of modification "towards more strongly-marked and permanent varieties, . . . [to] sub-species, and then to species."[24] On the other hand, to strengthen his theory and make it convincing, Darwin ideally needed series of similar but well-delineated species confined to geographically isolated areas to plug his argument for how one species might evolve into a different species. It was Galápagos's unique fauna and flora, which in many cases differs at the species level not only between the South American mainland and the archipelago but between the islands themselves, that would best answer this call.

For the next decade Galápagos excited Darwin. "[T]he more I go into the [Galápagos] Fauna, the more peculiar it is," he exclaimed to George Waterhouse in 1845. "[O]ut of 17 land shells 16 are new species confined to the group!"[25] Darwin's Galápagos collections were not perfect. They were limited, and in some areas, such as the reptiles, they were decidedly lacking. But collections made prior to Darwin's visit to Galápagos, few though they were, helped fill in some of the gaps.

With the help of a small assortment of reptile specimens brought to England on various ships and left to gather dust in museums, herpetologists decided there were two endemic species of tortoise and two endemic species of marine iguana in Galápagos. Hugh Cuming's collection of 90 seashells taken from the islands in 1829 also helped support Darwin's ideas, for they demonstrated a clear connection between Galápagos and neighboring lands. Darwin learned that over half the species were endemic to Galápagos and all belonged to genera found either along the western coast of South America or in the eastern Pacific. "I find in sea-shells the Galapagos is a point of union for two grand & otherwise most distinct concho-geographical divisions of the world," Darwin wrote in 1845.[26] Many of the marine shells Cuming collected have since been found outside Galápagos, so that today only a quarter of the marine mollusc fauna of Galápagos is considered endemic.[27] Nonetheless, Darwin was correct in identifying the places of origin of the shells. For Darwin it was important knowledge because it helped emphasize the fact that organisms of Galápagos had evolved from colonists from distant but feasibly accessible places, rather than having been created by God on site.

Best of all were the Galápagos plants. In 1843, Henslow, having sat upon Darwin's collection for seven years, handed the pressed specimens over to his future son-in-law, botanist Joseph Dalton Hooker. Hooker described Darwin's plants with speed and enthusiasm, and Darwin embraced the results as a desert welcomes rain. Galápagos was a veritable fountain of evolutionary fact. "What a wonderful spot it is!"[28] he wrote to Hooker in a fit of joy. In 1845, Hooker told Darwin that out of 185 species[29] he had examined so far, over half (100) were found exclusively in Galápagos and 85 were confined to just one island.[30] Genus after genus had multiple species that were "very *wonderfully* peculiar"[31] to the separate islands.

While based primarily on Darwin's plant collection, Hooker's *Enumeration of the plants of the Galapagos Archipelago*,[32] published in 1847, includes about 65 taxa (many duplicated by Darwin) collected by naturalists who visited Galápagos before Darwin: David Douglas and John Scouler in 1825, James Macrae in 1825, and Hugh Cuming in 1829.[33] Hooker found their collections in the Herbarium of the Royal Botanic Gardens at Kew and acquired the rest from the Horticultural

Society of London through Henslow. As all these collections had yet to be described, Hooker took it upon himself to do so, and included them in his Galápagos paper. At one point Darwin became worried that Hooker had based most of his findings on the other men's collections but Hooker was quick to assure his friend. "Your collection was the main material," he wrote, " [though] Macrae . . . formed the best part of the Albemarle Isld Collections."[34] Hooker also included 30 taxa collected by the first collectors to follow Darwin's 1835 visit: Abel du Petit-Thouars in 1838, and Thomas Edmonston and John Goodridge in 1846.[35] Edmonston's contribution was made partly thanks to Darwin. Darwin had written to him shortly before his trip on HMS *Herald*, urging him "to collect everything at the Galapagos, & attend particularly to the productions of the *different* islands."[36] Edmonston collected several Galápagos plants, but was accidentally and fatally shot in Peru just one week after leaving the islands.[37] Although his specimens made it back to England, many were in poor condition and had not been kept separated by island. Today several plant species, including the common Galápagos carpet weed *Sesuvium edmonstonei*, bear Edmonston's name.

When Hooker began describing the Galápagos plants he not only bolstered Darwin's evidence for evolution, he boosted Darwin's confidence, so much so that Darwin decided to divulge his brewing thoughts. On January 11, 1844, Darwin wrote his friend a letter in which he disclosed, with far less conviction than he felt, "I am almost convinced . . . that species are not (it is like confessing a murder) immutable."[38] To Darwin's delight Hooker, who was young and open to new ideas, took Darwin's "presumption"[39] in stride and asked to hear more.

With renewed enthusiasm Darwin rewrote his *Journal* to include all he had learned from analysis of his specimens. His Galápagos chapter grew by a third and hinted boldly of his transmutationist ideas:

> Seeing every height crowned with its crater, and the boundaries of most of the lava-streams still distinct, we are led to believe that within a period, geologically recent, the unbroken ocean was here spread out. Hence, both in space and time, we seem to be brought somewhat near to that great fact—that mystery of mysteries—the first appearance of new beings on this earth.

Although these words could be interpreted as meaning that Galá-pagos was a "centre of creation," a place where God had recently put down new species, Darwin no longer saw it this way. He continued with a lengthy discussion on the relationships of the various faunal and floral inhabitants of the islands that implied (rather than openly stated) his conviction that the "new beings" had evolved from other beings. He wrote:

> The distribution of the tenants of this archipelago would not be nearly so wonderful, if, for instance, one island had a mocking-thrush, and a second island some other quite distinct species; — if one island had its genus of lizard and a second island another distinct genus. . . . But it is the circumstance, that several of the islands possess their own species of the tortoise, mocking-thrush, finches, and numerous plants, these species having the same general habits, occupying analogous situations, and obviously filling the same place in the natural economy of this archipelago, that strikes me with wonder.[40]

This new edition was published in the summer of 1845. Two months later Hooker deflated Darwin's growing confidence. The two friends were discussing the works of French botanical writer Frédéric Gérard, when Hooker scoffed, "no one has hardly a right to examine the question of species who has not minutely described many."[41] Hooker's scorn was directed at Gérard, but Darwin took it to heart. He defended his position by reminding himself, and Hooker, that he had "dabbled in several branches of Nat. Hit: [*sic*, Natural History] & seen good specific men work out my species & know something of geology; (an indispensable union)."[42] Still, he recognized that he might get "more kicks than half-pennies"[43] for his views and the worry of this may well have inspired his next move.

Seeds across the Sea

In 1846 Darwin began examining some of the marine invertebrate specimens from the *Beagle* voyage that had not yet been described. He started with a new barnacle species, which he described and named *Arthrobalanus minutes*.[44] He became enthralled by its beauty, fascinated by its structure, and inspired to examine more barnacle species. Among the specimens he looked at were four species collected from Galápagos: *Conchoderma virgata, Balanus tintinnabulum, Platylepus decorata, and Tetraclita porosa* var *nigresens*.[45] He described and named two other species that can be found in Galápagos—*Megabalanus vinaceus* and *Balanus trigonus*—but did not include the Galápagos islands in their distribution.[46] It was the beginning of a systematic study of the entire order of barnacles (Cirripedia), an endeavor that lasted almost eight years. Darwin's barnacle study became such a steady occupation that one of Darwin's young sons, assuming all fathers worked alike, asked a neighbor's child, "Where does your father do his barnacles?"[47] Despite delaying the disclosure of his theory of evolution by means of natural selection, Darwin's barnacle career gave him valuable insight into the processes of evolution. For example, the barnacles showed him how gonochoristic (separate sex) species had evolved from hermaphroditic ones in insensibly small stages. The study also helped earn him, in 1853, the Royal Medal of the Royal Society. Throughout his barnacle years Darwin all but forgot Galápagos. But in 1855 he returned his mind to the field to address the problem of how oceanic islands might be colonized. Darwin was sceptical of Edward Forbes' then-popular sunken land bridge theory. He was sure that most oceanic islands, including Galápagos, had never been connected to the mainland, as Forbes believed, but had emerged from the sea as volcanoes, and then been colonized by plants and animals traveling over water from distant lands. During the first year of the voyage, and 60 miles off the coast of Monte Video (Montevideo, Uruguay), he had encountered thousands of aeronautic spiders being carried by the wind on silken parachutes; they had coated the *Beagle*'s rigging with a fine "Gossamer web."[48] On another occasion a beetle flew onto the deck when the ship was 45 miles from land.[49] Darwin had little doubt that other organisms were capable of traveling large distances over water, by drifting in the wind, or by floating alone or on rafts of vegetation. He suspected that many seeds were dispersed this way, but he had never actually put this idea to the test. It was time to do so.

The subject of seed dispersal had haunted Darwin for some time. As early as 1837 he had forecast to Henslow, "At some future time I shall want to know . . . whether seeds could probably endure floating on salt water"[50] for long distances. Nine years later, in 1846, Darwin was "delighted to hear" that Hooker had taken it upon himself to "systematically . . . [go] through the individual powers of transport of the Galápagos plants." Darwin applauded Hooker's initiative for he had "often wished to see this done [and had] never met with such a discussion."[51] He reminded Hooker of the migrant bobolink he had collected on James Island, implying that it might represent one kind of vector for the transport of seeds from the American continent to Galápagos.[52]

Now, in 1855, another nine years down the line, Darwin was ready to study the problem of seed dispersal himself. With his children's help, for by now he had fathered nine (with the tenth and last on the way) he set about proving his conviction with experiments. He floated a variety of seeds in tanks of salt water for long periods of time, after which he tried to germinate them. He also planted seeds taken from mud caked on ducks' feet and extracted from the feces of various zoo animals, and recorded which ones successfully grew into seedlings and which ones failed to sprout. Darwinians in every sense of the word, his children reveled in these mucky experiments. For Darwin himself dispersal turned into an obsessive hobby. It also grew to include other life forms. In 1857 he submerged land snails in seawater to see if they could potentially survive a long sea passage and professed the intent to do the same with lizard eggs, if only he could find some.[53] But most of his research was devoted to seeds, and it often took the form of eccentric requests to both unsuspecting acquaintances, and habituated family members alike. To African traveler James Lamont he wrote:

> As you are so great a sportsman perhaps you will kindly look to one very trifling point for me, as my neighbours here think it too absurd to notice —Namely whether the feet of birds are dirty, whether a few grains of dirt do not adhere occasionally to their feet. I especially want to know how this is in the case of birds like herons

& waders which stalk in the mud —You will guess that this relates
to dispersal of seeds —which is one of my greatest difficulties.[54]

To his eldest son, William, about to begin studies at the University
of Cambridge in 1858, Darwin requested: "If you go out shooting
look at Birds' feet & see if any dirt sticks to them: I want to collect
such dirt, & see if by any splendid chance a plant would come up, for
then could I not carry seeds across the sea!"[55]

The Book That Shook the World

All this time Darwin's private scribblings had kept pace with his work
on evolution, but they had remained private. In 1842 he had written
a 35-page sketch on natural selection, which grew into a 251-page
essay by 1844. Then, after the barnacle years, Darwin set to work
compiling his "huge pile of notes . . . in relation to the transmutation
of species"[56] and writing a lengthy book on natural selection.[57] By
1856 Darwin had a full story to tell. But was the world ready to receive it? Darwin decided to test the waters by consulting with Lyell.
Lyell was not fully convinced of transmutation (he drew the line at
human evolution), but urged Darwin to publish his ideas. Darwin
had spent enough time absorbed "in the application of the theory,"[58]
it was high time he published before someone beat him to it. Despite Lyell's encouragement Darwin hesitated. Perhaps he was loath
to distress his beloved but devoutly religious wife, Emma. His health,
chronically poor since a few years after his return to England, was
becoming worse and he could ill afford the stress that would surely
accompany publication of his ideas. He had not forgotten his reaction
when in October 1844 Robert Chambers, a middle-class essayist, had
published an evolutionary book called *Vestiges of the Natural History
of Creation*.[59] Darwin had recoiled at the controversy it spurned. He
had read a scathing attack by Sedgwick in the *Edinburgh Review* "with
fear & trembling."[60] While Darwin rejected most of Chambers arguments, for they were highly speculative ideas about a "law of development"[61] that connected all living things, he lamented how Chambers

had "done the subject [of Transmutation] harm."[62] Now, twelve years later, Chambers' downfall was still making it that much harder for Darwin to come out of the closet himself.

On June 18, 1858, Darwin received a bombshell in the mail, a time bomb that threatened to destroy any hope of receiving priority for his life's work if he did not publish and publish soon. A letter arrived from the Malay Archipelago, where the author, Alfred Russel Wallace, was on an eight-year collecting expedition. Wallace had hit upon the same concept of natural selection as an explanation for evolution that Darwin had come up with almost twenty years earlier. Darwin immediately wrote to Lyell, "Your words have come true with a vengeance that I sh^d. be forestalled. . . . I never saw a more striking coincidence, if Wallace had my M.S. sketch written out in 1842 he could not have made a better short abstract!"[63] Lyell and Hooker urged Darwin, as a first and honorable step, to go public with Wallace. Papers by both were consequently read out at a meeting of the Linnean Society on July 1, 1858. Things happened quickly after that. Spurred on by competition Darwin abstracted his unfinished treatise on natural selection into "the book that shook the world."[64] *On The Origin of Species* was published on November 24, 1859, and all 1250 copies sold out that very same day. Galápagos, the catalyst for Darwin's conception of evolution, features prominently in the two chapters on Geographical Distribution:

> The most striking and important fact for us in regards to the inhabitants of islands, is their affinity to those of the nearest mainland, without being actually the same species. . . . The law which causes the inhabitants of an archipelago, though specifically distinct, to be closely allied to those of the nearest continent, we sometimes see displayed on a small scale, yet in a most interesting manner, within the limits of the same archipelago. Thus the several islands of the Galapagos are tenanted, as I have elsewhere shown, in a quite marvellous [*sic*] manner, by very closely related species; so that the inhabitants of each separate island, though mostly distinct, are related in an incomparably closer degree to each other than to the inhabitants of any other part of the world.[65]

Championing Darwin's case were the Galápagos mockingbirds. For though he peppered the chapters with supporting facts about the flora and fauna of Galápagos, he did so in general terms only. He refrained from distinguishing the finches from the "twenty-six land birds" that Gould described from Galápagos. He did not even use the words tortoise and iguana, when referring to the reptiles of the archipelago. But he did specify the "three closely-allied species of mocking-thrush, each confined to its own island."[66]

As Darwin had expected, his book was greeted with a mixture of outrage and enthusiasm. It sparked a fire of debate that raged through the scientific community, and beyond. Darwin lost friends and gained others. Asa Gray, botany professor at Harvard University, was his staunchest supporter overseas. Darwin had corresponded with Asa Gray for several years and their communications had formed part of the papers submitted by Darwin and Wallace to the Linnean Society in 1858. Comparative zoologist Thomas Henry Huxley was Darwin's most passionate and vociferous advocate at home. Huxley coined the term "Darwinism" and became known as "Darwin's bulldog," for defending it. On June 30, 1860, Huxley's bark and bite were keenly felt at a meeting of the British Association for the Advancement of Science in Oxford. After a rather dry talk on the influence of Darwinian theory by William Draper of New York University, the meeting became inflamed as Huxley, Hooker, and Bishop Samuel Wilberforce hotly debated Darwinism for the next four hours. Robert FitzRoy was in the audience and stood up, waving his Bible and pleading with the audience to listen to the voice of God. At the end of the day, both sides claimed they had won "The Great Debate." But of course it was only the beginning. The fire still smolders today.

"The Origin of 'the Origin'"

In the flurry of reaction following the announcement of Darwin's theory, Galápagos was a word that was often heard. Anyone who had any interest in the theory wanted to know how Darwin had arrived at it in the first place. There was no longer any need for secrecy on Darwin's part. The "origin of the 'Origin'"[67] came out like a confession, and it always included Galápagos. Shortly before the publication of his book Darwin had confided to Alfred Wallace, "Geographical Distrib. & Geographical relations of extinct to recent inhabitants of S. America first led me to subject. Especially case of Galapagos Isl[ds]."[68]

In December of the same year Darwin reminded Charles Lyell that "the law of succession [of living organisms from fossil forms] with the Galapagos Distribution first turned my mind on origin of species."[69] But the clearest outline of initial facts that led to Darwin's theory of evolution was spelled out in a letter Darwin wrote to German biologist Ernst Haeckel in 1864:

> In South America three classes of facts were brought strongly before my mind: 1[stly] the manner in which closely allied species replace species in going Southward. 2[ndly] the close affinity of the species inhabiting the Islands near to S. America to those proper to the Continent. This struck me profoundly, especially the difference of the species in the adjoining islets in the Galapagos Archipelago. 3[rdly] the relation of the living Edentata & Rodentia to the extinct species. I shall never forget my astonishment when I dug out a gigantic piece of armour like that of the living Armadillo.[70]

Darwin did not close the book on Galápagos immediately after publication of *The Origin of Species*. In 1846 he had written, "The Galapagos seems a perennial source of new things."[71] The islands, he believed, still had volumes to tell. Though he had long since exhausted analyzing his own Galápagos observations there were always the potential findings of future expeditions. He was quick to offer his encouragement to scientists thinking of visiting Galápagos, offering suggestions on what to look for once they got there. In 1863 he penned a particularly persuasive letter to ornithologist Osbert Salvin who was hesitating over the worth of an expedition to Galápagos. Darwin wrote:

> I think it would be scarcely possible to exaggerate the interest of a good collection of every species rigorously kept separate from each island. It would throw much light on variation (& as I believe

on the origin of Species) & on geographical distribution. No doubt many curious facts could be observed on the habits of the Birds & Reptiles. Probably there would be curious facts on the naturalisation [*sic*] & spreading of introduced plants & animals.

Nor did he let the opportunity pass of requesting some very specific observations to be made. "If you go, it would well deserve your attention to ascertain how the marine Amblyrhynchus [marine iguana] breeds. Pray attend to presence of sea-borne seeds in drift on the beaches exposed to prevailing currents."[72] Hooker joked that Salvin would not go, for his traveling companion Frederick Godman was a "fine looking young man of means" who might well "be bagged by some pretty girl before a year is over."[73] As it turned out Salvin did not make it to Galápagos, but Darwin was not disappointed for long. In 1868, Dr. Simeon Habel, an ornithologist from New York, spent five and a half months visiting seven of the Galápagos Islands, and collected over 300 specimens, each well labeled for locality. Salvin used Habel's collection, and Darwin's founding collection, to write a comprehensive treatise on the avifauna of Galápagos. In it he fully endorsed Darwin's message that immigrants from the South American continent originally colonized the islands of Galápagos, and their descendents gradually became "modified by the different circumstances with which they became surrounded."[74]

After this Darwin was content to let the "enchanted isles" gradually fade from his life. He always retained a connection to Galápagos, answering the occasional query about his Galápagos observations and discussing the evolution of some of its species, but for the last two decades of his life Darwin concentrated primarily on evolutionary topics that could be investigated closer to home. There was plenty to occupy his time.

At Home, at Work

From the moment he embarked on the voyage of the *Beagle* to the end of his days Darwin was a practical, hands-on naturalist. He never ceased exploring out of doors, watching nature, and theorizing on all he observed. After his return to England he also indulged in a scientific activity that, because of time constraints, had been restricted on the voyage of the *Beagle*: rigorous experimentation. He conducted as many experiments as his overactive curiosity and compulsive theorizing conjured to mind, and it was rare when Darwin was not in the middle of several drawn out investigations designed to test an idea or develop a concept. Indeed, when Darwin wasn't writing, he was cross-pollinating flowers or examining orchids in his greenhouse, breeding pigeons in his aviary, floating seeds and snails in tubs of salt water, watching bees and worms in his garden, studying the behavior of zoo animals (and even that of his own children), and carefully measuring specimens borrowed from museums.

As age and illness increasingly turned against him, he relied more heavily on his correspondents to gather facts from far afield, while he focused his taxed energies into the wonderfully accessible realm of botany and anything else that could be studied in his own backyard. He gathered and hoarded facts in the same ambitious manner that he had once collected specimens on the voyage. The results of his efforts were numerous scientific discoveries, which he published as periodical contributions and books on orchid fertilization,[75] climbing plants,[76] insectivorous plants,[77] fertilization,[78] heterostyly,[79] movement in plants,[80] and the formation of vegetable mould by earthworms.[81] Darwin's various studies also gave rise to books on broader themes. His experiments with artificial selection led to a fascination with reproduction, sexual selection, out-crossing, and heritability, and inspired a large tome on variation.[82] He also addressed human evolution in a book on the descent of man[83] and in another on the expressions and emotions of animals.[84] He even penned a short biography of his grandfather Erasmus Darwin.[85]

Finally, a few years before his death in 1882, Darwin composed his autobiography. In it "dear Galapagos"[86] was recalled, like an afterthought, for a final moment of glory. Nothing emotional, nothing lyrical, just a simple statement that belies the monumental significance the islands had to his becoming "the greatest Revolutionist in natural history of [the 19th] century, if not of all centuries"[87]: "I . . . reflect with high satisfaction on some of my scientific work," he wrote as though closing up his study before turning in for the night, "such as solving the problem of coral-islands, and making out the geological structure of certain islands. . . . Nor must I pass over the discovery of the singular relations of the animals and plants inhabiting the several islands of the Galapagos archipelago, and of all of them to the inhabitants of South America."[88]

~~~~~~~~~~~~~~~~~~~~~~~~~

Island and Site Names in Galápagos

Island Names[1]

English names known to Darwin and Beagle *crew*	Additional names in 1835, but not used by Darwin or Beagle *crew*	Modern Spanish names
Abingdon	Quita Sueños	Pinta
Albemarle	Santa Isabel	Isabela
Barrington		Santa Fé
Bartholomew[a]		Bartolomé
Bindloe		Marchena
Brattle		Tortuga
Charles	Marcos, Floriana	Floreana, Santa María
Chatham	San Clemente, La Aguada	San Cristóbal
Culpepper		Darwin
Duncan		Pinzón
Hood		Española
Indefatigable	Bolivia	Santa Cruz
James	Carenero, Olmedo	Santiago, San Salvador
Jervis		Rábida
Narborough		Fernandina
		Seymour
Tower	Douwes	Genovesa
Wenman		Wolf

[a]Named by Captain FitzRoy.

Site Names

Names in use in 1835	Modern English names	Modern Spanish names
Chatham Island	*Chatham Island*	*Isla San Cristóbal*
—	Frigatebird Hill	Cerro Tijeretas, Cerro Chivo
St Stephens Harbour,[e] Stephens Bay[c,d,f]	Stephens Bay	Bahía Stephens
—	—	Pan de Azucar
Terrapin Road[f]	—	Bahía Tortuga de Agua Dulce
Fresh Water Bay[f]	Freshwater Bay	Bahía Agua Dulce, La Honda
Kicker Rock[b,c,d,f]	Kicker Rock	León Dormido
Finger Hill[b] Finger Point[b,f]	—	Cerro Brujo, Punta Dedo
Craterized District[b]	—	Los Crateres
Charles/Floriana Island	*Charles Island*	*Isla Floreana*
Post Office Bay[c,d,f]	Post Office Bay	Bahía de Correo
Round Hill[c]	—	Cerro Pajas
The Settlement[e]	Haven of Peace	Asilo de Paz
Black Beach Bay,[c] Black Beach Road[d,f]	Black Beach	Rada Black Beach
Albemarle Island	*Albemarle Island*	*Isla Isabela*
Tagus Cove,[c,d] Blonde Cove,[c] Banks Cove[b,d]	Tagus Cove	Caleta Tagus
Iguana Cove[f]	Iguana Cove	Caleta Iguana
Pt Christopher[c,f]	Point Christopher	Punta Cristóbal
Whale Hill[b]	Darwin Volcano	Volcan Darwin
Volcano Hill[b]	—	Sierra Negra
—	Beagle Crater	Crater Beagle
James Island	*James Island*	*Isla Santiago*
Salina[b,e]	Salt Mine	Mina de Sal
—	—	Cerro Cowan

(*continued*)

Names in use in 1835	Modern English names	Modern Spanish names
James Island	James Island	Isla Santiago
Freshwater Cove of the Buccaniers,[e] Freshwater Bay[b]	Buccaneer Cove	Caleta Bucanero
Sugar Loaf[a,c]	Sugar Loaf	Pan de Azucar
—	—	Jaboncillos
Puerto Grande,[b] James Bay[f]	James Bay	Bahía James
—	—	Puerto Egas
—	—	Playa Espumilla
—	Sulivan Bay	Bahía Sulivan

Sources: [a]Appendix to FitzRoy's *Narrative*,[2] [b]Darwin's *Geological Notes*,[3] [c]*Beagle* Log,[4] [d] Fitz-Roy's *Narrative*,[5] [e]*Beagle* Diary,[6] [f] British Admiralty charts of Galápagos.[7]

Bearing Darwin's Name in Galápagos

Darwin is a common name in Galápagos. Various businesses and organizations, capitalizing on Darwin's famous association with Galápagos, use his name to promote their respective interests. In the town of Puerto Ayora on Indefatigable Island (Isla Santa Cruz), for example, Darwin designates a street, a hotel, a yacht, a tour agency, a taxi, and a research station.

Not surprisingly Darwin's name is also attached to several plants, animals, and geological formations in Galápagos, many of which are listed here. Darwin did not see all the organisms and natural phenomena that bear his name. For instance, he never recorded the carpenter bee (*Xylocopa darwinii*), Darwin's rice rat (*Nesoryzomys darwini*), or red-lipped bat fish (*Ogcocephalus darwini*). Nor did he explore the land formations named after him. Darwin Volcano is a possible exception, as he did land on the flank of the volcano, but never climbed it. Darwin Lake is somewhat of a misnomer. There are two salt-water lakes in the vicinity of Tagus Cove on Albemarle Island—Darwin Lake and the lake within Beagle Crater. Both were unnamed in Darwin's day. Although Darwin saw and wrote about both lakes, he physically explored only the second. Therefore, if any Galápagos lake should bear his name, the one contained by Beagle Crater would be most appropriate.

Plants

Darwiniothamnus spp.
Lecocarpus darwini
Tiquilia darwini

Croton scouleri var darwini
Gossypium darwini
Pleuropetalum darwini
Scalesia darwini

Arthropods
Darwin's darklings (Coleoptera)
Phalacrus darwini (Coleoptera)
Nitela darwini (Sphecidae)
Capsus darwini (Hemiptera, Miridae)
Darwinysius marginalis (Hemiptera, Heteroptera)
Darwinysius wenmanensis (Hemiptera, Heteroptera)
Amblyomma darwini (Arachnida)
Darwinneon crypticus (Arachnida)
Xylocopa darwinii (carpenter bee)

Molluscs
Bulimulus darwini (land snail)

Marine Invertebrates
Cavernulina darwini (a sea pen)
Pacifigorgia darwinii (a sea fan)

Fish
Semicossyphus darwini (Pacific red sheephead)
Ogcocephalus darwini (red-lipped batfish)

Birds
Darwin's finches (*Geospizinae* spp.)

Mammals
Darwin's rice rat *Nesoryzomys darwini* (extinct rice rat from Indefatigable Island)

Reptiles
Geochelone nigra darwini (James Island subspecies of Galápagos
tortoise)

Land Formations
Darwin Lake (Albemarle Island/Isla Isabela)
Darwin Volcano (Albemarle Island/Isla Isabela)
Darwin Bay (Tower Island/Isla Genovesa)
Isla Darwin (Culpepper Island)
Darwin Arch (Culpepper Island/Isla Darwin)

Appendix 3

HMS Beagle's Complement[1]

Captain	**Robert FitzRoy**
First Lieutenant	John Clements Wickham
Second Lieutenant	Bartholomew James Sulivan
Master	Edward Main Chaffers
Master's assistant	Alexander Burns Usborne*
Surgeon	Robert Mac-Cormick* (returned to England April 1832; replaced by Benjamin Bynoe)
Assistant surgeon	Benjamin Bynoe (replaced by William Kent September 1833 after Bynoe was promoted to surgeon)
Purser	George Rowlett* (died June 1834; replaced by John Edward Dring)
Mate	Alexander Derbishire* (returned to England April 1832; replaced by Arthur Mellersh)
Midshipman	Charles Richardson Johnson (joined *Beagle* April 1832)
Mate	Peter Benson Stewart
Mate and assistant surveyor	John Lort Stokes
Midshipman	Arthur Mellersh (became mate April 1832)
Midshipman	Philip Gidley King (became ship's artist in 1835; left *Beagle* February 1836)

(*continued*)

Captain	**Robert FitzRoy**
Volunteer 1st class	Charles Musters* (died June 1832; replaced by Charles Forsyth*)
Carpenter	Jonathan May
Clerk	Edward H. Hellyer (died March 1833)*
Supernumeraries:	Charles Robert Darwin, naturalist
	George James Stebbing, instrument maker
	August Earle, draughtsman* (left *Beagle* August 1832; replaced by Conrad Martens* who then left *Beagle* in 1834)
	Richard Matthews, missionary (left *Beagle* December 1835)
	Fuegian Indians (York Minster,* Fuegia Basket,* and Jemmy Button*) (left *Beagle* January 1833)
	Harry Fuller, captain's steward
	Syms Covington, Darwin's servant
Seamen	Roughly 50 sailors

*Not on board the *Beagle* in Galápagos.

Acknowledgments

~~~~~~~~~~~~~~~~~~~~~~

While reading Darwin's accounts of the voyage of the *Beagle*, we were impressed by the number and variety of people who had assisted or otherwise influenced Darwin throughout his travels. Gauchos, native Indians, estate owners and other locals, ship captains, colonists, travelers and missionaries provided logistical assistance or shared important local natural history knowledge. In Galápagos, Nicholas Lawson, the inhabitants of Charles and James islands, and the *Beagle* crew were especially instrumental in Darwin's endeavors as a collector of specimens and facts. Darwin was an amazing observer and explorer in his own right, and an exceptional theoretician and synthesizer of information. But neither on the voyage nor after he returned to England did he operate alone. The eyes, hands and voices of many "little" people were behind his greatness. While Darwin recognized these helpers in his notes, letters and diary, he did not always acknowledge them in his publications.

When Darwin wrote up his *Journal* for publication, he unintentionally offended Captain FitzRoy by neglecting to acknowledge the officers of the *Beagle* for their help with his collections and explorations. FitzRoy's reaction was thunderous. "I was . . . astonished," he raged, "at the total omission of any notice of the officers—either particular —or general. —My memory is rather tenacious respecting a variety of transactions in which you were concerned with them; and others in the Beagle. Perhaps you are not aware that the ship which carried us safely was the first employed in exploring and surveying whose Officers were not ordered to collect—and were therefore at liberty to keep the best of all—nay, all their specimens for themselves. To their honour—they gave you the preference."[1]

Darwin's changed his preface to read, "Both to the Captain FitzRoy and to all the Officers of the Beagle, I shall ever feel most thankful for the undeviating kindness with which I was treated, during our long voyage."[2] Whether or not this final version satisfied FitzRoy is unknown, but the damage was done. It was the beginning of the end of a fine but fragile friendship.

Lest we repeat Darwin's omission and incur the wrath of our own "captain and crew," we offer our sincere thanks to all those who have helped in infinite and invaluable ways with our literary "voyage," our expedition to retrace Darwin's footsteps, and the writing of this book.

First we thank our children Olivia and Devon for their patience with our "obsession" with Darwin, as they sometimes perceive it. They have been enthusiastic and helpful participants on our most recent field trips to retrace Darwin's footsteps within the Galápagos Islands. And although they have long outgrown the stage of believing that Darwin and their maternal grandfather (Peter Grant) are one and the same—for the resemblance between the two men in photographs is uncanny—they, like us, consider Darwin as part of the family.

We are also eternally grateful for the advice and support of Peter and Rosemary Grant, who have helped in countless ways.

Our research trips within Galápagos would not have been possible without the sanction of the Charles Darwin Research Station and authorization from the Galápagos National Park Service. Members of both institutions provided invaluable logistical support and advice. We are also grateful to the numerous people who assisted with our transportation between sites and islands; these include the captain and crew of the *Queen Karen, North Star, Beagle V, Santa Cruz, Isabela II,* and *Queen Mabel.* We gratefully acknowledge the staff of EMETEBE, INGALA, and Tame Airlines for providing additional transport.

We thank the staff of the Cambridge University Library (including the Darwin Archive), the Public Records Office of London, the UK Hydrographic Office, the Royal Geographical Society of London, the Cambridge University Museum of Zoology, the Sedgwick Museum of Earth Sciences, and the Cambridge University Herbarium for providing access to Darwin's manuscripts and specimens and to FitzRoy's logs and maps of Galápagos. The library of the Charles Darwin Research Station was another important resource. The volunteer staff was consistently helpful in our literary pursuits, but one librarian deserves special acknowledgement for his assistance both while resident in Galápagos and after his return to England. John Simcox's generosity and professional attitude to library research helped us surmount the difficulty of being in one place while needing to look something up in an archive, sometimes thousands of miles away.

We also acknowledge the assistance of the various people, many of whom we list here, who helped with other aspects of the project. Some provided logistical assistance, others helped with scientific and historical advice and suggestions. A few assisted in an editorial capacity. Still others are simply to be thanked for some very enjoyable and enlightening discussions about Darwin and Galápagos:

David Asensio, Ken Atherton, Jill Awkerman, David Balfour, Robert Bensted-Smith, Ruth Boada, Diego Bonilla, Barry Boyce, Janet Browne, Frank Bungartz, Wilson Cabrera, Joe Cain, David Catling, Robin Catchpole, Gordon Chancellor, Rita Chancellor, Fuzz Crompton, Eliecer Cruz, Felipe Cruz, Robert Curry, Virginia Dunn, Gayle Davis-Merlen, Jacqueline De Roy, Steve Divine, Niles Eldredge, Sam Elworthy, Birgitte Fessel, Daniel Fitter, Robert Fleischer, Cliff and Dawn Frith, Brian Gardiner, Dennis Geist, James Gibbs, Sally Gibson, Stephen Gregory, Jack Grove, Anne Guezou, Bernardo Gutierrez, Karen Harpp, Donna Harris, Sandra Herbert, Paquita Hoeck, Mathew James, Dimitri Karetnikov, Lukas Keller, Rex Kenner, Ann Keynes, Quentin Keynes, Randal Keynes, Richard Darwin Keynes, Robert Kirk, David Kohn, Annie Lagarde, Edward Larson, Octavio Latorre , Steve Laurie, Ken Levenstein, Cathy Luchetti, Craig MacFarland, John Madunich, Cruz Marquez, Godfrey Merlen, Andrew Murray, Gina Murrell, David Norman, Andrew Nystrom, Christine Parent, John Parker, Brigitte Pelner, Adam Perkins, Ken Petren, Richard Polatty, Duncan Porter, Nigel Sitwell, Heidi Snell, Howard Snell, David Stern, Paul Stewart, Giok Stewart-Ong, Frank Sulloway, Ray Symonds, Washington Tapia, Keith Thompson, Alan Tye, Godfrey Waller, Graham Watkins, Jerry Wellington, Martin Wikelski, John Woram.

Finally, we are indebted to Charles Darwin for providing the footsteps through Galápagos for us to follow.

Notes

~~~~~~~~~~~~~~~~~~~~~~~~~~~~~~~~

Introduction: Darwin's Islands

1. Charles Robert Darwin. *Journal of researches into the natural history and geology of the countries visited during the voyage of H.M.S. Beagle round the world, under the Command of Capt. Fitz Roy, R.N.,* 2d edition. John Murray, London, 1845, p. 377.

2. The 13 principle islands of Galápagos each have an area greater than 10 square kilometers. The remaining islands are all 5 square kilometers or smaller.

3. J.F.W. Herschel to Charles Lyell. Letter 20 February 1836. In: *Charles Darwin's Notebooks.* Paul H. Barrett, Peter J. Gautrey, Sandra Herbert, David Kohn, and Sydney Smith, editors. Cornell University Press, New York, 1987, p. 413.

4. Julian Huxley. Introduction. In: William Irvine. *Apes, Angels and Victorians: Darwin, Huxley and Evolution.* Time-Life Books, Chicago, 1982, p. xviii.

5. Loren Eiseley. *Darwin's Century: Evolution and the Men Who Discovered It.* Anchor Books, New York, 1990, p. 2.

6. Douglas J. Futuyma. *Evolutionary Biology,* 3rd edition. Sinauer Associates, Massachusetts, 1998. pp. 4–23.

7. Ibid.

8. Julian Huxley. Introduction. In: William Irvine. *Apes, Angels, and Victorians. The Story of Darwin, Huxley and Evolution.* Time-Life Books, Chicago, 1982, p. xviii.

9. Ernst Mayr. *What Evolution Is.* Basic Books, New York, 2001, p. 9.

10. The word "On" was dropped from the title of the 6th edition, published in 1872, as well as from all subsequent editions.

11. Ernst Mayr. *What Makes Biology Unique?* Cambridge University Press, New York, 2004, p. 117.

12. Peter J. Bowler. *Charles Darwin: The Man and His Influence.* Cambridge University Press, Cambridge, UK, 1996, p. 128.

13. Ibid., pp. 145–146.

14. Ernst Mayr. *Darwin's Influence on Modern Thought.* Lecture given to the Royal Swedish Academy of Sciences, 1999. See also Ernst Mayr. *What Makes Biology Unique?* Cambridge University Press, New York, 2004, pp. 84 and 95.

15. Janet Browne. *Charles Darwin: Voyaging* (Vol. I of a biography). Knopf, New York, 1995. And Janet Browne. *Charles Darwin: The Power of Place* (Vol. II of a biography). Princeton University Press, Princeton, NJ, 2002.

16. Adrian Desmond and James Moore. *Darwin.* Penguin Books, New York, 1992.

17. Richard Keynes, *Fossils, Finches and Fuegians: Charles Darwin's Adventures and Discoveries on the Beagle, 1832–1836.* HarperCollins, London, 2002.

18. Randal Keynes. *Annie's Box: Charles Darwin, His Daughter and Human Evolution.* Fourth Estate, London, 2001.

19. Niles Eldredge. *Darwin: Discovering the Tree of Life.* WW Norton, New York, 2005.

20. David Quammen. *The Reluctant Mr. Darwin: An Intimate Portrait of Charles Darwin and the Making of His Theory of Evolution.* Great Discoveries. W.W Norton, New York, 2006.

21. Sandra Herbert. *Darwin, Geologist.* Cornell University Press, Ithaca, NY, 2005.

22. Edward J. Larson. *Evolution: The Remarkable History of a Scientific Theory.* Modern Library, New York, 2004.

23. Peter J. Bowler. *Charles Darwin: The Man and His Influence.* Cambridge University Press, Cambridge, UK, 1996. And Peter J. Bowler. *Evolution: The History of an Idea.* University of California Press, Los Angeles, 2003.

24. Julian Huxley. Charles Darwin: Galápagos and After. In: *The Galápagos: Proceedings of the Symposia of the Galápagos International Scientific Project,* Robert I. Bowman, editor. University of California Press, Berkeley, 1966.

25. Charles Darwin. *Journal 1837–1843* (DAR 158 in the Cambridge University Library), transcribed in *The Correspondence of Charles Darwin,* Vol. 2: *1837–1843,* Frederick Burkhardt and Sydney Smith, editors. Cambridge University Press, Cambridge, UK, 1987, Appendix II, p. 431.

26. Charles Darwin. Letter to A. R. Wallace, 6 April 1859. In: *The Correspondence of Charles Darwin,* Vol. 7: *1858–1859,* Frederick Burkhardt and Sydney Smith, editors. Cambridge University Press, Cambridge, UK, 1992, p. 279.

27. Edward J. Larson. *Evolution: The Remarkable History of a Scientific Theory.* Modern Library, New York, 2004, p. 68–69.

28. Victor Wolfgang Von Hagen. *Ecuador and the Galápagos Islands.* University of Oklahoma Press, 1949, p. 234.

29. Julian Huxley. Charles Darwin: Galápagos and After. In: *The Galápagos: Proceedings of the Symposia of the Galápagos International Scientific Project,* Robert I. Bowman, editor. University of California Press, Berkeley, 1966, p. 3.

30. Frank J. Sulloway. 1987. Darwin and the Galápagos: Three Myths. *Oceanus* 30(2): 79–85.

31. Frank J. Sulloway. 1984. Darwin and the Galápagos. *Biological Journal of the Linnean Society* 21: 29–59.

32. Frank J. Sulloway. 1982. Darwin and His Finches: The Evolution of a Legend. *Journal of the History of Biology* 15(1): 1–53. See also: Frank J. Sulloway. 1987. Darwin and the Galápagos: Three Myths. *Oceanus* 30(2), 79–85.

33. Sandra Herbert. *Darwin, Geologist.* Cornell University Press, Ithaca, NY, 2005, p. 312.

34. Ibid.

35. Richard Keynes, editor. *Charles Darwin's Zoology Notes & Specimen Lists from H.M.S. Beagle.* Cambridge University Press, Cambridge, UK, 2000.

36. Duncan Porter. 1980. The Vascular Plants of Joseph Dalton Hooker's *An enumeration of the plants of the Galápagos Archipelago; with descriptions of those which are new. Botanical Journal of the Linnean Society* 81:79–134.

37. Paul N. Pearson. 1996. Charles Darwin on the Origin and Diversity of Igneous Rocks. *Earth Sciences History* 15: 49–67.

38. Sandra Herbert. *Darwin, Geologist.* Cornell University Press, Ithaca, NY, 2005, p. 116–128.

39. David Kohn, Gina Murrell, John Parker, and Mark Whitehorn. August 4, 2005. What Henslow Taught Darwin. *Nature* 436: 643–645.

40. Charles Darwin. *Charles Darwin's Beagle Diary,* Richard D. Keynes, editor. Cambridge University Press, Cambridge, UK, 1988, p. 445.

41. Charles Darwin. 1835. *Beagle* field notebook *Galapagos Otaheite Lima.* Darwin Archive, Cambridge University Library Microfilm EH1.17. (The original notebook was stolen from Down House in the early 1980s). Transcribed from microfilm by Thalia Grant.

42. Charles Darwin. 1835. *Notes on the geology of places visited on the voyage.* Darwin Archive, Cambridge University Library DAR 37.2. Transcribed by Thalia Grant.

43. Charles Darwin. *Diary of observations on zoology of the places visited during the voyage.* Darwin Archive, Cambridge University Library (1832–1836) DAR 31.2. Transcribed by Richard Keynes.

44. Nora Barlow, editor. 1963. Darwin's ornithological notes. *Bulletin of the British Museum (Natural History).* Historical Series 2(7): 201–278. We also obtained a copy of the original manuscript (DAR 29.2) from the Darwin Archive of CUL.

45. Duncan M. Porter. 1987. Darwin's notes on *Beagle* plants. *Bulletin of the British Museum (Natural History).* Historical Series 14(2):145–233.

46. Charles Darwin. *Charles Darwin's Beagle Diary,* Richard D. Keynes, editor. Cambridge University Press, Cambridge, UK, 1988.

47. Charles Darwin. Specimens Down House Notebook 63.6 (*List of Beagle Specimens*) Printed Numbers 3345. Transcribed by Richard Keynes. Also geological specimen notebook DAR 236 (property of Sedgwick Museum), housed in Darwin Archive, Cambridge University Library, Cambridge, UK.

48. Ship's log of the *Beagle* 1831–1836. ADM53/236, Part 2. Public Records Office, Kew, UK. Transcribed by Thalia Grant. Also Captain's log of the *Beagle* ADM51/3055 Public Records Office, Kew, UK. Transcribed by Thalia Grant.

49. Robert FitzRoy. British Admiralty Charts L945, L946, L947, L948, L949, L950 from 1837 and Chart 1375 printed in 1841. *Galapagos Islands Surveyed by Capt. Robt. Fitz Roy R. N. and the Officers of H. M. S. Beagle, 1836.* Hydrographic Office of Taunton, UK.

50. Prior to the year 2000 the United States enabled "selective availability" for all nonmilitary GPS users. The result was purposeful, random errors of up to 100 meters. The GPS measurements that we recorded at the sites Darwin visited, and published in our paper (Gregory Estes, K. Thalia Grant, and Peter R. Grant. 2000. *Darwin in Galápagos: His Footsteps through the Archipelago.* Notes and Records of the Royal Society of London. 54(3): 343–368) were therefore low in resolution. Since the United States disabled "selective availability" in May 2000, we have been involved

in re-measuring the location of many of the sites, this time to a standard accuracy of ±15 meters. (WGS-84). Some of these new measurements are included in this book.

51. Frank J. Sulloway. 1984. Darwin and the Galápagos. *Biological Journal of the Linnean Society* 21:29–59.

52. Gregory Estes, K. Thalia Grant, and Peter R. Grant. 2000. Darwin in Galápagos: His Footsteps through the Archipelago. *Notes and Records of the Royal Society of London* 54(3): 343–368

53. Nora Barlow, editor. *The Autobiography of Charles Darwin 1809–1882* . WW Norton, New York, 1958, p. 141.

54. Charles Darwin. *On the Origin of Species by Means of Natural Selection, or the Preservation of Favoured Races in the Struggle for Life.* John Murray, London, 1859, p. 489.

55. Richard D. Keynes. *Fossils, Finches and Fuegians.* HarperCollins, London, 2002, p. xix.

56. Toby Green. *Saddled with Darwin: A Journey through South America on Horseback.* Orion, London, 1999, p. 38.

57. Ian Thornton. *Darwin's Islands: A Natural History of the Galápagos.* Natural History Press, New York, 1971.

58. John Milton. *Paradise Lost, A Poem in Twelve Books*, 2nd ed. S. Simmons, London, 1674 (Book 1, lines 254–255).

59. Robert I. Bowman. Contributions to Science from the Galápagos. In: *Galápagos: Key Environments*, Roger Perry, editor. Pergamon Press, Oxford, UK, 1984, p. 278.

60. Fray Tomás de Berlanga. 1535. *Letter to His Majesty from Fray Tomás de Berlanga, describing his Voyage from Panamá to Puerto Viejo, and the hardships he encountered in this navigation.* In: *Colección de Documentos Inéditos relativos al Descubrimiento, Conquista y Organización de las Antiguas Posesiones Españolas de América y Oceanía.* Tomo XLI, Cuaderno II. Imprenta de Manuel G. Hernández, Madrid, 1884, pp. 538–544.

61. William Ambrose Cowley. 1688. *Cowley's Voyage Round the World.* Sloane MS, 1050. British Library, London.

62. Woodes Rogers. 1712. *A Cruising Voyage Round the World.* 1928 reprint. Longmans, Green, New York.

63. William Ambrose Cowley. 1688. *Cowley's Voyage Round the World.* Sloane MS, 1050. British Library, London.

64. Lord George Anson Byron. *Voyage of H.M.S. Blonde to the Sandwich Islands, in the years 1824–25*, Maria Graham, editor. John Murray, London, 1826.

65. Herman Melville. The Encantadas; or Enchanted Isles. (Originally published in 1854) In: *Billy Budd and Other Tales.* Signet Classic, New York, 1979, pp. 233–234

66. Charles Robert Darwin. *Narrative of the surveying voyages of His Majesty's Ships Adventure and Beagle between the years 1826 and 1836, describing their examination of the southern shores of South America, and the Beagle's circumnavigation of the globe. Journal and remarks. 1832–1836.* Henry Colburn, London, 1839, p. 629 (addenda to p. 465).

67. For an overview of important research conducted in Galápagos since Darwin, we recommend reading Edward J. Larson. *Evolution's Workshop: God and Science on the Galápagos Islands*. Basic Books, New York, 2001.

68. Charles Robert Darwin. *Narrative of the surveying voyages of His Majesty's Ships Adventure and Beagle between the years 1826 and 1836, describing their examination of the southern shores of South America, and the Beagle's circumnavigation of the globe. Journal and remarks. 1832–1836*. Henry Colburn, London, 1839, p. 478.

69. The Galápagos Memorial Expedition led by Victor von Hagen in 1935 rallied international naturalists concerned about the future of the islands, and urged them to act. The Galápagos Committee of London, headed by Julian Huxley, took over this movement and successfully campaigned for the allotment of a portion of the archipelago as a natural reserve. From Edward J. Larson. *Evolution's Workshop, God and Science on the Galápagos Islands*. Basic Books, New York, 2001, p. 166. See also Victor Wolfgang von Hagen. *Ecuador and the Galápagos Islands*. University of Oklahoma Press, 1949, p. 215.

70. Some of the nongovernmental organizations working in Galápagos include World Wildlife Fund, Conservation International, and WildAid.

71. Gregory Estes, K. Thalia Grant, and Peter R. Grant. 2000. Darwin in Galápagos: His Footsteps through the Archipelago. *Notes and Records of the Royal Society of London* 54(3): 343–368.

72. Wolfe, Thomas. *Look Homeward, Angel*. Scribner's, New York, 1929.

73. James A. Secord. 1991 The Discovery of a Vocation: Darwin's Early Geology. *British Journal for the History of Science* 24: 133–157.

74. Loren Eiseley. *Darwin's Century: Evolution and the Men Who Discovered It*. Anchor Books, New York, 1990, pp. 148–149.

75. David Kohn, veteran Darwin scholar who has transcribed many pages of Darwin's holographic manuscripts, estimates that in addition to Darwin's published works Darwin left about 11.69 million words of unpublished material, in the form of notes, letters, and essays. In 2007, only 28% of this material had been officially transcribed and published to modern standards.

76. R. B. Freeman. *The Works of Charles Darwin: An Annotated Bibliographical Handlist*, 2nd ed. Dawsons, Folkstone, UK, 1977.

77. Douglas J. Futuyma. *Evolutionary Biology*, 3rd ed. Sinauer, Sunderland, MA, 1998, pp. 4–5.

78. Darwin used the term "evolution" for the first time in his book *The Descent of Man* published in 1871. In 1872 he inserted "evolution" into the sixth edition of *The Origin of Species*.

CHAPTER 1: CURIOUS BEGINNINGS

1. Charles Darwin. *Life: An Autobiographical Fragment*. Written August 1838. In: *The Correspondence of Charles Darwin*, Vol. 2: 1837–1843, Frederick Burkhardt and Sydney Smith, editors. Cambridge University Press, Cambridge, UK, 1986, p. 440.

2. Adrian Desmond and James Moore. *Darwin.* Penguin Books, New York, 1992, p. 5.

3. William Paley. *Natural Theology, or Evidences of the Existence and Attributes of the Deity, collected from the Appearances of Nature.* Faulder, London, 1802.

4. Ernst Mayr. *What Makes Biology Unique.* Cambridge University Press, New York, 2004, p. 16.

5. Janet Browne. *Charles Darwin. Voyaging.* AA Knopf, New York, 1995, p. 129.

6. The words "science" and "scientist" were not used until the 1830s. See Dorinda Outram. *The Enlightenment,* 2nd edition. Cambridge University Press, Cambridge, UK, 2005, pp. 94–95.

7. Dorinda Outram. *The Enlightenment,* 2nd edition. Cambridge University Press, Cambridge, UK, 2005, pp. 93–108.

8. Ernst Mayr. *What Makes Biology Unique.* Cambridge University Press, New York, 2004, p. 16.

9. Carl Linnaeus. *Systema naturae per regna tria naturae, secundum classes, ordines, genera, species, cum characteribus, differentiis, synonymis, locis* (System of nature through the three kingdoms of nature, according to classes, orders, genera and species, with characters, differences, synonyms, places). 1735.

10. Peter J. Bowler. *Evolution: The History of an Idea.* University of California Press, Berkeley, 2003, p. 3.

11. Ibid., p. 27.

12. Douglas J. Futuyma. *Evolutionary Biology.* Sinauer Associates, Sunderland, MA, 1979, p. 4.

13. Peter J. Bowler. *Evolution: The History of an Idea.* University of California Press, Berkeley, 2003, p. 4.

14. Paul-Henri Thiry (Baron d'Holbach), Denis Diderot, and Julien Offray de La Mettrie.

15. Peter J. Bowler. *Evolution: The History of an Idea.* University of California Press, Berkeley, 2003, p. 81–85.

16. Erasmus Darwin. *Zoonomia, or The laws of organic life.* 2 vols. Printed for J. Johnson, London, 1794–1796.

17. Loren Eiseley. *Darwin's Century: Evolution and the Men Who Discovered It.* Anchor Books, New York, 1990, p. 35.

18. Jean-Baptiste Lamarck. *Philosophie zoologique ou exposition des considérations relatives à l'histoire naturelle des animaux.* Dentu, Paris 1809.

19. Peter J. Bowler. *Evolution: The History of an Idea.* University of California Press, Berkeley, 2003, p. 236.

20. Stephen J. Gould. *The Panda's Thumb: More Reflections in Natural History.* WW Norton, New York, 1992.

21. Janet Browne. *Charles Darwin: Voyaging.* AA Knopf, New York, 1995, p. 5.

22. Charles Darwin. *Life: An autobiographical fragment.* Written August 1838. In: *The Correspondence of Charles Darwin,* Vol. 2: 1837–1843, Frederick Burkhardt and Sydney Smith, editors. Cambridge University Press, Cambridge, UK, 1986, p. 440.

23. Ibid.

24. Ibid., p. 22.

25. Ibid., p. 23. See also Charles Darwin. *Life. An autobiographical fragment.* Written August 1838. In: *The Correspondence of Charles Darwin*, Vol. 2: 1837–1843, Frederick Burkhardt and Sydney Smith, editors. Cambridge University Press, Cambridge, UK, 1986, p. 439.

26. Charles Darwin. *Life. An autobiographical fragment.* Written August 1838. In: *The Correspondence of Charles Darwin*, Vol. 2: 1837–1843, Frederick Burkhardt and Sydney Smith, editors. Cambridge University Press, Cambridge, UK, 1986, p. 439.

27. Charles Darwin. *The Autobiography of Charles Darwin 1809–1882*, Nora Barlow, editor. WW. Norton, New York, 1958, p. 22.

28. Ibid., p. 46.

29. Ibid., p. 27.

30. Ibid., p.141.

31. Ibid., p. 44.

32. Ibid., p. 23.

33. Ibid., p. 27.

34. Edna Healey. *Emma Darwin: The Inspirational Wife of a Genius.* Headline, London, 2001, p. 110.

35. Charles Darwin. *The Autobiography of Charles Darwin 1809–1882*, Nora Barlow, editor. WW Norton, New York, 1958, p. 46.

36. Ibid., p. 46.

37. Ibid., p. 28.

38. Ibid., pp. 47–48.

39. Ibid., p. 48.

40. James Hutton. *Theory of the Earth; or an Investigation of the laws observable in the Composition, Dissolution, and Restoration of Land upon the Globe, 1788.* Transactions of the Royal Society of Edinburgh, Vol. I, Part II, pp. 209–304.,

41. Charles Darwin. *The Autobiography of Charles Darwin, 1809–1882*, Nora Barlow, editor. WW Norton, New York, 1958, p. 53.

42. James A. Secord. 1991. The Discovery of a Vocation: Darwin's Early Geology. *British Journal for the History of Science* 24: 133–157.

43. Erasmus Darwin. *Zoonomia, or The laws of organic life*, 2 vols. London. Printed for J. Johnson, 1794–1796.

44. Charles Darwin. *The Autobiography of Charles Darwin, 1809–1882*, Nora Barlow, editor. WW Norton, New York, 1958, p. 49.

45. Ibid., p. 57.

46. Charles Darwin. Letter to Asa Gray, 5 September 1857. In: *The Correspondence of Charles Darwin*, Vol. 6: 1856–1857, Frederick Burkhardt and Sydney Smith, editors. Cambridge University Press, Cambridge, UK, 1990, p. 134.

47. Charles Darwin. *The Autobiography of Charles Darwin, 1809–1882*, Nora Barlow, editor. WW Norton, New York, 1958, p. 60.

48. Ibid., p. 62.

49. Ibid., p. 63.

50. K.G.V. Smith. 1987. Darwin's insects: Charles Darwin's entomological notes, with an introduction and comments by Kenneth G. V. Smith. *Bulletin of the British Museum (Natural History) Historical Series* 14(1): 1–143.

51. Charles Darwin. *The Autobiography of Charles Darwin, 1809–1882*, Nora Barlow, editor. WW Norton, New York, 1958, p. 82.

52. Ibid., p. 60.

53. Ibid., p. 64.

54. Ibid., p. 64.

55. D. Kohn, G. Murrell, J. Parker, and M. Whitehorn. 2005. What Henslow Taught Darwin. *Nature* 436: 643–645.

56. Charles Darwin. *The Autobiography of Charles Darwin, 1809–1882*, Nora Barlow, editor. WW Norton, New York, 1958, p. 64.

57. Ibid., p. 67–68.

58. Charles Darwin. Letter to C. S. Darwin, 28 April 1831. In: *The Correspondence of Charles Darwin*, Vol. 1: *1821–1836*. Frederick Burkhardt, Sydney Smith, David Kohn, and William Montgomery, editors. Cambridge University Press, Cambridge, UK, 1985, p. 122.

59. Charles Darwin. Letter to W. D. Fox, 9 July 1831. In: *The Correspondence of Charles Darwin*, Vol. 1: *1821–1836*. Frederick Burkhardt, Sydney Smith, David Kohn, and William Montgomery, editors. Cambridge University Press, Cambridge, UK, 1985, p. 124.

60. Ibid.

61. Sandra Herbert. Remembering Charles Darwin as a Geologist. In: *Charles Darwin 1809–1882: A Centennial Commemorative*, Chapman and Durval, editors. Nova Pacifica, 1892, p. 241.

62. James A. Secord. 1991. The Discovery of a Vocation: Darwin's Early Geology. *British Journal for the History of Science* 24: 143.

63. IIbid p. 142.

64. Francis Darwin, editor. *The life and letters of Charles Darwin, including an autobiographical chapter*, Vol. 1. John Murray, London, 1887, p. 57.

65. Martin J. S. Rudwick. Darwin and the World of Geology (Commentary). In: *The Darwinian Heritage*, David Kohn, editor. Princeton University Press, Princeton, NJ, 1985, p. 514.

66. Charles Darwin. *The Autobiography of Charles Darwin, 1809–1882*, Nora Barlow, editor. WW Norton, New York, 1958, p. 70.

67. Ibid., p. 52.

68. Janet Browne. *Charles Darwin: Voyaging*. AA Knopf, New York, 1995, p. 139.

69. Susan Darwin. Letter to Charles Darwin, 12 February 1836. In: *The Correspondence of Charles Darwin*, Vol. 1: *1821–1836*. Frederick Burkhardt, Sydney Smith, David Kohn, and William Montgomery, editors. Cambridge University Press, Cambridge, UK, 1985, p. 488.

70. Charles Darwin. In: *Charles Darwin's Beagle Diary*, Richard D. Keynes, editor. Cambridge University Press. Cambridge, UK, 1988, p. 27. See also Charles Darwin. Letter to J. S. Henslow, 18 May–16 June 1832. In: *The Correspondence of Charles Darwin*, Vol. 1: *1821–1836*. Frederick Burkhardt, Sydney Smith, David Kohn, and William Montgomery, editors. Cambridge University Press, Cambridge, UK, 1985, p. 237.

71. Charles Darwin. *The Autobiography of Charles Darwin, 1809–1882*, Nora Barlow, editor. WW Norton, New York, 1958, p. 68.

72. Sandra Herbert. Darwin the Young Geologist. In: *The Darwinian Heritage*, David Kohn, editor. Princeton University Press, Princeton, NJ, 1985, pp. 492–493.

73. James A. Secord. 1991. The Discovery of a Vocation: Darwin's Early Geology. *British Journal for the History of Science* 24: 148.

74. Josiah Wedgwood II. Letter to R. W. Darwin, 31 August 1831. In: *The Correspondence of Charles Darwin*, Vol. 1: *1821–1836*. Frederick Burkhardt, Sydney Smith, David Kohn, and William Montgomery, editors. Cambridge University Press, Cambridge, UK, 1985, p. 134.

CHAPTER 2: VOYAGE OF THE *BEAGLE*

1. Charles Darwin. *The Autobiography of Charles Darwin, 1809–1882*, Nora Barlow, editor. WW Norton, New York, 1958, p. 76.

2. Adrian Desmond and James Moore. *Darwin*. Penguin Books, London, 1992, p. 103.

3. For the Admiralty's instructions and plan of the voyage, see Robert FitzRoy. *Narrative of the surveying voyages of His Majesty's Ships Adventure and Beagle between the years 1826 and 1836, describing their examination of the southern shores of South America, and the Beagle's circumnavigation of the globe: Proceedings of the second expedition, 1831–36, under the command of Captain Robert Fitz-Roy, R.N.* Henry Colburn, London, 1839, pp. 17–40.

4. Richard D. Keynes. *Fossils, Finches and Fuegians*. HarperCollins, London, 2002, pp. 336–337.

5. Charles Darwin. Letter to J. S. Henslow, 9 September 1831. In: *The Correspondence of Charles Darwin*, Vol. 1: *1821–1836*. Frederick Burkhardt, Sydney Smith, David Kohn, and William Montgomery, editors. Cambridge University Press, Cambridge, UK, 1985, p. 149.

6. Charles Darwin. Letter to Charles Whitley, 15 November 1831. In: *The Correspondence of Charles Darwin*, Vol. 1: *1821–1836*. Frederick Burkhardt, Sydney Smith, David Kohn, and William Montgomery, editors. Cambridge University Press, Cambridge, UK, 1985, p. 181.

7. Charles Darwin. Letter to J. S. Henslow, 5 September 1831. In: *The Correspondence of Charles Darwin*, Vol. 1: *1821–1836*. Frederick Burkhardt, Sydney Smith, David Kohn, and William Montgomery, editors. Cambridge University Press, Cambridge, UK, 1985, p. 142.

8. Charles Darwin. Letter to R. W. Darwin, 31 August 1831. In: *The Correspondence of Charles Darwin*, Vol. 1: *1821–1836*. Frederick Burkhardt, Sydney Smith, David Kohn, and William Montgomery, editors. Cambridge University Press, Cambridge, UK, 1985, p. 133.

9. Josiah Wedgwood II. Letter to R. W. Darwin, 31 August 1831. In: *The Correspondence of Charles Darwin*, Vol. 1: *1821–1836*. Frederick Burkhardt, Sydney Smith, David Kohn, and William Montgomery, editors. Cambridge University Press, Cambridge, UK, 1985, p. 134.

10. Charles Darwin. Letter to Caroline Darwin, 25 April 1832. In: *The Correspondence of Charles Darwin*, Vol. 1: *1821–1836*. Frederick Burkhardt, Sydney Smith,

David Kohn, and William Montgomery, editors. Cambridge University Press, Cambridge UK, 1985, p. 225.

11. Charles Darwin. In: *Charles Darwin's Beagle Diary,* Richard D. Keynes, editor. Cambridge University Press, Cambridge, UK, 2001, p. xii.

12. Charles Darwin. Letter to Robert FitzRoy, 10 October 1831. In: *The Correspondence of Charles Darwin,* Vol. 1: *1821–1836.* Frederick Burkhardt, Sydney Smith, David Kohn, and William Montgomery, editors. Cambridge University Press, Cambridge, UK, 1985, p. 175.

13. Charles Darwin. Letter to J. S. Henslow, 3 December 1831. In: *The Correspondence of Charles Darwin,* Vol. 1: *1821–1836.* Frederick Burkhardt, Sydney Smith, David Kohn, and William Montgomery, editors. Cambridge University Press, Cambridge, UK, 1985, p. 186.

14. Fanny Owen. Letter to Charles Darwin, 2 December 1831. In: *The Correspondence of Charles Darwin,* Vol. 1: *1821–1836.* Frederick Burkhardt, Sydney Smith, David Kohn, and William Montgomery, editors. Cambridge University Press, Cambridge, UK, 1985, p. 184.

15. Charles Darwin. Chronology 1821–1836 (Appendix I). In: *The Correspondence of Charles Darwin,* Vol. 1: *1821–1836.* Frederick Burkhardt, Sydney Smith, David Kohn, and William Montgomery, editors. Cambridge University Press, Cambridge, UK, 1985, p. 539

16. Charles Darwin. Letter to W. D. Fox, 17 November 1831. In: *The Correspondence of Charles Darwin,* Vol. 1: *1821–1836.* Frederick Burkhardt, Sydney Smith, David Kohn, and William Montgomery, editors. Cambridge University Press, Cambridge, UK, 1985, pp. 182–183.

17. Charles Darwin. Letter to W. D. Fox, 26 August 1829. In: *The Correspondence of Charles Darwin,* Vol. 1: *1821–1836.* Frederick Burkhardt, Sydney Smith, David Kohn, and William Montgomery, editors. Cambridge University Press, Cambridge, UK, 1985, p. 92.

18. Charles Darwin. Letter to W. D. Fox, 24 December 1828. In: *The Correspondence of Charles Darwin,* Vol. 1: *1821–1836.* Frederick Burkhardt, Sydney Smith, David Kohn, and William Montgomery, editors. Cambridge University Press, Cambridge, UK, 1985, p. 72.

19. Charles Darwin. *The Autobiography of Charles Darwin, 1809–1882,* Nora Barlow, editor. WW Norton, New York, 1958, p. 79.

20. Keith Thompson. *HMS Beagle: The Story of Darwin's Ship.* WW Norton, New York, 1995, p. 133.

21. FitzRoy was promoted to the rank of full Captain in early 1835. Charles Darwin. In: *Charles Darwin's Beagle Diary*, Richard D. Keynes, editor. Cambridge University Press, Cambridge, UK, 1988, Footnote 1, p. 324.

22. Charles Darwin. Letter to J. S. Henslow, 15 November 1831. In: *The Correspondence of Charles Darwin,* Vol. 1: *1821–1836.* Frederick Burkhardt, Sydney Smith, David Kohn, and William Montgomery, editors. Cambridge University Press, Cambridge, UK, 1985, pp. 179–180.

23. Adrian Desmond and James Moore. *Darwin.* Penguin Books, London, 1992, p. 169.

24. Keith S. Thomson. *HMS Beagle: The Story of Darwin's Ship.* WW Norton, New York, 1995, pp. 122–123.

25. Charles Darwin. Letter to J. S. Henslow, 30 October 1831. In: *The Correspondence of Charles Darwin,* Vol. 1: *1821–1836.* Frederick Burkhardt, Sydney Smith, David Kohn, and William Montgomery, editors. Cambridge University Press, Cambridge, UK, 1985, p. 176.

26. For a description of the cabin see Keith S. Thomson. *HMS Beagle: The Story of Darwin's Ship.* WW Norton, New York, 1995, pp. 124–126. For a sketch of the cabin and a list of books on board the *Beagle* see Appendix IV. In: *The Correspondence of Charles Darwin,* Vol. 1: *1821–1836.* Frederick Burkhardt, Sydney Smith, David Kohn, and William Montgomery, editors. Cambridge University Press, Cambridge, UK, 1985, pp. 553–566.

27. Charles Darwin. Letter to R. W. Darwin, 08 February–01 March 1832. In: *The Correspondence of Charles Darwin,* Vol. 1: *1821–1836.* Frederick Burkhardt, Sydney Smith, David Kohn, and William Montgomery, editors. Cambridge University Press, Cambridge, UK, 1985, p. 202.

28. See Appendix 3 for a list of crew.

29. Robert FitzRoy. *Narrative of the surveying voyages of His Majesty's Ships Adventure and Beagle between the years 1826 and 1836, describing their examination of the southern shores of South America, and the Beagle's circumnavigation of the globe: Proceedings of the second expedition, 1831–36, under the command of Captain Robert Fitz-Roy, R.N.* Henry Colburn, London, 1839, p. 20.

30. Charles Darwin. In: *Charles Darwin's Beagle Diary,* Richard D. Keynes, editor. Cambridge University Press, Cambridge, UK, 1988, p. 64.

31. Robert FitzRoy. *Narrative of the surveying voyages of His Majesty's Ships Adventure and Beagle between the years 1826 and 1836, describing their examination of the southern shores of South America, and the Beagle's circumnavigation of the globe: Proceedings of the second expedition, 1831–36, under the command of Captain Robert Fitz-Roy, R.N.* Henry Colburn, London, 1839, p. 20.

32. Charles Darwin. In: *Charles Darwin's Beagle Diary,* Richard D. Keynes, editor. Cambridge University Press, Cambridge, UK, 1988, p. 78.

33. Charles Darwin. Letter to C. Whitley, 9 September 1831. In: *The Correspondence of Charles Darwin,* Vol. 1: *1821–1836.* Frederick Burkhardt, Sydney Smith, David Kohn, and William Montgomery, editors. Cambridge University Press, Cambridge, UK, 1985, p. 150.

34. Charles Darwin. Letter to Caroline Darwin, 12 November, 1831. In: *The Correspondence of Charles Darwin,* Vol. 1: *1821–1836.* Frederick Burkhardt, Sydney Smith, David Kohn, and William Montgomery, editors. Cambridge University Press, Cambridge, UK, 1985, p. 178.

35. Charles Darwin. Letter to Susan Darwin, 14 September 1831. In: *The Correspondence of Charles Darwin,* Vol. 1: *1821–1836.* Frederick Burkhardt, Sydney Smith, David Kohn, and William Montgomery, editors. Cambridge University Press, Cambridge, UK, 1985, pp. 154–155.

36. Charles Darwin. In: *Charles Darwin's Beagle Diary,* Richard D. Keynes, editor. Cambridge University Press, Cambridge, UK, 1988, p. 8.

37. Charles Lyell. *Principles of Geology, Being an Attempt to Explain the Former Changes of the Earth's Surface, by Reference to Causes Now in Operation,* Vol. 1. John Murray, London, 1830. Volume II was published in 1832 and Volume III in 1833.

38. Charles Darwin. *The Autobiography of Charles Darwin, 1809–1882,* Nora Barlow, editor. WW Norton, New York, 1958, p. 101.

39. Charles Darwin. Letter to Susan Darwin 14, September 1831. In: *The Correspondence of Charles Darwin,* Vol. 1: *1821–1836.* Frederick Burkhardt, Sydney Smith, David Kohn, and William Montgomery, editors. Cambridge University Press, Cambridge, UK, 1985, p. 155.

40. Charles Darwin. Letter to Caroline Darwin, 25–26 April 1832. In: *The Correspondence of Charles Darwin,* Vol. 1: *1821–1836.* Frederick Burkhardt, Sydney Smith, David Kohn, and William Montgomery, editors. Cambridge University Press, Cambridge, UK, 1985, p. 226.

41. Charles Darwin. Letter to Catherine Darwin, 20–29 July 1834. In: *The Correspondence of Charles Darwin,* Vol. 1: *1821–1836.* Frederick Burkhardt, Sydney Smith, David Kohn, and William Montgomery, editors. Cambridge University Press, Cambridge, UK, 1985, p. 393.

42. Charles Darwin. *The Autobiography of Charles Darwin, 1809–1882,* Nora Barlow, editor. WW Norton, New York, 1958, p. 74.

43. Robert FitzRoy. *Narrative of the surveying voyages of His Majesty's Ships Adventure and Beagle between the years 1826 and 1836, describing their examination of the southern shores of South America, and the Beagle's circumnavigation of the globe: Proceedings of the second expedition, 1831–36, under the command of Captain Robert Fitz-Roy, R.N.* Henry Colburn, London, 1839, pp. 61–62.

44. For an overview of the relationship between Darwin and FitzRoy during the voyage, see Introduction to Charles Darwin. In: *Charles Darwin's Beagle Diary,* Richard D. Keynes, editor. Cambridge University Press, Cambridge, UK, 2001, pp. xxii–xxiii.

45. Charles Darwin. *The Autobiography of Charles Darwin, 1809–1882,* Nora Barlow, editor. WW Norton, New York, 1958, p. 85.

46. Robert FitzRoy and Charles Robert Darwin. 1836. A letter, containing remarks on the moral state of Tahiti, New Zealand, &c. *South African Christian Recorder* 2(4, September): 221–238.

47. Robert FitzRoy. Letter to Charles Darwin, 16 November 1837. In: *The Correspondence of Charles Darwin,* Vol. 2: *1837–1843,* Frederick Burkhardt and Sydney Smith, editors. Cambridge University Press, Cambridge, UK, 1986, pp. 57–59.

48. Charles Darwin. *The Autobiography of Charles Darwin, 1809–1882,* Nora Barlow, editor. WW Norton, New York, 1958, p. 76.

49. Charles Darwin. In: *Charles Darwin's Beagle Diary,* Richard D. Keynes, editor. Cambridge University Press, Cambridge, UK, 1988, p. 244.

50. Francis Darwin. *The Life of Charles Darwin.* (First published by John Murray, London, 1902.) Senate, London, 1995, p. 126.

51. Charles Darwin. Journal 1837–1843 (Appendix II). In: *The Correspondence of Charles Darwin,* Vol. 2: *1837–1843,* Frederick Burkhardt and Sydney Smith, editors. Cambridge University Press, Cambridge, UK, 1986, p. 433.

52. Francis Darwin, editor. *The life and letters of Charles Darwin, including an autobiographical chapter*, Vol. 1. John Murray, London, 1887, p. 395.

53. Ibid., p. 223.

54. Charles Darwin. In: *Charles Darwin's Beagle Diary*, Richard D. Keynes, editor. Cambridge University Press, Cambridge, UK, 1988, p. 49.

55. Charles Darwin. *The Autobiography of Charles Darwin, 1809–1882*, Nora Barlow, editor. WW Norton, New York, 1958, p. 75.

56. Francis Darwin. *The Life of Charles Darwin*. (First published by John Murray, London, 1902.) Senate, London, 1995, p. 126.

57. Charles Darwin. Letter to Catherine Darwin, 20–99 July 1834. In: *The Correspondence of Charles Darwin*, Vol. 1: *1821–1836*. Frederick Burkhardt, Sydney Smith, David Kohn, and William Montgomery, editors. Cambridge University Press, Cambridge, UK, 1985, p. 393.

58. Charles Darwin. Letter to Caroline Darwin, 30 March–12 April 1833. In: *The Correspondence of Charles Darwin*, Vol. 1: *1821–1836*. Frederick Burkhardt, Sydney Smith, David Kohn, and William Montgomery, editors. Cambridge University Press, Cambridge, UK, 1985, p. 305.

59. Charles Darwin. Letter to Catherine Darwin, 20–29 July 1834. In: *The Correspondence of Charles Darwin*, Vol. 1: *1821–1836*. Frederick Burkhardt, Sydney Smith, David Kohn, and William Montgomery, editors. Cambridge University Press, Cambridge, UK, 1985, p. 393.

60. Ibid.

61. Francis Darwin. *The Life of Charles Darwin*. (First published by John Murray, London, 1902.) Senate, London, 1995, p. 126.

62. Charles Darwin. Letter to Caroline Darwin, 25–26 April 1832. In: *The Correspondence of Charles Darwin*, Vol. 1: *1821–1836*. Frederick Burkhardt, Sydney Smith, David Kohn, and William Montgomery, editors. Cambridge University Press, 1985, Cambridge, UK, p. 226.

63. Charles Darwin. In: *Charles Darwin's Beagle Diary*, Richard D. Keynes, editor. Cambridge University Press, Cambridge, UK, 1988, p. 84.

64. Charles Darwin. Letter to Catherine Darwin, 20–29 July 1834. In: *The Correspondence of Charles Darwin*, Vol. 1: *1821–1836*. Frederick Burkhardt, Sydney Smith, David Kohn, and William Montgomery, editors. Cambridge University Press, Cambridge, UK, 1985, p. 392.

65. Charles Darwin. *The Autobiography of Charles Darwin, 1809–1882*, Nora Barlow, editor. WW Norton, New York, 1958, p. 79.

66. Charles Darwin. In: *Charles Darwin's Beagle Diary*, Richard D. Keynes, editor. Cambridge University Press, Cambridge, UK, 1988, p. 81.

67. Richard D. Keynes, editor. *Charles Darwin's Zoology Notes & Specimen Lists from H.M.S. Beagle*. Cambridge University Press, Cambridge, UK, 2000, p. ix.

68. Janet Browne. *Charles Darwin: Voyaging*. AA Knopf, New York, 1995, p. 227.

69. Charles Darwin. Letter to R. W. Darwin, 10 February 1832. In: *The Correspondence of Charles Darwin*, Vol. 1: *1821–1836*. Frederick Burkhardt, Sydney Smith, David Kohn, and William Montgomery, editors. Cambridge University Press, Cambridge, UK, 1985, p. 206.

70. Charles Darwin. In: *Charles Darwin's Beagle Diary*, Richard D. Keynes, editor. Cambridge University Press, Cambridge, UK, 1988, p. 18. See also Charles Darwin. Letter to R. W. Darwin, 08 February–01 March 1832. In: *The Correspondence of Charles Darwin*, Vol. 1: *1821–1836*. Frederick Burkhardt, Sydney Smith, David Kohn, and William Montgomery, editors. Cambridge University Press, Cambridge, UK, 1985, p. 201.

71. Caroline Darwin. Letter to Sarah Elizabeth Wedgwood, 05 October 1836. In: *The Correspondence of Charles Darwin*, Vol. 1: *1821–1836*. Frederick Burkhardt, Sydney Smith, David Kohn, and William Montgomery, editors. Cambridge University Press, Cambridge, UK, 1985, pp. 504–505.

72. Robert FitzRoy. *Narrative of the surveying voyages of His Majesty's Ships Adventure and Beagle between the years 1826 and 1836, describing their examination of the southern shores of South America, and the Beagle's circumnavigation of the globe: Proceedings of the second expedition, 1831–36, under the command of Captain Robert Fitz-Roy, R.N.* Henry Colburn, London, 1839, p. 125.

73. Charles Darwin. Letter to Caroline Darwin, 30 March–12 April 1833. In: *The Correspondence of Charles Darwin*, Vol. 1: *1821–1836*. Frederick Burkhardt, Sydney Smith, David Kohn, and William Montgomery, editors. Cambridge University Press, Cambridge, UK, 1985, p. 303.

74. Charles Darwin. In: *Charles Darwin's Beagle Diary*, Richard D. Keynes, editor. Cambridge University Press, Cambridge, UK, 1988, p. 148.

75. Robert FitzRoy. *Narrative of the surveying voyages of His Majesty's Ships Adventure and Beagle between the years 1826 and 1836, describing their examination of the southern shores of South America, and the Beagle's circumnavigation of the globe: Proceedings of the second expedition, 1831–36, under the command of Captain Robert Fitz-Roy, R.N.* Henry Colburn, London, 1839, p. 125.

76. Charles Darwin. Letter to W. D. Fox, May 1832. In: *The Correspondence of Charles Darwin*, Vol. 1: *1821–1836*. Frederick Burkhardt, Sydney Smith, David Kohn, and William Montgomery, editors. Cambridge University Press, Cambridge, UK, 1985, p. 232.

77. Charles Darwin. Letter to Caroline Darwin, 2–6 April 1832. In: *The Correspondence of Charles Darwin*, Vol. 1: *1821–1836*. Frederick Burkhardt, Sydney Smith, David Kohn, and William Montgomery, editors. Cambridge University Press, Cambridge, UK, 1985, p. 219.

78. Charles Darwin. In: *Charles Darwin's Beagle Diary*, Richard D. Keynes, editor. Cambridge University Press, Cambridge, UK, 1988, p. 442.

79. Richard D. Keynes, editor. *Charles Darwin's Zoology Notes & Specimen Lists from H.M.S. Beagle.* Cambridge University Press, Cambridge, UK, 2000, pp. x–xi.

80. Charles Darwin. In: *Charles Darwin's Beagle Diary*, Richard D. Keynes, editor. Cambridge University Press, Cambridge, UK, 1988, p. 442.

81. Charles Darwin. Letter to Susan Darwin, 4 August 1836. In: *The Correspondence of Charles Darwin*, Vol. 1: *1821–1836*. Frederick Burkhardt, Sydney Smith, David Kohn, and William Montgomery, editors. Cambridge University Press, Cambridge, UK, 1985, p. 503.

CHAPTER 3: GEARING FOR GALÁPAGOS

1. Charles Darwin. In: *Charles Darwin's Beagle Diary*, Richard D. Keynes, editor. Cambridge University Press, Cambridge, UK, 1988, p. 446.

2. Sandra Herbert. *Darwin, Geologist*. Cornell University Press, Ithaca, NY, 2005, p. 311.

3. John Bowlby. Charles Darwin: A New Life. WW Norton, New York, 1991 p. 179.

4. Charles Darwin. *The Autobiography of Charles Darwin, 1809–1882*, Nora Barlow, editor. WW Norton, New York, 1958, p. 79.

5. Charles Darwin. In: *Charles Darwin's Beagle Diary*, Richard D. Keynes, editor. Cambridge University Press, Cambridge, UK, 1988, p. 19.

6. Charles Darwin. Letter to R. W. Darwin, 8 February–1 March 1832. In: *The Correspondence of Charles Darwin*, Vol. 1: *1821–1836*. Frederick Burkhardt, Sydney Smith, David Kohn, and William Montgomery, editors. Cambridge University Press, Cambridge, UK, 1985, p. 201.

7. Charles Darwin. In: *Charles Darwin's Beagle Diary*, Richard D. Keynes, editor. Cambridge University Press, Cambridge, UK, 1988, p. 20.

8. Ibid., p. 23.

9. Robert FitzRoy. Letter to Francis Beaufort, 5 March 1832. In: *The Correspondence of Charles Darwin*, Vol. 1: *1821–1836*. Frederick Burkhardt, Sydney Smith, David Kohn, and William Montgomery, editors. Cambridge University Press, Cambridge, UK, 1985, p. 205.

10. Charles Darwin. In: *Charles Darwin's Beagle Diary*, Richard D. Keynes, editor. Cambridge University Press, Cambridge, UK, 1988, p. 25.

11. Richard D. Keynes, editor. *Charles Darwin's Zoology Notes & Specimen Lists from H.M.S. Beagle*. Cambridge University Press, Cambridge, UK, 2000, p. ix.

12. Charles Darwin. *The Autobiography of Charles Darwin, 1809–1882*, Nora Barlow, editor. WW Norton, New York, 1958, p. 81.

13. Ibid.

14. Charles Robert Darwin. *Narrative of the surveying voyages of His Majesty's Ships Adventure and Beagle between the years 1826 and 1836, describing their examination of the southern shores of South America, and the Beagle's circumnavigation of the globe: Journal and remarks, 1832–1836*. Henry Colburn, London, 1839. Darwin's volume was republished on its own, also in 1839, under the title *Journal of Researches into the Geology and Natural History of the Countries Visited during the Voyage of HMS Beagle round the World, under the Command of Capt. Fitz Roy*. In 1845 a second edition with several changes to the text was printed as Charles Darwin. *Journal of Researches into the Natural History and Geology of the Countries Visited during the Voyage of HMS Beagle round the World, under the Command of Capt. Fitz Roy, R.N.*, 2nd edition. John Murray, London, 1845. Later editions became known as *Voyage of the Beagle*.

15. Charles Darwin. *The Structure and Distribution of Coral Reefs. Being the First Part of The Geology of the Voyage of the Beagle, Under the Command of Capt. FitzRoy, R.N. During the Years 1832 to 1836*. Smith Elder, London, 1842.

16. Charles Darwin. *Geological Observations on the Volcanic Islands visited during the Voyage of H. M. S. Beagle, together with some brief notices of the Geology of Australia and the Cape of Good Hope. Being the second part of the Geology of the Voyage of the Beagle, under the Command of Capt. FitzRoy, R.N. during the years 1832 to 1836.* Smith Elder, London, 1844.

17. Charles Darwin. *Geological Observations on South America. Being the third part of the Geology of the Voyage of the Beagle, under the Command of Capt. FitzRoy, R.N. during the years 1832 to 1836.* Smith Elder, London, 1846.

18. Charles Robert Darwin, editor. *The Zoology of The Voyage of H.M.S. Beagle, under the command of Captain FitzRoy, R.N., During the Years 1832 to 1836.* Smith Elder, London. 1838–1843.

19. Duncan M. Porter. The Beagle Collector and His Collections. In: *The Darwinian Heritage*, David Kohn, editor. Princeton University Press, Princeton, NJ, 1985, p. 984.

20. A. C. Harker, c. 1907. Catalogue of the "Beagle" Collection of Rocks, made by Charles Darwin during the voyage of H.M.S. "Beagle," 1832–6.

21. Charles Darwin. *On the Origin of Species by Means of Natural Selection, or the Preservation of Favoured Races in the Struggle for Life.* John Murray, London, 1859, p. 398.

22. Francis Darwin, editor. *The foundations of The origin of species. Two essays written in 1842 and 1844.* Cambridge University Press, Cambridge, UK, 1909, p. 31.

23. J. D. Hooker. Letter to Charles Darwin, 23 March 1845. In: *The Correspondence of Charles Darwin*, Vol. 3: *1844–1846*, Frederick Burkhardt and Sydney Smith, editors. Cambridge University Press, Cambridge, UK, 1987, pp. 163–164.

24. Charles Darwin. *On the Origin of Species by Means of Natural Selection, or the Preservation of Favoured Races in the Struggle for Life.* John Murray, London, 1859, p. 398.

25. Charles Darwin. In: *Charles Darwin's Beagle Diary*, Richard D. Keynes, editor. Cambridge University Press, Cambridge, UK, 1988, pp. 42–43.

26. Ibid., p. 23.

27. Ibid., p. 444.

28. Charles Darwin. Letter to J. S. Henslow 18 May–16 June 1832. In: *The Correspondence of Charles Darwin*, Vol. 1: *1821–1836*. Frederick Burkhardt, Sydney Smith, David Kohn, and William Montgomery, editors. Cambridge University Press, Cambridge, UK, 1985, p. 237.

29. Charles Darwin. In: *Charles Darwin's Beagle Diary*, Richard D. Keynes, editor. Cambridge University Press, Cambridge, UK, 1988, p. 42.

30. Caroline Darwin. Letter to Charles Darwin, 28 October 1833. In: *The Correspondence of Charles Darwin*, Vol. 1: *1821–1836*. Frederick Burkhardt, Sydney Smith, David Kohn, and William Montgomery, editors. Cambridge University Press, Cambridge, UK, 1985, p. 345.

31. Charles Darwin. Letter to Susan Darwin, 4 August 1836. In: *The Correspondence of Charles Darwin*, Vol. 1: *1821–1836*. Frederick Burkhardt, Sydney Smith, David Kohn, and William Montgomery, editors. Cambridge University Press, Cambridge, UK, 1985, p. 503.

32. Charles Darwin. *The Autobiography of Charles Darwin, 1809–1882*, Nora Barlow, editor. WW Norton, New York, 1958, p. 44.

33. Ibid., p. 139.

34. Charles Darwin. Letter to J. S. Henslow, 26 October–24 November 1832. In: *The Correspondence of Charles Darwin*, Vol. 1: *1821–1836*. Frederick Burkhardt, Sydney Smith, David Kohn, and William Montgomery, editors. Cambridge University Press, Cambridge, UK, 1985, p. 280.

35. Charles Darwin. In: *Charles Darwin's Beagle Diary*, Richard D. Keynes, editor. Cambridge University Press, Cambridge, UK, 1988, p. 109.

36. Niles Eldredge. *Darwin: Discovering the Tree of Life*. WW Norton, New York, 2005, pp. 117–119.

37. Charles Darwin. In: *Charles Darwin's Beagle Diary*, Richard D. Keynes, editor. Cambridge University Press, Cambridge, UK, 1988, p. 106.

38. Janet Browne. *Charles Darwin: Voyaging*. AA Knopf, New York, 1995 p. 225.

39. Robert FitzRoy. *Narrative of the surveying voyages of His Majesty's Ships Adventure and Beagle between the years 1826 and 1836, describing their examination of the southern shores of South America, and the Beagle's circumnavigation of the globe: Proceedings of the second expedition, 1831–36, under the command of Captain Robert Fitz-Roy, R.N.* Henry Colburn, London, 1839, p. 107.

40. Charles Darwin. Letter to Catherine Darwin, 20–29 July 1834. In: *The Correspondence of Charles Darwin*, Vol. 1: *1821–1836*. Frederick Burkhardt, Sydney Smith, David Kohn, and William Montgomery, editors. Cambridge University Press, Cambridge, UK, 1985, p. 393.

41. Charles Darwin. In: *Charles Darwin's Beagle Diary*, Richard D. Keynes, editor. Cambridge University Press, Cambridge, UK, 1988, p. 134.

42. Ibid., p. 444.

43. Charles Robert Darwin. *Narrative of the surveying voyages of His Majesty's Ships Adventure and Beagle between the years 1826 and 1836, describing their examination of the southern shores of South America, and the Beagle's circumnavigation of the globe: Journal and remarks, 1832–1836*. Henry Colburn, London, 1839, p. 228.

44. Janet Browne. *Charles Darwin: Voyaging*. AA. Knopf, New York, 1995, p. 249.

45. Ibid., pp. 249–250.

46. Charles Darwin. In: *Charles Darwin's Beagle Diary*, Richard D. Keynes, editor. Cambridge University Press, Cambridge, UK, 1988, p. 137.

47. Ibid., p. 143.

48. Ibid., pp. 226–227.

49. Ibid., p. 144.

50. Ibid., p. 145.

51. Charles Robert Darwin. *Narrative of the surveying voyages of His Majesty's Ships Adventure and Beagle between the years 1826 and 1836, describing their examination of the southern shores of South America, and the Beagle's circumnavigation of the globe: Journal and remarks, 1832–1836*. Henry Colburn, London, 1839, p. 254.

52. Ibid., p. 256.

53. Charles Darwin. In: *Charles Darwin's Beagle Diary*, Richard D. Keynes, editor. Cambridge University Press, Cambridge, UK, 1988, p. 148.

54. Robert FitzRoy. *Narrative of the surveying voyages of His Majesty's Ships Adventure and Beagle between the years 1826 and 1836, describing their examination of the southern shores of South America, and the Beagle's circumnavigation of the globe: Proceedings of the second expedition, 1831–36, under the command of Captain Robert Fitz-Roy, R.N.* Henry Colburn, London, 1839, p. 273.

55. Darwin sometimes spelled the captain's name Lowe.

56. Robert FitzRoy. *Narrative of the surveying voyages of His Majesty's Ships Adventure and Beagle between the years 1826 and 1836, describing their examination of the southern shores of South America, and the Beagle's circumnavigation of the globe: Proceedings of the second expedition, 1831–36, under the command of Captain Robert Fitz-Roy, R.N.* Henry Colburn, London, 1839, p. 331.

57. Charles Robert Darwin, editor. *Birds Part 3 No. 5 of The zoology of the voyage of H.M.S. Beagle, by John Gould. Edited and superintended by Charles Darwin.* Smith Elder, London, 1841, p. 140.

58. Robert FitzRoy. *Narrative of the surveying voyages of His Majesty's Ships Adventure and Beagle between the years 1826 and 1836, describing their examination of the southern shores of South America, and the Beagle's circumnavigation of the globe: Proceedings of the second expedition, 1831–36, under the command of Captain Robert Fitz-Roy, R.N.* Henry Colburn, London, 1839, p. 192.

59. Ibid., p. 192.

60. Charles Darwin. Letter to Catherine Darwin, 8 November 1834. In: *The Correspondence of Charles Darwin,* Vol. 1: *1821–1836.* Frederick Burkhardt, Sydney Smith, David Kohn, and William Montgomery, editors. Cambridge University Press, Cambridge, UK, 1985, p. 418.

61. Sandra Herbert. Remembering Charles Darwin as a Geologist. In: *Charles Darwin, 1809–1882: A Centennial Commemorative,* Chapman and Durval, editors. Nova Pacifica, 1982, p. 241.

62. Charles Darwin. In: *Charles Darwin's Beagle Diary,* Richard D. Keynes, editor. Cambridge University Press, Cambridge, UK, 1988, p. 147.

63. Ibid.

64. Richard D. Keynes, editor. *Charles Darwin's Zoology Notes & Specimen Lists from H.M.S. Beagle.* Cambridge University Press, Cambridge, UK, 2000, pp. 209–210.

65. See Charles Robert Darwin. *Narrative of the surveying voyages of His Majesty's Ships Adventure and Beagle between the years 1826 and 1836, describing their examination of the southern shores of South America, and the Beagle's circumnavigation of the globe: Journal and remarks, 1832–1836.* Henry Colburn, London, 1839, p. 250; and Richard D. Keynes (editor) *Charles Darwin's Zoology Notes & Specimen Lists from H.M.S. Beagle.* Cambridge University Press, Cambridge, UK, 2000, p. 210.

66. Nora Barlow, editor. 1963. Darwin's Ornithological Notes. *Bulletin of the British Museum (Natural History).* Historical Series. 2, (7): 201–278 (p. 262).

67. Richard D. Keynes. 1997. Steps on the Path to the *Origin of Species. Journal of Theoretical Biology* 187 (4): 461–467.

68. Charles Robert Darwin. *Narrative of the surveying voyages of His Majesty's Ships Adventure and Beagle between the years 1826 and 1836, describing their exami-*

nation of the southern shores of South America, and the Beagle's circumnavigation of the globe: *Journal and remarks, 1832–1836.* Henry Colburn, London, 1839, p. 399.

69. Richard D. Keynes. 1997. Steps on the Path to the *Origin of Species. Journal of Theoretical Biology* 187 (4): 461–467.

70. Charles Robert Darwin, editor. *Birds Part 3 No. 5 of The zoology of the voyage of H.M.S. Beagle, by John Gould. Edited and superintended by Charles Darwin.* Smith Elder, London, 1841, p. 124.

71. Richard D. Keynes, editor. *Charles Darwin's Zoology Notes & Specimen Lists from H.M.S. Beagle.* Cambridge University Press, Cambridge, UK, 2000, p. 102.

72. Charles Darwin. Letter to J. S. Henslow, 26 October–24 November 1832. In: *The Correspondence of Charles Darwin,* Vol. 1: *1821–1836.* Frederick Burkhardt, Sydney Smith, David Kohn, and William Montgomery, editors. Cambridge University Press, Cambridge, UK, 1985, p. 280.

73. Charles Robert Darwin, editor. *Birds Part 3 No. 5 of The zoology of the voyage of H.M.S. Beagle, by John Gould. Edited and superintended by Charles Darwin.* Smith Elder, London, 1841, p. 125.

74. Sandra Herbert. 1995. From Charles Darwin's Portfolio: An Early Essay on South American Geology and Species. *Earth Sciences History* 14 (1): 23–36.

75. Ibid.

76. Robert FitzRoy. *Narrative of the surveying voyages of His Majesty's Ships Adventure and Beagle between the years 1826 and 1836, describing their examination of the southern shores of South America, and the Beagle's circumnavigation of the globe: Proceedings of the second expedition, 1831–36, under the command of Captain Robert Fitz-Roy, R.N.* Henry Colburn, London, 1839, pp. 216–217.

77. Charles Darwin. In: *Charles Darwin's Beagle Diary,* Richard D. Keynes, editor. Cambridge University Press, Cambridge, UK, 1988, p. 225. See also Charles Darwin. Letter to Catherine Darwin, 6 April 1834. In: *The Correspondence of Charles Darwin,* Vol. 1: *1821–1836.* Frederick Burkhardt, Sydney Smith, David Kohn, and William Montgomery, editors. Cambridge University Press, Cambridge, UK, 1985, p. 381.

78. Robert FitzRoy. *Narrative of the surveying voyages of His Majesty's Ships Adventure and Beagle between the years 1826 and 1836, describing their examination of the southern shores of South America, and the Beagle's circumnavigation of the globe: Proceedings of the second expedition, 1831–36, under the command of Captain Robert Fitz-Roy, R.N.* Henry Colburn, London, 1839, pp. 319–320.

79. Charles Darwin. In: *Charles Darwin's Beagle Diary,* Richard D. Keynes, editor. Cambridge University Press, Cambridge, UK, 1988, p. 263.

80. Catherine Darwin. Letter to Charles Darwin, 28 January 1835. In: *The Correspondence of Charles Darwin,* Vol. 1: *1821–1836.* Frederick Burkhardt, Sydney Smith, David Kohn, and William Montgomery, editors. Cambridge University Press, Cambridge, UK, 1985, p. 424.

81. Charles Darwin. Letter to Catherine Darwin, 8 November 1834. In: *The Correspondence of Charles Darwin,* Vol. 1: *1821–1836.* Frederick Burkhardt, Sydney Smith, David Kohn, and William Montgomery, editors. Cambridge University Press, Cambridge, UK, 1985, pp. 418–419.

82. Charles Darwin. Letter to Caroline Darwin, 10–13 March 1835. In: *The Correspondence of Charles Darwin*, Vol. 1: *1821–1836*. Frederick Burkhardt, Sydney Smith, David Kohn, and William Montgomery, editors. Cambridge University Press, Cambridge, UK, 1985, p. 433.

83. Ibid., p. 434.

84. Charles Darwin. Letter to Catherine Darwin, 8 November 1834. In: *The Correspondence of Charles Darwin*, Vol. 1: *1821–1836*. Frederick Burkhardt, Sydney Smith, David Kohn, and William Montgomery, editors. Cambridge University Press, Cambridge, UK, 1985, pp. 418–419.

85. Caroline Darwin. Letter to Charles Darwin, 30 March 1835. In: *The Correspondence of Charles Darwin*, Vol. 1: *1821–1836*. Frederick Burkhardt, Sydney Smith, David Kohn, and William Montgomery, editors. Cambridge University Press, Cambridge, UK, 1985, p. 438.

86. Charles Darwin. In: *Charles Darwin's Beagle Diary*, Richard D. Keynes, editor. Cambridge University Press, Cambridge, UK, 1988, p. 308.

87. Charles Darwin. Letter to Susan Darwin, 23 April 1835. In: *The Correspondence of Charles Darwin*, Vol. 1: *1821–1836*. Frederick Burkhardt, Sydney Smith, David Kohn, and William Montgomery, editors. Cambridge University Press, Cambridge, UK, 1985, p. 445.

88. Charles Darwin. Letter to J. S. Henslow, 18 April 1835. In: *The Correspondence of Charles Darwin*, Vol. 1: *1821–1836*. Frederick Burkhardt, Sydney Smith, David Kohn, and William Montgomery, editors. Cambridge University Press, Cambridge, UK, 1985, p. 440.

89. Charles Darwin. In: *Charles Darwin's Beagle Diary*, Richard D. Keynes, editor. Cambridge University Press, Cambridge, UK, 1988, p. 292.

90. Ibid., p. 296.

91. Robert FitzRoy. *Narrative of the surveying voyages of His Majesty's Ships Adventure and Beagle between the years 1826 and 1836, describing their examination of the southern shores of South America, and the Beagle's circumnavigation of the globe: Proceedings of the second expedition, 1831–36, under the command of Captain Robert Fitz-Roy, R.N.* Henry Colburn, London, 1839, pp. 413–414.

92. Charles Darwin. In: *Charles Darwin's Beagle Diary*, Richard D. Keynes, editor. Cambridge University Press, Cambridge, UK, 1988, p. 300.

93. Ibid., p. 302.

94. Robert FitzRoy. *Narrative of the surveying voyages of His Majesty's Ships Adventure and Beagle between the years 1826 and 1836, describing their examination of the southern shores of South America, and the Beagle's circumnavigation of the globe: Proceedings of the second expedition, 1831–36, under the command of Captain Robert Fitz-Roy, R.N.* Henry Colburn, London, 1839, p. 484.

95. Charles Darwin. In: *Charles Darwin's Beagle Diary*, Richard D. Keynes, editor. Cambridge University Press, Cambridge, UK, 1988, p. 350.

96. Robert FitzRoy. *Narrative of the surveying voyages of His Majesty's Ships Adventure and Beagle between the years 1826 and 1836, describing their examination of the southern shores of South America, and the Beagle's circumnavigation of the globe: Proceedings of the second expedition, 1831–36, under the command of Captain Robert Fitz-Roy, R.N.* Henry Colburn, London, 1839, p. 484.

97. Charles Darwin. In: *The Correspondence of Charles Darwin*, Vol. 2: *1837–1843*, Frederick Burkhardt and Sydney Smith, editors. Cambridge University Press, Cambridge, UK 1986, p. 153.

98. Charles Darwin. In: *Charles Darwin's Beagle Diary*, Richard D. Keynes, editor. Cambridge University Press, Cambridge, UK, 1988, p. 148.

99. Robert FitzRoy. *Narrative of the surveying voyages of His Majesty's Ships Adventure and Beagle between the years 1826 and 1836, describing their examination of the southern shores of South America, and the Beagle's circumnavigation of the globe: Proceedings of the second expedition, 1831–36, under the command of Captain Robert Fitz-Roy, R.N.* Henry Colburn, London, 1839, p. 331.

100. Ibid., p. 448.

101. Books on the *Beagle*. In: *The Correspondence of Charles Darwin*, Vol. 1: *1821–1836*. Frederick Burkhardt, Sydney Smith, David Kohn, and William Montgomery, editors. Cambridge University Press, Cambridge, UK 1985, Appendix IV, pp. 553–556.

102. James Colnett. *A Voyage to the South Atlantic and Round Cape Horn into the Pacific Ocean, for the Purpose of Extending the Spermaceti Whale Fisheries, and other objects of commerce, by ascertaining the ports, bays, harbours, and anchoring births, in certain islands and coasts in those seas at which the ships of the British merchants might be refitted.* W Bennett, London, 1798.

103. Lord George Anson Byron. *Voyage of H.M.S. Blonde to the Sandwich Islands, in the years 1824–25*, Maria Graham, editor. John Murray, London, 1826.

104. British author Maria Graham created the book by compiling and editing the official papers and journals of several of the ship's company, including Robert Dampier and naturalist Andrew Bloxam.

105. Basil Hall. *Extracts from a journal written on the coasts of Chili, Peru and Mexico for the years 1820, 1821, 1822*, 2 vols. E. Littell, Philadelphia, William Brown, printer, 1824.

106. William Dampier. *A New Voyage Round the World. Describing particularly, The Isthmus of America, several Coasts and Islands in the West Indies, the Isles of Cape Verd, the Passage by Terra del Fuego, the South Sea Coasts of Chili, Peru, and Mexico; the Isle of Guam one of the Ladrones, Mindanao, and other Philippine and East-India Islands near Cambodia, China, Formosa, Luconia, Celebes, &c. New Holland, Sumatra, Nicobar Isles; the Cape of Good Hope, and Santa Hellena. THEIR Soil, Rivers, Harbours, Plants, Fruits, Animals, and Inhabitants. THEIR Customs, Religion, Government, Trade, &c. Illustrated with Particular Maps and Draughts.* James Knapton, London, 1697.

107. Basil Hall. *Extracts from a journal written on the coasts of Chili, Peru and Mexico for the years 1820, 1821, 1822*, 2 vols. E Littell, Philadelphia, William Brown, printer, 1824, p. 80.

108. Ibid.

109. Amasa Delano. *A Narrative of Voyages and Travels, in the Northern and Southern Hemispheres: Comprising Three Voyages Round the World; Together with a Voyage of Survey and Discovery, in the Pacific Ocean and Oriental Islands.* E. G. House, Boston, 1817.

110. Adrian Desmond and James Moore. *Darwin.* Penguin Books, London, 1992, pp. 42–43.

111. For further information about Lonesome George and his subspecies see Peter C. H. Pritchard. 2005. The Pinta Tortoise: Globalization and the Extinction of Island Species. *Worldviews for the 21st Century: A Monograph Series* 3 (2): 46 pages.

112. Charles Darwin. Letter to Caroline Darwin, 19 July–12 August 1835. In: *The Correspondence of Charles Darwin*, Vol. 1: *1821–1836*. Frederick Burkhardt, Sydney Smith, David Kohn, and William Montgomery, editors. Cambridge University Press, Cambridge, UK, 1985, p. 458.

113. Charles Darwin. Letter to W. D. Fox, 9–12 August 1835. In: *The Correspondence of Charles Darwin*, Vol. 1: *1821–1836*. Frederick Burkhardt, Sydney Smith, David Kohn, and William Montgomery, editors. Cambridge University Press, Cambridge, UK, 1985, p. 460.

114. Charles Darwin. Letter to Alexander Burns Usborne, 1–5 September 1835. In: *The Correspondence of Charles Darwin*, Vol. 1: *1821–1836*. Frederick Burkhardt, Sydney Smith, David Kohn, and William Montgomery, editors. Cambridge University Press, Cambridge, UK, 1985, p. 464.

115. Charles Darwin. Letter to J. S. Henslow, 12 August 1835. In: *The Correspondence of Charles Darwin*, Vol. 1: *1821–1836*. Frederick Burkhardt, Sydney Smith, David Kohn, and William Montgomery, editors. Cambridge University Press, Cambridge, UK, 1985, p. 461.

116. Charles Robert Darwin. *Narrative of the surveying voyages of His Majesty's Ships Adventure and Beagle between the years 1826 and 1836, describing their examination of the southern shores of South America, and the Beagle's circumnavigation of the globe: Journal and remarks, 1832–1836.* Henry Colburn, London, 1839, p. 453.

117. Charles Darwin. Letter to Caroline Darwin, 27 December 1835. In: *The Correspondence of Charles Darwin*, Vol. 1: *1821–1836*. Frederick Burkhardt, Sydney Smith, David Kohn, and William Montgomery, editors. Cambridge University Press, Cambridge, UK, 1985, p. 471.

118. Roughly 88 pages of DAR 37.2 are filled with written text. An additional 17 pages contain side notes that are referenced in the text, additional comments of various lengths, and light pencil sketches. There are 115 foolscap pages in all, including memorandums, references to other works, and comments added at a later date.

119. Paul N. Pearson. 1996. Charles Darwin on the Origin and Diversity of Igneous Rocks. *Earth Sciences History* 15(1): 49–67 (p. 54).

CHAPTER 4: CHATHAM ISLAND (ISLA SAN CRISTÓBAL)

1. Charles Robert Darwin. *Narrative of the surveying voyages of His Majesty's Ships Adventure and Beagle between the years 1826 and 1836, describing their examination of the southern shores of South America, and the Beagle's circumnavigation of the globe: Journal and remarks, 1832–1836.* Henry Colburn, London, 1839, p. 455.

1. Robert FitzRoy. *Narrative of the surveying voyages of His Majesty's Ships Adventure and Beagle between the years 1826 and 1836, describing their examination of the southern shores of South America, and the Beagle's circumnavigation of the globe:*

Proceedings of the second expedition, 1831–36, under the command of Captain Robert Fitz-Roy, R.N. Henry Colburn, London, 1839, p. 485.

2. Ibid.

3. Ibid., p. 486.

4. Ship's log, HMS *Beagle.* ADM 53/236. Public Records Office, London, September 16, 1835.

5. Robert FitzRoy. *Narrative of the surveying voyages of His Majesty's Ships Adventure and Beagle between the years 1826 and 1836, describing their examination of the southern shores of South America, and the Beagle's circumnavigation of the globe: Proceedings of the second expedition, 1831–36, under the command of Captain Robert Fitz-Roy, R.N.* Henry Colburn, London, 1839, p. 486.

6. Ibid.

7. Charles Darwin. *The Autobiography of Charles Darwin, 1809–1882,* Nora Barlow, editor. WW Norton, New York, 1958, p. 85.

8. John Milton. *Paradise Lost: A Poem in Twelve Books,* 2nd ed. S Simmons, London, 1674 (Book 1, line 756).

9. Robert FitzRoy. *Narrative of the surveying voyages of His Majesty's Ships Adventure and Beagle between the years 1826 and 1836, describing their examination of the southern shores of South America, and the Beagle's circumnavigation of the globe: Proceedings of the second expedition, 1831–36, under the command of Captain Robert Fitz-Roy, R.N.* Henry Colburn, London, 1839, p. 486.

10. Richard Charles Darwin. In: *Charles Darwin's Beagle Diary,* Richard D. Keynes, editor. Cambridge University Press, Cambridge, UK, 1988, p. 352.

11. Robert FitzRoy. *Narrative of the surveying voyages of His Majesty's Ships Adventure and Beagle between the years 1826 and 1836, describing their examination of the southern shores of South America, and the Beagle's circumnavigation of the globe: Proceedings of the second expedition, 1831–36, under the command of Captain Robert Fitz-Roy, R.N.* Henry Colburn, London, 1839, p. 486.

12. Ibid., pp. 486–487.

13. Charles Darwin. *Notes on the geology of places visited on the voyage.* Darwin Archive, Cambridge University Library DAR 37.2 (1835) Folio 758.

14. Ibid., Folios 732–733.

15. Charles Darwin. *Geological Observations on the Volcanic Islands visited during the Voyage of H. M. S. Beagle, together with some brief notices of the Geology of Australia and the Cape of Good Hope. Being the second part of the Geology of the Voyage of the Beagle, under the Command of Capt. FitzRoy, R.N. during the years 1832 to 1836.* Smith Elder, London, 1844. p. 114.

16. Charles Darwin. *The Autobiography of Charles Darwin, 1809–1882,* Nora Barlow, editor. WW Norton, New York, 1958, p. 141.

17. Tom Simkin. Geology of Galápagos Islands. In: Roger Perry (editor). *Galápagos: Key Environments,* Roger Perry, editor. Pergamon Press, Oxford, UK, 1984, pp. 15–41.

18. Charles Darwin. In: *Charles Darwin's Beagle Diary,* Richard D. Keynes, editor. Cambridge University Press, Cambridge, UK, 1988, p. 353.

19. Ibid.

20. Richard D. Keynes, editor. *Charles Darwin's Zoology Notes & Specimen Lists from H.M.S. Beagle.* Cambridge University Press, Cambridge, UK, 2000, p. 294.

21. Marine iguanas rid their bodies of excess salt by this method.

22. Richard D. Keynes, editor. *Charles Darwin's Zoology Notes & Specimen Lists from H.M.S. Beagle.* Cambridge University Press, Cambridge, UK, 2000, p. 294.

23. Charles Darwin. In: *Charles Darwin's Beagle Diary,* Richard D. Keynes, editor. Cambridge University Press, Cambridge, UK, 1988, p. 196.

24. Ibid., p. 315.

25. Charles Robert Darwin. *Narrative of the surveying voyages of His Majesty's Ships Adventure and Beagle between the years 1826 and 1836, describing their examination of the southern shores of South America, and the Beagle's circumnavigation of the globe: Journal and remarks, 1832–1836.* Henry Colburn, London, 1839, p. 468.

26. Ibid.

27. Ibid., p. 467.

28. Charles Robert Darwin. *Narrative of the surveying voyages of His Majesty's Ships Adventure and Beagle between the years 1826 and 1836, describing their examination of the southern shores of South America, and the Beagle's circumnavigation of the globe: Journal and remarks, 1832–1836.* Henry Colburn, London 1839, p. 468.

29. Ibid.

30. Iguanas have only rarely been found in the stomachs of sharks. Moray eels and aggressive reef fish such as the hieroglyphic hawkfish (*Cirrhitus rivulatus*) pose more of a threat for young iguanas. Sea lions often play with iguanas in the water by pulling their tails and sometimes injure the iguanas in the process, though the wounds are rarely fatal. (Personal communication with Dr. Martin Wikelski, Princeton University.)

31. Richard D. Keynes, editor. *Charles Darwin's Zoology Notes & Specimen Lists from H.M.S. Beagle.* Cambridge University Press, Cambridge, UK, 2000, p. 294.

32. Ibid., p. 293.

33. Lord George Anson Byron. *Voyage of H.M.S. Blonde to the Sandwich Islands, in the years 1824–25:* March 27, 1825. Maria Graham, editor. John Murray, London, 1826.

34. Richard D. Keynes, editor. *Charles Darwin's Zoology Notes & Specimen Lists from H.M.S. Beagle.* Cambridge University Press, Cambridge, UK, 2000, p. 296.

35. Charles Robert Darwin. *Narrative of the surveying voyages of His Majesty's Ships Adventure and Beagle between the years 1826 and 1836, describing their examination of the southern shores of South America, and the Beagle's circumnavigation of the globe: Journal and remarks, 1832–1836.* Henry Colburn, London, 1839, p. 466.

36. Ibid., p. 628 (addenda to p. 465).

37. K. Rassman, T. Tautz, F. Trillmich, and C. Gliddon. 1997. The Microevolution of the Galápagos Marine Iguana *Amblyrhynchus cristatus* Assessed by Nuclear and Mitochondrial Genetic Analyses. *Molecular Ecology* 6:437–452.

38. Linda J. Cayot, Kornelia Rassmann, and Fritz Trillmich. April 1994. Are Marine Iguanas Endangered on Islands with Introduced Predators? *Noticias de Galápagos* 53:13.

39. Martin Wikelski and Karin Nelson. 2004. Conservation of Galápagos Marine Iguanas (*Amblyrhynchus cristatus*). *Iguana* 11(4):191–197.

40. Charles Robert Darwin. *Narrative of the surveying voyages of His Majesty's Ships Adventure and Beagle between the years 1826 and 1836, describing their examination of the southern shores of South America, and the Beagle's circumnavigation of the globe: Journal and remarks, 1832–1836.* Henry Colburn, London, 1839, pp. 468–469.

41. Charles Darwin. Letter to Osbert Salvin,11 May 1863. *The Correspondence of Charles Darwin,* Vol. 11: *1863,* Frederick Burkhardt, Duncan M. Porter, Sheila Ann Dean, Jonathan R. Topham, and Sarah Wilmot, editors. Cambridge University Press, Cambridge, UK, 1999, pp. 404–405.

42. Michael H. Jackson. *Galápagos: A Natural History.* University of Calgary Press, Canada, 1993, pp. 124–125.

43. Irenaus Eibl-Eibesfeldt. The Large Iguanas of the Galápagos Islands. In: *Galápagos: Key Environments,* Roger Perry, editor. Pergamon Press, Oxford, UK, 1984, p. 164.

44. Charles Robert Darwin. *Narrative of the surveying voyages of His Majesty's Ships Adventure and Beagle between the years 1826 and 1836, describing their examination of the southern shores of South America, and the Beagle's circumnavigation of the globe: Journal and remarks, 1832–1836.* Henry Colburn, London, 1839, p. 454.

45. Robert FitzRoy. *Narrative of the surveying voyages of His Majesty's Ships Adventure and Beagle between the years 1826 and 1836, describing their examination of the southern shores of South America, and the Beagle's circumnavigation of the globe: Proceedings of the second expedition, 1831–36, under the command of Captain Robert Fitz-Roy, R.N.* Henry Colburn, London, 1839, p. 487.

46. Charles Darwin. In: *Charles Darwin's Beagle Diary,* Richard D. Keynes, editor. Cambridge University Press, Cambridge, UK, 1988, p. 353.

47. Ibid.

48. Richard D. Keynes, editor. *Charles Darwin's Zoology Notes & Specimen Lists from H.M.S. Beagle.* Cambridge University Press, Cambridge, UK, 2000, p. 300.

49. Charles Robert Darwin, editor. *Fish Part 4 No. 4 of The zoology of the voyage of H.M.S. Beagle. By Leonard Jenyns. Edited and superintended by Charles Darwin.* Smith Elder, London, 1842, p. VI.

50. The fish Darwin collected are listed here. The scientific names ascribed by Leonard Jenyns (Charles Robert Darwin, editor. *Fish Part 4 No. 4 of The zoology of the voyage of H.M.S. Beagle. By Leonard Jenyns. Edited and superintended by Charles Darwin.* Smith Elder, London, 1842) are followed by modern nomenclature in brackets (Jack Grove and Robert Lavenberg's *The Fishes of the Galápagos Islands.* Standford University Press, 1997). The first 5 species are endemic to Galápagos. *Serranus albo-maculatus* (Whitespotted sand bass *Paralabrax albomaculatus*), *Pristipoma cantharinum* (Sheephead grunt *Orthopristis cantharinus*), *Tetrodon angusticeps* (Concave puffer *Sphoeroides angusticeps*), *Gobiesox pœcilophthalmos* (Red clingfish *Arcos poecilophthalmus*), *Prionotus Miles* (Orangethroat searobin *Prionotus miles*), *Serranus labriformis* (Flag cabrilla grouper *Epinephelus labriformis*), *Serranus olfax* (Bacalao *Mycteroperca olfax*), *Scorpæna Histrio* (Bandfin scorpionfish *Scorpaena histrio*), *Prionodes fasciatus* (Misty grouper *Epinephelus mystacinus*), *Latilus princeps*

(Ocean whitefish *Caulolatilus princeps*), *Chrysophrys taurina* (Galápagos porgy *Calamus taurinus*), *Gobius lineatus* (Southern frillfin *Bathygobius lineatus*), *Cossyphus Darwini* (Pacific red sheephead *Semicossyphus darwini*), *Muræna lentiginosa* (Lentil moray *Muraena lentiginosa*), *Tetrodon annulatus* (Bullseye puffer *Sphoeroides annulatus*).

51. Charles Robert Darwin, editor. *Fish Part 4 No. 4 of The zoology of the voyage of H.M.S. Beagle. By Leonard Jenyns. Edited and superintended by Charles Darwin.* Smith Elder, London, 1842, p. 102.

52. Richard D. Keynes, editor. *Charles Darwin's Zoology Notes & Specimen Lists from H.M.S. Beagle.* Cambridge University Press, Cambridge, UK, 2000, pp. 360–361.

53. Ibid.

54. Charles Darwin was often frustrated by his inability to draw well, and urged other naturalists to learn the skill so that they might illustrate what they observed. For comments regarding Darwin's lack of artistic talent see Charles Darwin. *The Autobiography of Charles Darwin, 1809–1882,* Nora Barlow, editor. WW Norton, New York, 1958, p. 47; and Francis Darwin. *The Life of Charles Darwin.* (First published by John Murray, London, 1902.) Studio Editions, 1995, p. 67. Also Charles Darwin. Letter to R. W. Darwin, 8 February–1 March 1832. In: *The Correspondence of Charles Darwin,* Vol. 1: *1821–1836.* Frederick Burkhardt, Sydney Smith, David Kohn, and William Montgomery, editors. Cambridge University Press, Cambridge, UK, 1985, p. 204

55. Robert FitzRoy. *Narrative of the surveying voyages of His Majesty's Ships Adventure and Beagle between the years 1826 and 1836, describing their examination of the southern shores of South America, and the Beagle's circumnavigation of the globe: Proceedings of the second expedition, 1831–36, under the command of Captain Robert Fitz-Roy, R.N.* Henry Colburn, London, 1839, p. 487.

56. Darwin never mentioned visiting Finger Hill in his diary or in his field notebook. Nor did he anywhere record spending the night on shore with FitzRoy on September 17. This contrasts with his specification (in his diary, field notebook, and later in his *Journal*) that he slept one night on a beach on September 21. The only indications that Darwin explored Finger Hill at all come from his geology notes (DAR 37.2). There, his description of Finger Hill follows his descriptions of Pan de Azucar (which he explored on September 18) and the Craterized District (September 21–22), suggesting that he examined Finger Hill several days after FitzRoy's visit to the hill. Darwin collected some of his geological specimens from Finger Hill by boat (probably on September 21), as this is the only way to access the dikes. The others, and specifically the sample of basalt from the interior of Finger Hill, were most likely collected on the way back to the ship after Darwin and Covington were picked up from the Craterized District on September 22. It is even conceivable that Darwin walked to Finger Hill from the Craterized District (a distance of 2.5 km) on September 22, before being picked up by boat.

57. Robert FitzRoy. *Narrative of the surveying voyages of His Majesty's Ships Adventure and Beagle between the years 1826 and 1836, describing their examination of the southern shores of South America, and the Beagle's circumnavigation of the globe: Appendix to Volume II.* Henry Colburn, London, 1839, p. 336.

58. Charles Darwin. *Beagle* field notebook *Galapagos Otaheite Lima* (1835). Darwin Archive, Cambridge University Library Microfilm EH1.17, p. 31b.

59. Charles Darwin. In: *Charles Darwin's Beagle Diary*, Richard D. Keynes, editor. Cambridge University Press, Cambridge, UK, 1988, p. 353.

60. Duncan Porter, editor. 1987. Darwin's Notes on *Beagle* Plants. *Bulletin of the British Museum (Natural History) Historical Series* 14 (2): 182.

61. Ibid.

62. Ibid.

63. Duncan Porter. 1980. The Vascular Plants of Joseph Dalton Hooker's *An enumeration of the plants of the Galapagos Archipelago; with descriptions of those which are new. Botanical Journal of the Linnean Society* 81: 129.

64. James Colnett. *A Voyage to the South Atlantic and Round Cape Horn into the Pacific Ocean, for the Purpose of Extending the Spermaceti Whale Fisheries, and other objects of commerce, by ascertaining the ports, bays, harbours, and anchoring births, in certain islands and coasts in those seas at which the ships of the British merchants might be refitted.* W Bennett, London, 1798, p. 157.

65. Ira L. Wiggins and Duncan M. Porter. *Flora of the Galápagos Islands.* Stanford University Press, California, 1971, pp. 673–674.

66. Charles Darwin. Letter to J. D. Hooker, 11–12 July 1845. In: *The Correspondence of Charles Darwin*, Vol. 3: *1844–1846*, Frederick Burkhardt and Sydney Smith, editors. Cambridge University Press, Cambridge, UK, 1987, p. 219.

67. Charles Darwin. Letter to J. D. Hooker, 28–29 January 1836. In: *The Correspondence of Charles Darwin*, Vol. 1: *1821–1836*. Frederick Burkhardt, Sydney Smith, David Kohn, and William Montgomery, editors. Cambridge University Press, Cambridge, UK, 1985, p. 485.

68. Darwin did not label all his plant specimens according to island at the time of their collection. However, it has been pointed out that because he must have used a plant press, which stores specimens in chronological order, his plants were naturally kept separated by island. In contrast Darwin mixed up the finches and several other specimens he collected from Chatham and Charles islands. See Frank J. Sulloway. 1982. Darwin and His Finches: The Evolution of a Legend. *Journal of the History of Biology* 15 (1, Spring): 29.

69. Charles Darwin. Letter to J. D. Hooker, 31 March 1844. In: *The Correspondence of Charles Darwin*, Vol. 3: *1844–1846*, Frederick Burkhardt and Sydney Smith, editors. Cambridge University Press, Cambridge, UK 1987, p. 23.

70. David Kohn. *Darwin's Garden: An Evolutionary Adventure.* Catalog of an exhibition at the New York Botanical Garden, April 25–July 20, 2008 and the Huntington Library and Botanical Gardens, San Marino, California, October 4, 2008–January 5, 2009. New York Botanical Garden publication, 2008.

71. Charles Darwin. Letter to J. S. Henslow, 1 November, 1836. In: *The Correspondence of Charles Darwin*, Vol. 1: *1821–1836*. Frederick Burkhardt, Sydney Smith, David Kohn, and William Montgomery, editors. Cambridge University Press, Cambridge, UK, 1985, p. 515.

72. Charles Darwin. Letter to J.D. Hooker, 3 September, 1846. In: *Origins: Selected Letters of Charles Darwin 1822–1859*, Frederick Burkhardt, editor. Cambridge University Press, Cambridge, UK, 2008, p. 94.

73. Charles Robert Darwin. *Narrative of the surveying voyages of His Majesty's Ships Adventure and Beagle between the years 1826 and 1836, describing their examination*

of the southern shores of South America, and the Beagle's circumnavigation of the globe: Journal and remarks, 1832–1836. Henry Colburn, London, 1839, p. 629 (addenda to p. 465).

74. Duncan Porter. 1980. The Vascular Plants of Joseph Dalton Hooker's An enumeration of the plants of the Galapagos Archipelago; with descriptions of those which are new. Botanical Journal of the Linnean Society 81: 79–134.

75. Joseph Hooker. Letter to Charles Darwin, after 12 July 1845. In: The Correspondence of Charles Darwin, Vol. 3: 1844–1846, Frederick Burkhardt and Sydney Smith, editors. Cambridge University Press, Cambridge, UK, 1987, p. 221.

76. Charles Robert Darwin. Narrative of the surveying voyages of His Majesty's Ships Adventure and Beagle between the years 1826 and 1836, describing their examination of the southern shores of South America, and the Beagle's circumnavigation of the globe: Journal and remarks, 1832–1836. Henry Colburn, London, 1839, p. 629 (addenda to p. 465).

77. Charles Darwin. Letter to J. S. Henslow, 3 November 1838. The Correspondence of Charles Darwin, Vol. 7: 1858–1859, Frederick Burkhardt and Sydney Smith, editors. Cambridge University Press, Cambridge, UK, 1992, p. 470.

78. Charles Darwin. In: Charles Darwin's Beagle Diary, Richard D. Keynes, editor. Cambridge University Press, Cambridge, UK, 1988, p. 353.

79. Charles Robert Darwin. Narrative of the surveying voyages of His Majesty's Ships Adventure and Beagle between the years 1826 and 1836, describing their examination of the southern shores of South America, and the Beagle's circumnavigation of the globe: Journal and remarks, 1832–1836. Henry Colburn, London, 1839, p. 455.

80. Ibid., p. 475.

81. Ibid., p. 476.

82. Charles Darwin. In: Charles Darwin's Beagle Diary, Richard D. Keynes, editor. Cambridge University Press, Cambridge, UK, 1988, p. 353.

83. Charles Robert Darwin. Narrative of the surveying voyages of His Majesty's Ships Adventure and Beagle between the years 1826 and 1836, describing their examination of the southern shores of South America, and the Beagle's circumnavigation of the globe: Journal and remarks, 1832–1836. Henry Colburn, London, 1839, p. 476.

84. Ibid., p. 478.

85. Charles Robert Darwin. Narrative of the surveying voyages of His Majesty's Ships Adventure and Beagle between the years 1826 and 1836, describing their examination of the southern shores of South America, and the Beagle's circumnavigation of the globe: Journal and remarks, 1832–1836. Henry Colburn, London, 1839, p. 474.

86. Charles Darwin. Beagle field notebook Galapagos Otaheite Lima (1835). Darwin Archive, Cambridge University Library Microfilm EH1.17, p. 30b. Darwin answered his botany question with a "yes," but only after he got back to England. In March 1839, after Darwin (or perhaps Henslow) showed David Don his Galápagos plant specimens, Darwin jotted in a notebook, "[Professor] Don would have known the Composites of Galapagos were South American." Notebook E (100) In: Charles Darwin's Notebooks 1836–1844, Paul H. Barrett, Peter J. Gautrey, Sandra Herbert, David Kohn, and Sydney Smith, editors. Cornell University Press, Ithaca, NY 1987.

87. Richard D. Keynes, editor. *Charles Darwin's Zoology Notes & Specimen Lists from H.M.S. Beagle.* Cambridge University Press, Cambridge, UK, 2000, p. 298.

88. Charles Darwin. *Journal of Researches into the natural history and geology of the countries visited during the voyage of H.M.S. Beagle round the world, under the Command of Capt. Fitz Roy, R.N.,* 2d ed. John Murray, London, 1845, p. 393.

89. Charles Darwin. *Notes on the geology of places visited on the voyage.* Darwin Archive, Cambridge University Library DAR 37.2 (1835) Folio 750(4).

90. Ibid.

91. Charles Darwin. *Geological Observations on the Volcanic Islands visited during the Voyage of H. M. S. Beagle, together with some brief notices of the Geology of Australia and the Cape of Good Hope. Being the second part of the Geology of the Voyage of the Beagle, under the Command of Capt. FitzRoy, R.N. during the years 1832 to 1836.* Smith Elder, London, 1844, p. 100.

92. Charles Darwin. *Notes on the geology of places visited on the voyage.* Darwin Archive, Cambridge University Library DAR 37.2 (1835) Folio 749(3).

93. Charles Robert Darwin. *Narrative of the surveying voyages of His Majesty's Ships Adventure and Beagle between the years 1826 and 1836, describing their examination of the southern shores of South America, and the Beagle's circumnavigation of the globe: Journal and remarks, 1832–1836.* Henry Colburn, London, 1839, p. 463.

94. Richard D. Keynes, editor. *Charles Darwin's Zoology Notes & Specimen Lists from H.M.S. Beagle.* Cambridge University Press, Cambridge, UK, 2000, p. 292.

95. Charles Darwin. *Notes on the geology of places visited on the voyage.* Darwin Archive, Cambridge University Library DAR 37.2 (1835) Folio 727(2).

96. Charles Darwin. In: *Charles Darwin's Beagle Diary,* Richard D. Keynes, editor. Cambridge University Press, Cambridge, UK, 1988, p. 353.

97. Charles Darwin. *Notes on the geology of places visited on the voyage.* Darwin Archive, Cambridge University Library DAR 37.2 (1835) Folio 749(3).

98. Ibid., Folio 750 (4).

99. Charles Darwin. *Notes on the geology of places visited on the voyage.* Darwin Archive, Cambridge University Library DAR 37.2 (1835) Folio 749(3).

100. Charles Darwin. *Geological Observations on the Volcanic Islands visited during the Voyage of H. M. S. Beagle, together with some brief notices of the Geology of Australia and the Cape of Good Hope. Being the second part of the Geology of the Voyage of the Beagle, under the Command of Capt. FitzRoy, R.N. during the years 1832 to 1836.* Smith Elder, London, 1844, p. 99.

101. Charles Darwin. *Notes on the geology of places visited on the voyage.* Darwin Archive, Cambridge University Library DAR 37.2 (1835) Folio 751(5).

102. Ibid., Folio 750(4).

103. Ibid., Folio 787(41).

104. Charles Darwin. *Geological Observations on the Volcanic Islands visited during the Voyage of H. M. S. Beagle, together with some brief notices of the Geology of Australia and the Cape of Good Hope. Being the second part of the Geology of the Voyage of the Beagle, under the Command of Capt. FitzRoy, R.N. during the years 1832 to 1836.* Smith Elder, London, 1844, p. 112.

105. Ibid., p. 99.

106. Charles Darwin. In: *Charles Darwin's Beagle Diary*, Richard D. Keynes, editor. Cambridge University Press, Cambridge, UK, 2001, p. 353.

107. Robert FitzRoy. *Narrative of the surveying voyages of His Majesty's Ships Adventure and Beagle between the years 1826 and 1836, describing their examination of the southern shores of South America, and the Beagle's circumnavigation of the globe: Proceedings of the second expedition, 1831–36, under the command of Captain Robert Fitz-Roy, R.N.* Henry Colburn, London, 1839, p. 488.

108. Charles Darwin. In: *Charles Darwin's Beagle Diary*, Richard D. Keynes, editor. Cambridge University Press, Cambridge, UK, 1988, p. 362.

109. Robert FitzRoy. *Narrative of the surveying voyages of His Majesty's Ships Adventure and Beagle between the years 1826 and 1836, describing their examination of the southern shores of South America, and the Beagle's circumnavigation of the globe: Proceedings of the second expedition, 1831–36, under the command of Captain Robert Fitz-Roy, R.N.* Henry Colburn, London, 1839, p. 498.

110. Tortoises from this same population can be viewed in captivity at an additional visiting site (also called La Galapaguera) in the highlands of Chatham Island.

111. Robert FitzRoy. *Narrative of the surveying voyages of His Majesty's Ships Adventure and Beagle between the years 1826 and 1836, describing their examination of the southern shores of South America, and the Beagle's circumnavigation of the globe: Proceedings of the second expedition, 1831–36, under the command of Captain Robert Fitz-Roy, R.N.* Henry Colburn, London, 1839, p. 498.

112. For an abstract of the meteorological journal kept on board the *Beagle* see Robert FitzRoy. *Narrative of the surveying voyages of His Majesty's Ships Adventure and Beagle between the years 1826 and 1836, describing their examination of the southern shores of South America, and the Beagle's circumnavigation of the globe: Appendix to Volume II.* Henry Colburn, London, 1839, p. 45.

113. Charles Darwin. In: *Charles Darwin's Beagle Diary*, Richard D. Keynes, editor. Cambridge University Press, Cambridge, UK, 1988, p. 354.

114. Charles Darwin. *Beagle* field notebook *Galapagos Otaheite Lima* (1835). Darwin Archive, Cambridge University Library Microfilm EH1.17, p. 19b.

115. Charles Darwin. In: *Charles Darwin's Beagle Diary*, Richard D. Keynes, editor. Cambridge University Press, Cambridge, UK, 1988, p. 354.

116. Robert FitzRoy. *Narrative of the surveying voyages of His Majesty's Ships Adventure and Beagle between the years 1826 and 1836, describing their examination of the southern shores of South America, and the Beagle's circumnavigation of the globe: Proceedings of the second expedition, 1831–36, under the command of Captain Robert Fitz-Roy, R.N.* Henry Colburn, London, 1839, p. 488.

117. Robert FitzRoy. *Remarks on Galapagos Islands, the NE Coast of Tierra del Fuego, and Magellan Strait. H.M.S. Beagle 1835.* In: A list of Documents Sent to the Hydrographer of the Admiralty Sept 16, 1846, and April 14, 1845. Sailing Directions and Nautical Remarks referring to the Coasts of South America, and the Galapagos Islands, intended to be incorporated with the Directions and Remarks published by the Admiralty on the authority of Captain Phillip Parker King.

118. During the authors' expedition to retrace Darwin's footsteps through Galá-pagos, swells breaking on the steeply inclined pebble beach of La Honda made land-ing impossible by boat. We had to swim ashore.

119. Robert FitzRoy. *Remarks on Galapagos Islands, the NE Coast of Tierra del Fuego, and Magellan Strait. H.M.S. Beagle 1835.* In: A list of Documents Sent to the Hydrographer of the Admiralty Sept 16, 1846, and April 14, 1845. Sailing Directions and Nautical Remarks referring to the Coasts of South America, and the Galapagos Islands, intended to be incorporated with the Directions and Remarks published by the Admiralty on the authority of Captain Phillip Parker King.

120. Ship's log, HMS *Beagle*. ADM 53/236. Public Records Office, London. Sep-tember 20, 1835.

121. Ibid.

122. See Frank Sulloway. 1982. Darwin's Conversion: The Beagle Voyage and Its Aftermath. *Journal of the History of Biology* 15(3): 325–396. (p. 344).

123. Charles Darwin. *Notes on the geology of places visited on the voyage.* Darwin Archive, Cambridge University Library DAR 37.2 (1835) Folio 731(6).

124. Ship's log, HMS *Beagle*. ADM 53/236. Public Records Office, London. Sep-tember 22, 1835.

125. George Poulett Scrope. *Considerations of Volcanoes: the probable causes of their phenomena, the laws which determine their march, the disposition of their prod-ucts, and their connexion with the present state and past history of the globe; leading to the establishment of a new theory of the Earth.* W. Philips, London, 1825.

126. Charles Darwin. *Notes on the geology of places visited on the voyage.* Darwin Archive, Cambridge University Library DAR 37.2 (1835) Folio 731(6) and 754(8).

127. Ibid., Folios 730(5) and 752(6).

128. Richard D. Keynes, editor. *Charles Darwin's Zoology Notes & Specimen Lists from H.M.S. Beagle.* Cambridge University Press, Cambridge, UK, 2000, p. 300.

129. Ibid.

130. Charles Darwin. *Notes on the geology of places visited on the voyage.* Darwin Archive, Cambridge University Library DAR 37.2 (1835) Folio 727(2).

131. Charles Darwin. *Geological Observations on the Volcanic Islands visited dur-ing the Voyage of H. M. S. Beagle, together with some brief notices of the Geology of Australia and the Cape of Good Hope. Being the second part of the Geology of the Voy-age of the Beagle, under the Command of Capt. FitzRoy, R.N. during the years 1832 to 1836.* Smith Elder, London, 1844, p. 100.

132. Charles Darwin. *Notes on the geology of places visited on the voyage.* Darwin Archive, Cambridge University Library DAR 37.2 (1835) Folio 729(4).

133. Charles Darwin. *Beagle* field notebook *Galapagos Otaheite Lima* (1835). Darwin Archive, Cambridge University Library Microfilm EH1.17, p. 19b.

134. Charles Darwin. *Notes on the geology of places visited on the voyage.* Darwin Archive, Cambridge University Library DAR 37.2 (1835) Folio 754(8).

135. Ibid., Folio 755(9)

136. Charles Darwin. *Beagle* field notebook *Galapagos Otaheite Lima* (1835). Darwin Archive, Cambridge University Library Microfilm EH1.17, p. 23b.

137. Charles Lyell. *Principles of Geology, being an attempt to explain the former changes of the Earth's surface, by reference to causes now in operation,* Vol. 1. John Murray, London, 1830. Reprinted by The University of Chicago Press, Chicago, 1990, Woodcut number 12, p. 336.

138. Charles Darwin. Letter to J. D. Hooker, 11–12 July 1845. In: *The Correspondence of Charles Darwin,* Vol. 3: *1844–1846,* Frederick Burkhardt and Sydney Smith, editors. Cambridge University Press, Cambridge, UK, 1987, p. 218.

139. Charles Darwin. In: *Charles Darwin's Beagle Diary,* Richard D. Keynes, editor. Cambridge University Press, Cambridge, UK, 1988, p. 354.

140. James Colnett. *A Voyage to the South Atlantic and Round Cape Horn into the Pacific Ocean, for the Purpose of Extending the Spermaceti Whale Fisheries, and other objects of commerce, by ascertaining the ports, bays, harbours, and anchoring births, in certain islands and coasts in those seas at which the ships of the British merchants might be refitted.* W Bennett, London, 1798, p. 51.

141. Charles Darwin. *Beagle* field notebook *Galapagos Otaheite Lima* (1835). Darwin Archive, Cambridge University Library Microfilm EH1.17, pp. 19b–20b.

142. James Colnett. *A Voyage to the South Atlantic and Round Cape Horn into the Pacific Ocean, for the Purpose of Extending the Spermaceti Whale Fisheries, and other objects of commerce, by ascertaining the ports, bays, harbours, and anchoring births, in certain islands and coasts in those seas at which the ships of the British merchants might be refitted.* W Bennett, London, 1798, p. 51.

143. Charles Darwin. *Beagle* field notebook *Galapagos Otaheite Lima* (1835). Darwin Archive, Cambridge University Library Microfilm EH1.17, p. 33b.

144. Charles Darwin. *Notes on the geology of places visited on the voyage.* Darwin Archive, Cambridge University Library DAR 37.2 (1835) Folio 755(9).

145. Charles Robert Darwin. *Narrative of the surveying voyages of His Majesty's Ships Adventure and Beagle between the years 1826 and 1836, describing their examination of the southern shores of South America, and the Beagle's circumnavigation of the globe: Journal and remarks, 1832–1836.* Henry Colburn, London, 1839, p. 456.

146. Charles Darwin. In: *Charles Darwin's Beagle Diary,* Richard D. Keynes, editor. Cambridge University Press, Cambridge, UK, 1988, p. 354.

147. Charles Robert Darwin. *Narrative of the surveying voyages of His Majesty's Ships Adventure and Beagle between the years 1826 and 1836, describing their examination of the southern shores of South America, and the Beagle's circumnavigation of the globe: Journal and remarks, 1832–1836.* Henry Colburn, London, 1839, p. 456.

148. Charles Darwin. *Beagle* field notebook *Galapagos Otaheite Lima* (1835). Darwin Archive, Cambridge University Library Microfilm EH1.17, p. 20b.

149. Charles Darwin. In: *Charles Darwin's Beagle Diary,* Richard D. Keynes, editor. Cambridge University Press, Cambridge, UK, 1988, p. 354.

150. Charles Robert Darwin. *Narrative of the surveying voyages of His Majesty's Ships Adventure and Beagle between the years 1826 and 1836, describing their examination of the southern shores of South America, and the Beagle's circumnavigation of the globe: Journal and remarks, 1832–1836.* Henry Colburn, London, 1839, p. 456.

151. John Milton. *Paradise Lost,* A Poem in Twelve Books. Second edition. S. Simmons, London 1674 (Book 1, lines 670-674).

152. Charles Darwin. *Notes on the geology of places visited on the voyage.* Darwin Archive, Cambridge University Library DAR 37.2 (1835) Folio 757(11).

153. Charles Darwin. *Beagle* field notebook *Galapagos Otaheite Lima* (1835). Darwin Archive, Cambridge University Library Microfilm EH1.17, pp. 31b–32b.

154. There is nothing in the literature to indicate Darwin swam in Galápagos. However, he must have bathed somehow in those five weeks. The sea temperature was recorded at 6 p.m. in Stephens Bay by the *Beagle* crew. For a complete list of temperatures taken in Galápagos, see Robert FitzRoy. *Narrative of the surveying voyages of His Majesty's Ships Adventure and Beagle between the years 1826 and 1836, describing their examination of the southern shores of South America, and the Beagle's circumnavigation of the globe: Appendix to Volume II.* Henry Colburn, London, 1839, p. 45.

155. Charles Darwin. In: *Charles Darwin's Beagle Diary*, Richard D. Keynes, editor. Cambridge University Press, Cambridge, UK, 1988, p. 354.

156. Ibid.

157. Charles Robert Darwin. *Narrative of the surveying voyages of His Majesty's Ships Adventure and Beagle between the years 1826 and 1836, describing their examination of the southern shores of South America, and the Beagle's circumnavigation of the globe: Journal and remarks, 1832–1836.* Henry Colburn, London, 1839, p. 460.

158. Charles Robert Darwin, editor. *Mammalia Part 2 No. 4 of The zoology of the voyage of H.M.S. Beagle. By George R. Waterhouse. Edited and superintended by Charles Darwin.* Smith Elder, London, 1839, p. 65.

159. Richard D. Keynes, editor. *Charles Darwin's Zoology Notes & Specimen Lists from H.M.S. Beagle.* Cambridge University Press, Cambridge, UK 2000, p. 414.

160. Donna B. Harris and David W. Macdonald. 2007. Population Ecology of the Endemic Rodent *Nesoryxomys swarthi* in the Tropical Desert of the Galápagos Islands. *Journal of Mammalogy* 88(1): 208–219.

161. Robert C. Dowler, Darin S. Carroll, and Cody W. Edwards. 2000. Rediscovery of rodents (Genus *Nesoryzomys*) considered extinct in the Galápagos Islands. *Oryx* 34(2): 109–118.

162. In 1996 the authors found a rare Galápagos tomato plant (*Lycopersicon cheesmanii*) growing on the floor of one pit crater and a cut leaf daisy (*Lecocarpus darwinii*) clinging to the side of another. Specimens of both species were collected by Darwin from unspecified sites on Chatham Island.

163. Charles Darwin. *Beagle* field notebook *Galapagos Otaheite Lima* (1835). Darwin Archive, Cambridge University Library Microfilm EH1.17, p. 20b.

164. Peter. R. Grant. *Ecology and Evolution of Darwin's Finches.* Princeton University Press, Princeton, NJ, 1986, p. 52.

165. Richard D. Keynes, editor. *Charles Darwin's Zoology Notes & Specimen Lists from H.M.S. Beagle.* Cambridge University Press, Cambridge, UK, 2000, p. 416.

166. Charles Robert Darwin, editor. *Birds Part 3 No. 3 of The zoology of the voyage of H.M.S. Beagle, by John Gould. Edited and superintended by Charles Darwin.* Smith Elder, London, 1839, p. 46.

167. Richard D. Keynes, editor. *Charles Darwin's Zoology Notes & Specimen Lists from H.M.S. Beagle.* Cambridge University Press, Cambridge, UK, 2000, p. 297.

168. Ibid.

169. Sandra Herbert. *Darwin, Geologist.* Cornell University Press. Ithaca, NY, 2005. p 313. Gordon Chancellor (personal communication) has since suggested that Darwin wrote some of his field notes retrospectively, and that he may have penned this passage about mockingbirds when on Charles Island, and therefore after he had seen more than one island.

Box: Surveying Galápagos

1. Charles Darwin. In: *Charles Darwin's Beagle Diary*, Richard D. Keynes, editor. Cambridge University Press, Cambridge, UK, 1988, p. 95.

2. Aaron Arrowsmith, *Chart of the Galapagos surveyed by Merchant Ship Rattler, and drawn by Capt. James Colnett of the Royal Navy in 1793 1794: Additions & Corrections to 1817.* Hydrographical Office of Taunton, Taunton, UK, Shelf PB H297.

3. Basil Hall. *Extracts from a Journal written on the Coasts of Chili, Peru, and Mexico, in the years 1820, 1821, 1822,* Vol. II. E. Littell, Philadelphia, William Brown, printer, 1824, p. 83 and Appendix 1.

4. Robert FitzRoy. *Narrative of the surveying voyages of His Majesty's Ships Adventure and Beagle between the years 1826 and 1836, describing their examination of the southern shores of South America, and the Beagle's circumnavigation of the globe: Proceedings of the second expedition, 1831–36, under the command of Captain Robert Fitz-Roy, R.N.* Henry Colburn, London, 1839, p. 32.

5. Robert FitzRoy. *Narrative of the surveying voyages of His Majesty's Ships Adventure and Beagle between the years 1826 and 1836, describing their examination of the southern shores of South America, and the Beagle's circumnavigation of the globe: Appendix to Volume II.* Henry Colburn, London, 1839, p. 319.

6. At the end of the voyage Captain FitzRoy concluded that the movement of the ship was not a problem for chronometers, but that changes in temperature, such as those experienced when a timepiece was taken ashore or in a surveying boat, did alter, slightly, the rate of a chronometer. Robert FitzRoy. *Narrative of the surveying voyages of His Majesty's Ships Adventure and Beagle between the years 1826 and 1836, describing their examination of the southern shores of South America, and the Beagle's circumnavigation of the globe: Appendix to Volume II.* Henry Colburn, London, 1839, p. 326.

7. Charles Darwin. In: *Charles Darwin's Beagle Diary*, Richard D. Keynes, editor. Cambridge University Press, Cambridge, UK, 1988, p. 48.

8. William Ambrose Cowley. *Cowley's Voyage Round the World.* Sloane MS. 1050. British Library, London, 1688.

9. Gregory Estes, K. Thalia Grant, and Peter R. Grant. 2000. Darwin in Galápagos: His Footsteps through the Archipelago. *Notes and Records of the Royal Society of London* 54(3): 343–368.

10. Robert FitzRoy. *Narrative of the surveying voyages of His Majesty's Ships Adventure and Beagle between the years 1826 and 1836, describing their examination of the southern shores of South America, and the Beagle's circumnavigation of the globe:*

Proceedings of the second expedition, 1831–36, under the command of Captain Robert Fitz-Roy, R.N. Henry Colburn, London, 1839, p. 35.

11. Ibid.

12. Janet Browne. *Charles Darwin: Voyaging.* AA Knopf, New York, 1995, p. 247.

13. Charles Darwin. *Notes on the geology of places visited on the voyage.* Darwin Archive, Cambridge University Library DAR 37.2 (1835) Folio 746 (1)

14. Robert FitzRoy. *Narrative of the surveying voyages of His Majesty's Ships Adventure and Beagle between the years 1826 and 1836, describing their examination of the southern shores of South America, and the Beagle's circumnavigation of the globe: Appendix to Volume II.* Henry Colburn, London, 1839, pp. 203–204.

15. Ibid.

16. Charles Darwin. *Notes on the geology of places visited on the voyage.* Darwin Archive, Cambridge University Library DAR 37.2 (1835) Folio 747(1) .

17. Charles Darwin. *Notes on the geology of places visited on the voyage.* Darwin Archive, Cambridge University Library DAR 37.2 (1835) Folio 795 (46)

18. A. R. McBirney and H. Williams. 1969. Geology and Petrology of the Galápagos Islands. *Geological Society of America Memoirs* 118: 1–197 (p. 4).

19. Charles Darwin. *Notes on the geology of places visited on the voyage.* Darwin Archive, Cambridge University Library DAR 37.2 (1835) Folio 795(46) .

20. Michael H. Jackson. *Galápagos: A Natural History.* University of Calgary Press, Calgary, 2004, pp. 9–19.

Box: Golden Age of Whaling

1. James Colnett. *A Voyage to the South Atlantic and Round Cape Horn into the Pacific Ocean, for the Purpose of Extending the Spermaceti Whale Fisheries, and other objects of commerce, by ascertaining the ports, bays, harbours, and anchoring births, in certain islands and coasts in those seas at which the ships of the British merchants might be refitted.* W Bennett, London, 1798, pp. 147–158.

2. Robert FitzRoy. *Narrative of the surveying voyages of His Majesty's Ships Adventure and Beagle between the years 1826 and 1836, describing their examination of the southern shores of South America, and the Beagle's circumnavigation of the globe: Proceedings of the second expedition, 1831–36, under the command of Captain Robert Fitz-Roy, R.N.* Henry Colburn, London, 1839, p. 494.

3. Edouard A. Stackpole. *Whales and Destiny: The Rivalry between America, France, and Britain for Control of the Southern Whale Fishery, 1785–1825.* University of Massachusetts Press, Amherst, MA, 1972, p.127.

4. James Colnett. *A Voyage to the South Atlantic and Round Cape Horn into the Pacific Ocean, for the Purpose of Extending the Spermaceti Whale Fisheries, and other objects of commerce, by ascertaining the ports, bays, harbours, and anchoring births, in certain islands and coasts in those seas at which the ships of the British merchants might be refitted.* W Bennett, London, 1798, p. 157.

5. Charles Darwin. In: *Charles Darwin's Beagle Diary*, Richard D. Keynes, editor. Cambridge University Press, Cambridge, UK, 1988, p. 148.

6. Ibid.

7. Ibid., p. 363.

8. Charles Haskins Townsend. 1925. The Galápagos Tortoises in Their Relation to the Whaling Industry: A Study of Old Logbooks. *Zoologica* 4(3): 55–135.

9. Fritz Trillmich. Natural History of the Galápagos Fur Seal (*Arctocephalus galapagoensis*, Heller). In: *Galápagos: Key Environments*, Roger Perry, editor. Pergamon, Oxford, UK, 1984, p. 217.

10. H. E. Corley Smith. 1990. Early Attempts at Galapagos Conservation. *Noticias de Galápagos* 49: 6–7.

11. S. Salazar, S. Banks, and B. Milstead 2007. *Health and Population Status of the Galápagos Sea Lion and Fur Seal (Zalophus wollebaeki and Arctocephalus galapagoensis)*. Report submitted by the Charles Darwin Research Station to the Heinz Sielmann Foundation. December 2007. For earlier estimates of fur seal population size, see T. Dellinger and F. Trillmich. 1999. Fish Prey of the Sympatric Galapagos Fur Seals and Sea lions: Seasonal Variation and Niche Separation. *Canadian Journal of Zoology* 77: 1204–1216.

12. Robert FitzRoy *Remarks on Galapagos Islands, the NE Coast of Tierra del Fuego, and Magellan Strait.*

(A list of Documents Sent to the Hydrographer of the Admiralty Sept 16, 1846 and April 14, 1845. Sailing Directions and Nautical Remarks referring to the Coasts of South America, and the Galapagos Islands, intended to be incorporated with the Directions and Remarks published by the Admiralty on the authority of Captain Phillip Parker King), p. 7.

13. S. Salazar, S. Banks, and B. Milstead 2007. *Health and Population Status of the Galapagos Sea Lion and Fur Seal (Zalophus wollebaeki and Arctocephalus galapagoensis)*. Report submitted by the Charles Darwin Research Station to the Heinz Sielmann Foundation. December 2007.

14. Edouard A. Stackpole. *Whales and Destiny: The Rivalry between America, France, and Britain for Control of the Southern Whale Fishery, 1785–1825.* University of Massachusetts Press, Amherst, MA,1972, p. 350.

15. Charles Haskins Townsend. 1928. The Galapagos Islands Revisited. *Bulletin of the New York Zoological Society* 31(5): p. 159.

16. Charles Haskins Townsend. 1925. The Galapagos Tortoises in Their Relation to the Whaling Industry: A Study of Old Logbooks. *Zoologica* 4(3): 55–135.

17. J. Wood (Lieutenant Commander of H.M.Surveying vessel Pandora). Letter to Rear Admiral Sir G.F. Seymour, 29 September, 1847. *RGS Archives Journal* ms. No. 97.

18. Charles Haskins Townsend. 1925. The Galapagos Tortoises in Their Relation to the Whaling Industry: A Study of Old Logbooks. *Zoologica* 4(3): 55–135.

19. Octavio Latorre. *El Hombre en Las Islas Encantadas: La Historia Humana de Galápagos.* Producción Gráfica, Quito, Ecuador, 1999, p. 92.

20. J. N. Reynolds. *Voyage of the United States Frigate Potomac under the Command of Commodore John Downes, during the Circumnavigation of the Globe in the years 1831,1832,1833, and 1834.* Harper & Brothers, New York, 1835, Chapter 27, p. 470.

21. Charles Haskins Townsend. 1925. The Galápagos Tortoises in Their Relation to the Whaling Industry: A Study of Old Logbooks. *Zoologica* 4(3): 55–135.

22. James Gibbs, personal communication, 2007. See also: *Galápagos Giant Tortoises*. Charles Darwin Foundation Fact Sheet. 2006 www.darwinfoundation.org

23. *Journal of the ship Phoenix 1834–1837*. Log 194 Reel 18. Nantucket Historical Association, Nantucket, MA.

CHAPTER 5: CHARLES ISLAND (ISLA FLOREANA)

1. Charles Robert Darwin. *Narrative of the surveying voyages of His Majesty's Ships Adventure and Beagle between the years 1826 and 1836, describing their examination of the southern shores of South America, and the Beagle's circumnavigation of the globe: Journal and remarks, 1832–1836*. Henry Colburn, London, 1839. pp. 456–457.

2. Charles Darwin. Letter to W. D. Fox. May 1832. In: *The Correspondence of Charles Darwin*, Vol. 1: *1821–1836*. Frederick Burkhardt, Sydney Smith, David Kohn, and William Montgomery, editors. Cambridge University Press, Cambridge, UK, 1985, p. 232.

3. The assumption that Colnett erected the post office barrel is based on the fact that the fourth edition of his map of Galápagos, published by Aaron Arrowsmith in 1820, shows the words *Post Office* marked on Charles Island. However, that particular edition incorporates "Additions & Corrections" from ships that visited the islands after Colnett. The first three editions of Colnett's map (which were published by Aaron Arrowsmith in 1798, 1805, and 1808) do not show Post Office Bay at all. (see www.galapagos.to). Furthermore, in *A Voyage to the South Atlantic and Round Cape Horn into the Pacific Ocean* Colnett wrote that he was unable to land on Charles because of strong currents that bore his ship, the *Rattler*, away from shore.

4. David Porter. *Journal of a Cruise made to the Pacific Ocean in the United States Frigate Essex, in the years 1812 1813, and 1814. Containing descriptions of the Cape Verd Islands, Coasts of Brazil, Patagonia, Chili, and Peru, and of the Gallapagos Islands. Two volumes in one.* Bradford & Inskeep, Philadelphia, 1815, Chapter V.

5. Robert FitzRoy. *Narrative of the surveying voyages of His Majesty's Ships Adventure and Beagle between the years 1826 and 1836, describing their examination of the southern shores of South America, and the Beagle's circumnavigation of the globe: Proceedings of the second expedition, 1831–36, under the command of Captain Robert Fitz-Roy, R.N.* Henry Colburn, London, 1839, p. 490.

6. Ibid.

7. Charles Darwin. In: *Charles Darwin's Beagle Diary*, Richard D. Keynes, editor. Cambridge University Press, Cambridge, UK, 1988, p. 362.

8. Lawson reported the medicinal properties of the incense trees to Abel Du Petit-Thouars in 1838. See Abel Du Petit-Thouars. *Voyage Autour Du Monde Sur La Frégate Vénus Pendant Les Annees 1836–1839*, Vol. 2. Gide (editor), Paris, 1841. Translated from French by K. Thalia Grant and Anne Guezou, p. 287.

9. John Gray, Assistant Surgeon of the H.M.B. *Pandora*, and Thomas Borrowman, 2nd class engineer of the H.M.Surv. *Sampson. Accounts of the investigation of Charles and Chatham Islands of the Galapagos made by the order of G. Seymour, Rear Admiral.* 1847 Royal Geographical Society archives MS. No. 97.

10. Duncan M. Porter, editor. 1987. Darwin Notes on *Beagle Plants. Bulletin of the British Museum (Natural History).* Historical Series. 14(2): 182.

11. Charles Darwin. In: *Charles Darwin's Beagle Diary*, Richard D. Keynes, editor. Cambridge University Press, Cambridge, UK, 1988, p. 355.

12. Ibid., p. 356.

13. Robert FitzRoy. *Narrative of the surveying voyages of His Majesty's Ships Adventure and Beagle between the years 1826 and 1836, describing their examination of the southern shores of South America, and the Beagle's circumnavigation of the globe: Proceedings of the second expedition, 1831–36, under the command of Captain Robert Fitz-Roy, R.N.* Henry Colburn, London, 1839, p. 491.

14. Charles Darwin. In: *Charles Darwin's Beagle Diary*, Richard D. Keynes, editor. Cambridge University Press, Cambridge, UK, 1988, p. 355.

15. Robert FitzRoy. *Narrative of the surveying voyages of His Majesty's Ships Adventure and Beagle between the years 1826 and 1836, describing their examination of the southern shores of South America, and the Beagle's circumnavigation of the globe: Appendix to Volume II.* Henry Colburn, London, 1839, pp. 46 and 64.

16. Charles Darwin. In: *Charles Darwin's Beagle Diary*, Richard D. Keynes, editor. Cambridge University Press, Cambridge, UK, 1988, p. 347.

17. Ibid., p. 44.

18. Ibid., p. 355.

19. Ibid., p. 355.

20. Robert FitzRoy. Remarks on Galapagos Islands, the NE Coast of Tierra del Fuego, and Magellan Strait. H.M.S. Beagle 1835. In: A list of Documents Sent to the Hydrographer of the Admiralty Sept 16, 1846 and April 14, 1845. Sailing Directions and Nautical Remarks referring to the Coasts of South America, and the Galápagos Islands, intended to be incorporated with the Directions and Remarks published by the Admiralty on the authority of Captain Phillip Parker King.

21. Robert FitzRoy. *Narrative of the surveying voyages of His Majesty's Ships Adventure and Beagle between the years 1826 and 1836, describing their examination of the southern shores of South America, and the Beagle's circumnavigation of the globe: Proceedings of the second expedition, 1831–36, under the command of Captain Robert Fitz-Roy, R.N.* Henry Colburn, London, 1839, p. 491.

22. Charles Darwin. In: *Charles Darwin's Beagle Diary*, Richard D. Keynes, editor. Cambridge University Press, Cambridge, UK, 1988, p. 356.

23. Ibid., p. 354.

24. Ibid., p. 354.

25. Robert FitzRoy. *Narrative of the surveying voyages of His Majesty's Ships Adventure and Beagle between the years 1826 and 1836, describing their examination of the southern shores of South America, and the Beagle's circumnavigation of the globe: Proceedings of the second expedition, 1831–36, under the command of Captain Robert Fitz-Roy, R.N.* Henry Colburn, London, 1839, p. 491.

26. Ibid.

27. Charles Darwin. *Beagle* field notebook *Galapagos Otaheite Lima* (1835). Darwin Archive, Cambridge University Library Microfilm EH1.17, p. 34b.

28. Robert FitzRoy. *Narrative of the surveying voyages of His Majesty's Ships Adventure and Beagle between the years 1826 and 1836, describing their examination of the southern shores of South America, and the Beagle's circumnavigation of the globe: Proceedings of the second expedition, 1831–36, under the command of Captain Robert Fitz-Roy, R.N.* Henry Colburn, London, 1839, p. 491.

29. Ibid., p. 493.

30. There is no account of what was eaten at the feast. The food items have been compiled from the cultivated plants mentioned as growing on Charles by both Fitz-Roy and Darwin, and the domestic and native animals that were used as food. Darwin mentioned using a tortoise shell as a water pitcher in his *Journal*, but did not describe drinking tea on Charles. However, Abel du Petit-Thouars, during his visit to Asilo de Paz in 1838, noted that both the settlers and the visiting whalers habitually drank a bush tea made from the leaves of a lowland shrub. The identity of this shrub is not known but several factors indicate it was either Galápagos croton (*Croton scouleri*) or thin-leafed Darwin's shrub (*Darwiniothamnus tenufiolius*). 1. Petit-Thouars described the shrub as having "aromatic leaves" and the leaves of both *Croton* and *Darwiniothamnus* emit a pleasant odor when crushed. 2. The early 20th-century Norwegian settlers of Indefatigable Island reportedly drank tea made from croton leaves, until they suspected quite correctly that the tannins might not be good for them (early colonist Jacqueline De Roy, personal communication). 3. Croton tea, made from *Croton flavens*, is traditionally drunk in the West Indies. 4. Fleabane tea made from *Erigeron* species was a traditional drink in North America. *Darwiniothamnus* shrub, with its daisy like flowers, was originally thought to belong to the genus *Erigeron*.

31. Lawson's character has been derived from Du Petit-Thouars, Abel. *Voyage Autour Du Monde Sur La Frégate Vénus Pendant Les Annees 1836–1839*, Vol. 2. Gide (editor), Paris, 1841. Translated from French by K. Thalia Grant and Anne Guezou, p. 284.

32. Nora Barlow, editor. 1963. Darwin's Ornithological Notes. *Bulletin of the British Museum (Natural History)*. Historical Series. 2(7): 201–278 (p. 262).

33. Charles Darwin. *Journal of Researches* into *the natural history and geology of the countries visited during the voyage of H.M.S. Beagle round the world, under the Command of Capt. Fitz Roy, R.N.*, 2d ed. John Murray, London 1845, p. 379.

34. Nora Barlow, editor. 1963. Darwin's Ornithological Notes. *Bulletin of the British Museum (Natural History)*. Historical Series. 2(7): 201–278 (p. 262).

35. Richard D. Keynes, editor. *Charles Darwin's Zoology Notes & Specimen Lists from H.M.S. Beagle.* Cambridge University Press, Cambridge, UK, 2000, p. 298.

36. Charles Robert Darwin, editor. *Birds Part 3 No. 4 of The zoology of the voyage of H.M.S. Beagle, by John Gould. Edited and superintended by Charles Darwin.* Smith Elder, London, 1839, p. 64.

37. Ibid., p. 63.

38. Richard D. Keynes, editor. *Charles Darwin's Zoology Notes & Specimen Lists from H.M.S. Beagle.* Cambridge University Press, Cambridge, UK, 2000, p. 298.

39. Mr. George Robert Gray, ornithological assistant in the Zoological department of the British Museum, changed *Orpheus* to *Mimus*, after Gould left on a trip to Australia. See Charles Robert Darwin, editor. *Birds Part 3 No. 1 of The zoology of the voyage of H.M.S. Beagle, by John Gould. Edited and superintended by Charles Darwin.* Smith Elder, London, 1838, p. i.

40. Charles Robert Darwin, editor. *Birds Part 3 No. 1 of The zoology of the voyage of H.M.S. Beagle, by John Gould. Edited and superintended by Charles Darwin.* Smith Elder, London, 1838, p. 62.

41. The Galápagos group of mockingbirds comprise a clade known as *Nesomimus*. In 2007, the American Ornithologists' Union South American Checklist Committee officially merged *Nesomimus* with *Mimus*, based on the findings of Arbogast et al. 2006 (see endnote 42).

42. Brian S. Arbogast, Sergei V. Drovetski, Robert L. Curry, Peter T. Boag, Gilles Seutin, Peter R. Grant, Rosemary B. Grant, and David J. Anderson. 2006. The Orgin and Diversification of Galápagos Mockingbirds. *Evolution* 60(2): 370–382.

43. Richard D. Keynes, editor. *Charles Darwin's Zoology Notes & Specimen Lists from H.M.S. Beagle.* Cambridge University Press, Cambridge, UK, 2000, p. 298.

44. Nora Barlow, editor. 1963. Darwin's Ornithological Notes. *Bulletin of the British Museum (Natural History).* Historical Series. 2(7): 201–278 (p. 262).

45. Brian S. Arbogast, Sergei V. Drovetski, Robert L. Curry, Peter T. Boag, Gilles Seutin, Peter R. Grant, Rosemary B. Grant, and David J. Anderson. 2006. The Origin and Diversification of Galápagos Mockingbirds. *Evolution* 60(2): 370.

46. Francis Darwin, editor. 1909. *The foundations of The origin of species: Two essays written in 1842 and 1844.* Cambridge University Press, Cambridge, UK, 1909, p.161.

47. Sandra Herbert. *Darwin, Geologist.* Cornell University Press, Ithaca, 2005, pp. 315–317.

48. Lukas Keller (personal communication). See also: Brian S. Arbogast, Sergei V. Drovetski, Robert L. Curry, Peter T. Boag, Gilles Seutin, Peter R. Grant, Rosemary B. Grant, and David J. Anderson. 2006. The Origin and Diversification of Galápagos Mockingbirds. *Evolution* 60(2): 379.

49. P. R. Grant, R. L. Curry, and B. R. Grant. 2000. A Remnant Population of the Floreana Mockingbird on Champion Island, Galápagos. *Biological Conservation* 2: 285–290.

50. Robert L. Curry. 1986. Whatever Happened to the Floreana Mockingbird? *Noticias de Galápagos* 43: 13–15.

51. Lukas Keller and Paquita Hoeck, personal communication.

52. Akie Sako, Herbert Tichy, Colm O'hUigin, Peter R. Grant, B. Rosemary Grant, and Jan Klein. 2001. On the Origin of Darwin's Finches. *Molecular Biology and Evolution* 18(3): 299–311.

53. P. R. Grant and B. R. Grant. 2003. What Finches Can Teach Us About the Evolutionary Origin and Regulation of Biodiversity. *Bioscience* 53(10):965–975.

54. Brian S. Arbogast, Sergei V. Drovetski, Robert L. Curry, Peter T. Boag, Gilles Seutin, Peter R. Grant, Rosemary B. Grant, and David J. Anderson. 2006. The Origin and Diversification of Galápagos Mockingbirds. *Evolution* 60(2): 377.

55. Charles Darwin. In: *Charles Darwin's Beagle Diary*, Richard D. Keynes, editor. Cambridge University Press, Cambridge, UK, 1988, p. 356.

56. Francis Warriner. *Cruise of the United States Frigate Potomac Round the World, During the Years 1831–34. Embracing The Attack on Quallah Battoo, with Notices of Scenes, Manners, Etc., in Different Parts of Asia, South America, and the Islands of the Pacific.* Leavitt, Lord, New York, 1835, pp. 321–326.

57. Charles Darwin. In: *Charles Darwin's Beagle Diary*, Richard D. Keynes, editor. Cambridge University Press, Cambridge, UK, 1988, p. 356.

58. Peter C. H. Pritchard. *The Galápagos Tortoises: Nomenclatural and Survival Status.* Chelonian Research Monographs., No. 1, July 1996, pp. 62–65.

59. Octavio Latorre. El Hombre en las Islas Encantadas: La Historia Humana de Galápagos. (Man in the Enchanted Isles: The Human History of Galápagos). Producción Gráfica, printer, Quito, Ecuador, 1999, pp. 88–89.

60. Charles Darwin. In: *Charles Darwin's Beagle Diary*, Richard D. Keynes, editor. Cambridge University Press, Cambridge, UK, 1988, p. 356.

61. Nora Barlow, editor. 1963. Darwin's Ornithological Notes. *Bulletin of the British Museum (Natural History).* Historical Series. 2(7): 201–278 (p. 262).

62. Ibid.

63. Charles Darwin. Specimen catalogue. DAR 29.3:40, MS p. 7v Darwin Archive, Cambridge University Library.

64. Frank Sulloway. 1982. Darwin's Conversion: The *Beagle* Voyage and Its Aftermath. *Journal of the History of Biology* 15(3): 325–396 (p. 344).

65. Robert FitzRoy. *Narrative of the surveying voyages of His Majesty's Ships Adventure and Beagle between the years 1826 and 1836, describing their examination of the southern shores of South America, and the Beagle's circumnavigation of the globe: Proceedings of the second expedition, 1831–36, under the command of Captain Robert Fitz-Roy, R.N.* Henry Colburn, London, 1839, p. 504.

66. Charles Darwin. *Journal of Researches* into *the natural history and geology of the countries visited during the voyage of H.M.S. Beagle round the world, under the Command of Capt. Fitz Roy, R.N.* 2nd ed. John Murray, London, 1845, p. 394.

67. Georges Cuvier, 1827–35. *The animal kingdom arranged in conformity with its organization, with additional descriptions of all the species hitherto named, and of many not before noticed, by Edward Griffith and others,* Vol. 9 (of 16). London, 1831, pp. 63–64.

68. Georges Cuvier, 1827–35. *The animal kingdom arranged in conformity with its organization, with additional descriptions of all the species hitherto named, and of many not before noticed, by Edward Griffith and others,* Vol. 9 (of 16). London, 1831, p. 3.

69. Charles Robert Darwin. *Narrative of the surveying voyages of His Majesty's Ships Adventure and Beagle between the years 1826 and 1836, describing their examination of the southern shores of South America, and the Beagle's circumnavigation of the globe: Journal and remarks, 1832–1836.* Henry Colburn, London, 1839, p. 629 (addenda to p. 465).

70. Charles Darwin. Notebook Zed (DAR 118). In: *Charles Darwin's Notebooks, 1836–1844,* Paul Barrett, Peter Gautrey, Sandra Herbert, David Kohn, and Sydney Smith, editors. Cornell University Press, Ithaca, 1987, p. 482.

71. Paul Chambers. *A sheltered Life: The Unexpected History of the Giant Tortoise.* John Murray, London, 2005, pp. 62–65.

72. David Porter. *Journal of a Cruise made to the Pacific Ocean, by Captain David Porter, in the United States Frigate ESSEX , in the years 1812, 1813, and 1814. Containing Descriptions of the Cape Verde Islands, Coasts of Brazil, Patagonia, Chili, and Peru, and of the Gallapagos Islands; also, A full account of the Washington Groupe [sic] of Islands, the Manners, Customs, and Dress of the Inhabitants, &c. &c. Two volumes in one.* Bradford & Inskeep, Philadelphia, 1815, Chapter IX.

73. Darwin misquoted Porter in the 1845 edition of his *Journal,* by having him say that the James Island tortoises were better tasting than the others.

74. In 1874 Günther also spearheaded a move to protect the Aldabra tortoise (*Geochelone gigantea*) from extinction. Charles Darwin added his name to the petition. See Paul Chambers. *A Sheltered Life: The Unexpected History of the Giant Tortoise.* John Murray, London, 2005, p. 124.

75. Paul Chambers. *A Sheltered Life: The Unexpected History of the Giant Tortoise.* John Murray, London, 2005, p. 139.

76. A total of 15 taxa have been proposed, but whether the extinct Barrington Island (Isla Santa Fé) subspecies ever truly existed is questionable. The few bones that were found on Barrington were probably the remains of a tortoise from another island brought to Barrington by whalers or early settlers. (Adalgisa Caccone, Gabriele Gentile, James P. Gibbs, Thomas H. Fritts, Howard L. Snell, Jessica Betts, and Jeffrey R. Powell. 2002. Phylogeography and History of Giant Galápagos Tortoises. *Evolution* 56(10): 2052.) Peter Pritchard recognizes only 10 subspecies of Gálapagos tortoise and proposes subspecies status for 4 additional populations. (Peter C. H. Pritchard. *The Galápagos Tortoises: Nomenclatural and Survival Status.* Chelonian Research Monographs, No., 1 July 1996, pp. 50–51.)

77. It has been contended that there are 5 different subspecies on Albemarle Island (Michael H. Jackson. *Galápagos A Natural History.* University of Calgary Press, Calgary, 1993, p. 103). However, Peter Pritchard recognizes only 2 subspecies from Albemarle Island. (Peter C. H. Pritchard. *The Galápagos Tortoises: Nomenclatural and Survival Status.* Chelonian Research Monographs, No., 1 July 1996, p. 50.)

78. Adalgisa Caccone, James Gibbs, Valerio Ketmaier, Elizabeth Suatoni, and Jeffrey Powell. 1999. Origin and Evolutionary Relationships of Giant Galápagos Tortoises. *Proceedings of the National Academy of Sciences of the United States of America* 96(23):13223–13228.

79. Peter C. H. Pritchard. *The Galápagos Tortoises: Nomenclatural and Survival Status.* Chelonian Research Monographs, No., 1 July 1996, p. 19.

80. Robert FitzRoy. *Narrative of the surveying voyages of His Majesty's Ships Adventure and Beagle between the years 1826 and 1836, describing their examination of the southern shores of South America, and the Beagle's circumnavigation of the globe: Proceedings of the second expedition, 1831–36, under the command of Captain Robert Fitz-Roy, R.N.* Henry Colburn, London, 1839, p. 492.

81. Ibid.

82. Abel du Petit-Thouars, captain of *La Venús,* spent several days visiting the settlement of Asilo de Paz in June 1838. He wrote that several of the huts at the first

spring were then occupied, but made no mention of flower pots or tortoise shells. From Abel du Petit-Thouars. *Voyage Autour Du Monde Sur La Frégate Vénus Pendant Les Annees 1836–1839*. Gide, Paris, 1841, pp. 279–323.

83. Peter C. H. Pritchard. *The Galápagos Tortoises: Nomenclatural and Survival Status*. Chelonian Research Monographs, No. 1, July 1996, p. 44.

84. The tortoises in the south of Chatham Island may also have been larger than their northern counterparts. FitzRoy noted that the tortoises collected from Fresh Water Bay were elephantine, whereas the ones taken from Pan de Azucar were not very large.

85. Richard D. Keynes, editor. *Charles Darwin's Zoology Notes & Specimen Lists from H.M.S. Beagle*. Cambridge University Press, Cambridge, UK, 2000, p. xx.

86. Frank J. Sulloway. 1894. Darwin and the Galápagos. *Biological Journal of the Linnean Society of London* 21: 35. See also Paul Chambers. *A Sheltered Life: The Unexpected History of the Giant Tortoise*. John Murray, London, 2005, p. 49.

87. Robert FitzRoy. *Narrative of the surveying voyages of His Majesty's Ships Adventure and Beagle between the years 1826 and 1836, describing their examination of the southern shores of South America, and the Beagle's circumnavigation of the globe: Proceedings of the second expedition, 1831–36, under the command of Captain Robert Fitz-Roy, R.N.* Henry Colburn, London, 1839, pp. 504–505.

88. Ibid., p. 505.

89. William Dampier. *A New Voyage Round the World. Describing particularly, The Isthmus of America, seceral Coasts and Islands in the West Indies, the Isles of Cape Verd, the Passage by Terra del Fuego, the South Sea Coasts of Chili, Peru, and Mexico; the Isle of Guam one of the Ladrones, Mindanao, and other Philippine and East-India Islands near Cambodia, China, Formosa, Luconia, Celebes, &c. New Holland, Sumatra, Nicobar Isles; the Cape of Good Hope, and Santa Hellena. Their Soil, Rivers, Harbours, Plants, Fruits, Animals, and Inhabitants. Their Customs, Religion, Government, Trade, &c. Illustrated with Particular Maps and Draughts.* James Knapton, London, 1697, pp. 102–103.

90. Richard D. Keynes, editor. *Charles Darwin's Zoology Notes & Specimen Lists from H.M.S. Beagle*. Cambridge University Press, Cambridge, UK, 2000, p. xx.

91. Ibid., p. 416.

92. Ibid., p. 291.

93. Charles Darwin. *Journal of Researches into the natural history and geology of the countries visited during the voyage of H.M.S. Beagle round the world, under the Command of Capt. Fitz Roy, R.N.* 2nd ed. John Murray, London, 1845, p. 385.

94. Lord George Anson Byron. *Voyage of H.M.S. Blonde to the Sandwich Islands, in the years 1824–25*, Maria Graham, editor. John Murray, London (March 27, 1825), 1826.

95. F. Burkhardt and S. Smith, editors. *The Books on board the Beagle*. In: *The Correspondence of Charles Darwin*, Vol. 1: *1821–1836*. Frederick Burkhardt, Sydney Smith, David Kohn, and William Montgomery, editors. Cambridge University Press, Cambridge, UK, 1985, p. 559.

96. Charles Robert Darwin. *Narrative of the surveying voyages of His Majesty's Ships Adventure and Beagle between the years 1826 and 1836, describing their examination of the southern shores of South America, and the Beagle's circumnavigation*

of the globe: Journal and remarks, 1832–1836. Henry Colburn, London, 1839, pp. 465–466.

97. Charles Darwin. Letter to Catherine Darwin, 20–29 July 1834. In: *The Correspondence of Charles Darwin,* Vol. 1: *1821–1836.* Frederick Burkhardt, Sydney Smith, David Kohn, and William Montgomery, editors. Cambridge University Press, Cambridge, UK, 1985, p. 393.

98. Frank Sulloway. 1982. Darwin's Conversion: The *Beagle* Voyage and Its Aftermath. *Journal of the History of Biology* 15(3): 325–396 (p. 344).

99. Paul Chambers. *A Sheltered Life: The Unexpected History of the Giant Tortoise.* John Murray, London, 2005, pp. 177–198.

100. Octavio Latorre, personal communication.

101. Michael H. Jackson. *Galápagos: A Natural History.* University of Calgary Press, Calgary, 1993, p. 105.

102. BBC World News article, April 7, 2004. *Timmy the tortoise dies aged 160.*

103. BBC World News article, May 23, 2006. *'Clive of India's' tortoise dies.*

104. Charles Darwin. In: *Charles Darwin's Beagle Diary,* Richard D. Keynes, editor. Cambridge University Press, Cambridge, UK, 2001, p. 356.

105. Charles Robert Darwin. *Narrative of the surveying voyages of His Majesty's Ships Adventure and Beagle between the years 1826 and 1836, describing their examination of the southern shores of South America, and the Beagle's circumnavigation of the globe: Journal and remarks, 1832–1836.* Henry Colburn, London, 1839, p. 473.

106. Charles Darwin. In: *Charles Darwin's Beagle Diary,* Richard D. Keynes, editor. Cambridge University Press, Cambridge, UK, 1988, p. 355.

107. Ibid.

108. G. R. Waterhouse. Letter to Charles Darwin, 11 July 1845. In: *The Correspondence of Charles Darwin,* Vol. 3: *1844–1846,* Frederick Burkhardt and Sydney Smith, editors. Cambridge University Press, Cambridge, UK, 1987, p. 216.

109. Richard D. Keynes, editor. *Charles Darwin's Zoology Notes & Specimen Lists from H.M.S. Beagle.* Cambridge University Press, Cambridge, UK, 2000, p. 412.

110. Terrie L. Finston and Stewart B. Peck. 2004. Speciation in Darwin's Darklings: Taxonomy and Evolution of Stomion Beetles in the Galápagos Islands, Ecuador (Insecta: Coleoptera: Tenebrionidae). *Zoological Journal of the Linnean Society* 141: 135–152.

111. Richard D. Keynes, editor. *Charles Darwin's Zoology Notes & Specimen Lists from H.M.S. Beagle.* Cambridge University Press, Cambridge, UK, 2000, p. 361.

112. David W. Steadman. 1986. Holocene vertebrate fossils from Isla Floreana, Galápagos. *Smithsonian Contributions to Zoology,* No. 413.

113. Charles Darwin. In: *Charles Darwin's Beagle Diary,* Richard D. Keynes, editor. Cambridge University Press, Cambridge, UK, 1988, p. 356.

114. J. N. Reynolds. *Voyage of the United States Frigate Potomac under the command of Commodore John Downes, during the Circumnavigation of the Globe, in the years 1831, 1832, 1833, and 1834; including a particular account of the engagement at Quallah-Battoo, on the coast of Sumatra; with all the official documents relating to the same.* Harper & Brothers, New York, 1835, pp. 467–468.

115. Alan Tye, personal communication.

116. Peter R. Grant, B. Rosemary Grant, and Kenneth Petren. 2000. The Allopatric Phase of Speciation: The Sharp-beaked Ground Finch (*Geospiza difficilis*) on the Galápagos Islands. *Biological Journal of the Linnean Society* 69: 287–317.

117. Lord George Anson Byron. *Voyage of H.M.S. Blonde to the Sandwich Islands, in the years 1824–25*, Maria Graham, editor. John Murray, London, 1826.

118. J. N. Reynolds. *Voyage of the United States Frigate Potomac under the command of Commodore John Downes, during the Circumnavigation of the Globe, in the years 1831, 1832, 1833, and 1834; including a particular account of the engagement at Quallah-Battoo, on the coast of Sumatra; with all the official documents relating to the same.* Harper & Brothers, New York, 1835, pp. 465–467. See also Francis Warriner. *Cruise of the United States Frigate Potomac Round the World, During the Years 1831–34. Embracing The Attack on Quallah Battoo, with Notices of Scenes, Manners, Etc., in Different Parts of Asia, South America, and the Islands of the Pacific.* Leavitt, Lord, New York, 1835, pp. 319–320.

119. Charles Darwin. *Notes on the geology of places visited on the voyage.* Darwin Archive, Cambridge University Library DAR 37.2 (1835), Folio 747 (1).

120. Darwin used the word *cinder* when describing some of the pyroclastic material that he found at Buccaneer Cove on James Island and a rock specimen that was collected by Chaffers from Abingdon Island. See: Charles Darwin. *Notes on the geology of places visited on the voyage.* Darwin Archive, Cambridge University Library DAR 37.2 (1835), Folios 71 (1)–776(30).

121. According to the *Beagle*'s meteorological journal, Charles Island experienced a mix of "blue skies," "detached passing clouds," and "gloomy" weather on September 27. Robert FitzRoy. *Narrative of the surveying voyages of His Majesty's Ships Adventure and Beagle between the years 1826 and 1836, describing their examination of the southern shores of South America, and the Beagle's circumnavigation of the globe: Appendix to Volume II.* Henry Colburn, London, 1839, pp. 46 and 64.

122. Charles Darwin. *Notes on the geology of places visited on the voyage.* Darwin Archive, Cambridge University Library DAR 37.2 (1835), Folio 748(2).

123. Ibid.

124. William Healey Dall and Washington Henry Ochsner. June 22, 1928. Tertiary and Pleistocene Mollusca from the Galápagos Islands. *Proceedings of the California Academy of Sciences, Fourth Series.* 17(4): 97.

125. Godfrey Merlen, personal communication.

126. Fragments were found by Greg Estes in 2005. See also E. P. Vicenzi, A. R. McBirney, W. M. White, and M. Hamilton. 1990. The Geology and Geochemistry of Isla Marchena, Galápagos Archipelago: and Ocean Island Adjacent to a Mid-Ocean Ridge. *Journal of Volcanology and Geothermal Research* 40: 295; with additional communication from geologist Dennis Geist regarding the exact location of the shells found.

127. Charles Darwin. *Notes on the geology of places visited on the voyage.* Darwin Archive, Cambridge University Library DAR 37.2 (1835), Folio 785(39).

128. Abel Du Petit-Thouars. *Voyage Autour Du Monde Sur La Frégate Vénus Pendant Les Annees 1836–1839,* Vol. 2. Gide (editor), Paris, 1841. Translated from French by K. Thalia Grant and Anne Guezou, p. 296

129. In a letter to his sister Caroline, dated December 27, 1835, Darwin wrote, "My last letter was written from the Galápagos, since which time I have had no opportunity of sending another." (Charles Darwin. Letter to Caroline Darwin, 27 December 1835. In: *The Correspondence of Charles Darwin*, Vol. 1: *1821–1836*. Frederick Burkhardt, Sydney Smith, David Kohn, and William Montgomery, editors. Cambridge University Press, Cambridge, UK, 1985, p. 471.) A footnote by the editors reads, "No letter by CD from the Galápagos has been found."

Box: Early Human Colonization

1. Charles Darwin. *Journal of Researches into the natural history and geology of the countries visited during the voyage of H.M.S. Beagle round the world, under the Command of Capt. Fitz Roy, R.N.*, 2nd ed. John Murray, London, 1845, p. 375.

2. Charles Darwin. *Charles Darwin's Beagle Diary*, Richard D. Keynes, editor. Cambridge University Press, Cambridge, UK, 1988, p. 35.

3. Dortel de Tessan. 1842. *Carte d'une Partie de L'Archipel des Galápagos, Levée et dressée à bord de La Vénus, sous les ordres de Mr. A. Du-Petit-Thouars, . . . par Mr. De Tessan. (Chart of Part of the Galapagos Archipelago, prepared and drawn aboard the Venus, by order of Mr. A. Du-Petit-Thouars, by Mr. De Tessan).*

4. Personal communication with historian Octavio Latorre, who has read many letters and documents dating from and concerning the colonization of Galápagos in the Historical Archives of Quito, Ecuador.

5. For more information on the human history of Galápagos we recommend the following website run by John Woram: www.galapagos.to

6. Robert FitzRoy. *Narrative of the surveying voyages of His Majesty's Ships Adventure and Beagle between the years 1826 and 1836, describing their examination of the southern shores of South America, and the Beagle's circumnavigation of the globe: Proceedings of the second expedition, 1831–36, under the command of Captain Robert Fitz-Roy, R.N.* Henry Colburn, London, 1839, p. 491.

7. David Porter. *Journal of a Cruise made to the Pacific Ocean, by Captain David Porter, in the United States Frigate ESSEX , in the years 1812, 1813, and 1814. Containing Descriptions of the Cape Verde Islands, Coasts of Brazil, Patagonia, Chili, and Peru, and of the Gallapagos Islands; also, A full account of the Washington Groupe [sic] of Islands, the Manners, Customs, and Dress of the Inhabitants, &c. &c. Two volumes in one.* Bradford & Inskeep, Philadelphia, 1815, Chapter V.

8. Robert FitzRoy. *Narrative of the surveying voyages of His Majesty's Ships Adventure and Beagle between the years 1826 and 1836, describing their examination of the southern shores of South America, and the Beagle's circumnavigation of the globe: Proceedings of the second expedition, 1831–36, under the command of Captain Robert Fitz-Roy, R.N.* Henry Colburn, London, 1839, p. 491.

9. Octavio Latorre. *El Hombre en Las Islas Encantadas: La Historia Humana de Galápagos. (Man in the Enchanted Isles: The Human History of Galápagos).* Producción Gráfica, Quito, Ecuador, 1999. See also John Coulter, MD. *Adventures in the Pacific; with Observations on the Natural Productions, Manners and Customs of the*

Natives of the Various Islands; together with Remarks on Missionaries, British and Other Residents, Etc., Etc. William Curry, Jun., Dublin, 1845, Chapter V.

10. Robert FitzRoy. *Narrative of the surveying voyages of His Majesty's Ships Adventure and Beagle between the years 1826 and 1836, describing their examination of the southern shores of South America, and the Beagle's circumnavigation of the globe: Proceedings of the second expedition, 1831–36, under the command of Captain Robert Fitz-Roy, R.N.* Henry Colburn, London, 1839, p. 492.

11. John Coulter, MD. *Adventures in the Pacific; with Observations on the Natural Productions, Manners and Customs of the Natives of the Various Islands; together with Remarks on Missionaries, British and Other Residents, Etc., Etc.* Dublin: William Curry, Jun., Dublin, 1845, Chapter V.

12. Abel de Petit-Thouars. *Voyage Around the World on the Frigate La Vénus, during the years 1836–1839.* Gide et Cie., Paris, 1841, Chapter XIV (translation by K. Thalia Grant and Anne Guézou), p. 297.

13. Octavio Latorre. *El Hombre en Las Islas Encantadas: La Historia Humana de Galápagos. (Man in the Enchanted Isles: The Human History of Galápagos).* Producción Gráfica, Quito, Ecuador, 1999.

14. J. P. Lundh. 2001. *The Galápagos Islands: A Brief History.* www.galapagos.to

BOX: ALIEN INVADERS—THE HUMAN FACTOR

1. Charles Darwin. *On the Origin of Species by Means of Natural Selection, or the Preservation of Favoured Races in the Struggle for Life.* John Murray, London, 1959, p. 322.

2. The exception is Baltra, which does not have a fresh water source. Baltra has been altered ecologically through its use as a military base. It was first established as a U.S. Air Force base in World War II, and was later transformed into an Ecuadorian military base. The airport on Baltra is the main entryway to Galápagos for both visitors and residents.

3. Charles Darwin Foundation fact sheets (www.darwinfoundation.org) and Alan Tye, personal communication.

4. William Ambrose Cowley. *Cowley's Voyage Round the World.* Sloane MS. 1050. British Library, London, 1688, p. 16.

5. David Porter. *Journal of a Cruise made to the Pacific Ocean, by Captain David Porter, in the United States Frigate ESSEX, in the years 1812, 1813, and 1814. Containing Descriptions of the Cape Verde Islands, Coasts of Brazil, Patagonia, Chili, and Peru, and of the Gallapagos Islands; also, A full account of the Washington Groupe [sic] of Islands, the Manners, Customs, and Dress of the Inhabitants, &c. &c. Two volumes in one.* Bradford & Inskeep, Philadelphia, 1815, Chapter X.

6. Ibid., Chapter IX.

7. *Master's log of HMS TAGUS* July 1814. Public Records Office, London, ADM 52/4629.

8. John Coulter, MD. *Adventures in the Pacific; with Observations on the Natural Productions, Manners and Customs of the Natives of the Various Islands; together with*

Remarks on Missionaries, British and Other Residents, Etc., Etc. William Curry, Jun., Dublin, 1845, Chapter XI.

9. Peter C. H. Pritchard. *The Galápagos Tortoises: Nomenclatural and Survival Status.* Chelonian Research Monographs, Number 1, July 1966. p. 66.

10. Alexander Beatson. *Tracts Relative to the Island of St. Helena; written during a residence of five years. Illustrated with Views engraved by Mr. William Daniell, from the Drawings of Samuel Davis, Esq.* W Bulmer, London, 1816.

11. Charles Darwin. *Journal of Researches into the natural history and geology of the countries visited during the voyage of H.M.S. Beagle round the world, under the Command of Capt. Fitz Roy, R.N.,* 2nd ed. John Murray, London, 1845, p. 489.

12. R. C. Stauffer, editor. *Charles Darwin's Natural Selection; being the second part of his big species book written from 1836 to 1858.* Cambridge University Press, Cambridge, UK, 1975, p. 193.

13. The official establishment of the settlement on Charles Island occurred in 1832. However, in 1835 Darwin referred to the Charles Island settlement as being "five to 6 years" old, suggesting that it was unofficially started in 1829 or 1830. (Charles Darwin. In: *Charles Darwin's Beagle Diary,* Richard D. Keynes, editor. Cambridge University Press, Cambridge, UK, 1988, p. 356.) It is known that at least one person (John Johnston) was living on the island at the beginning of the decade, and it is likely that there were a few others.

14. David Porter. *Journal of a Cruise made to the Pacific Ocean, by Captain David Porter, in the United States Frigate ESSEX , in the years 1812, 1813, and 1814. Containing Descriptions of the Cape Verde Islands, Coasts of Brazil, Patagonia, Chili, and Peru, and of the Gallapagos Islands; also, A full account of the Washington Groupe [sic] of Islands, the Manners, Customs, and Dress of the Inhabitants, &c. &c. Two volumes in one.* Bradford & Inskeep, Philadelphia, 1815, Chapter VI.

15. Abel du Petit-Thouars. *Voyage Around the World on the Frigate La Vénus, during the years 1836–1839.* Gide et Cie., Paris, 1841, Chapter XIV, p. 309 (translation by K. Thalia Grant and Anne Guézou).

16. John Coulter, MD. *Adventures in the Pacific; with Observations on the Natural Productions, Manners and Customs of the Natives of the Various Islands; together with Remarks on Missionaries, British and Other Residents, Etc., Etc.* William Curry, Jun, Dublin, 1845, Chapter VI.

17. Ibid.

18. Abel du Petit-Thouars. *Voyage Around the World on the Frigate La Vénus, during the years 1836–1839.* Gide et Cie, Paris, 1841, Chapter XIV, p. 286 (translation by K. Thalia Grant and Anne Guézou).

19. Ibid., p. 309.

20. Robert FitzRoy. *Narrative of the surveying voyages of His Majesty's Ships Adventure and Beagle between the years 1826 and 1836, describing their examination of the southern shores of South America, and the Beagle's circumnavigation of the globe. Proceedings of the second expedition, 1831–36, under the command of Captain Robert Fitz-Roy, R.N.* Henry Colburn, London, 1839, pp. 499–500.

21. Charles Darwin. In: *Charles Darwin's Beagle Diary,* Richard D. Keynes, editor. Cambridge University Press, Cambridge, UK, 1988, p. 356.

22. Robert FitzRoy. *Narrative of the surveying voyages of His Majesty's Ships Adventure and Beagle between the years 1826 and 1836, describing their examination of the southern shores of South America, and the Beagle's circumnavigation of the globe. Proceedings of the second expedition, 1831–36, under the command of Captain Robert Fitz-Roy, R.N.* Henry Colburn, London, 1839, p. 492.

23. Abel du Petit-Thouars. *Voyage Around the World on the Frigate La Vénus, during the years 1836–1839.* Gide, Paris, 1841, Chapter XIV, p. 303 (translation by K. Thalia Grant and Anne Guézou).

24. Ibid.

25. See Charles Darwin. In: *Charles Darwin's Beagle Diary,* Richard D. Keynes, editor. Cambridge University Press, Cambridge, UK, 1988, p. 355; Robert FitzRoy. *Narrative of the surveying voyages of His Majesty's Ships Adventure and Beagle between the years 1826 and 1836, describing their examination of the southern shores of South America, and the Beagle's circumnavigation of the globe. Proceedings of the second expedition, 1831–36, under the command of Captain Robert Fitz-Roy, R.N.* Henry Colburn, London, 1839, pp. 491–493; and Octavio Latorre. *El Hombre en Las Islas Encantadas. La Historia Humana de Galápagos.* Producción Gráfica, Quito, Ecuador, 1999, p. 97.

26. John Coulter, MD. *Adventures in the Pacific; with Observations on the Natural Productions, Manners and Customs of the Natives of the Various Islands; together with Remarks on Missionaries, British and Other Residents, Etc., Etc.* William Curry, Jun, Dublin, 1845, Chapter VIII.

27. Commander William Edgar Cookson. *Report of Visit by Her Majesty's Ship "Peterel" to the Galapagos Islands, in July 1875. Printed for the use of the Foreign Office December 23, 1875.* Public Records Office, London, FO 881/2729

28. Jerry Guo. 15 September 2006. The Galápagos Islands Kiss Their Goat Problem Goodbye. *Science* 313: 1567.

BOX: A CONFUSION OF FINCHES

1. Michael Harris. *A Field Guide to the Birds of Galápagos.* Collins, London, 1974, p. 137.

2. Akie Sato, Herbert Tichy, Colm O'hUigin, Peter R. Grant, B. Rosemary Grant, and Jan Klein. 2001. On the Origin of Darwin's Finches. *Molecular Biology and Evolution* 18(3):299–311.

3. Jonathan Weiner. *The Beak of the Finch.* Vintage Books, New York, 1994, p. 134.

4. Charles Darwin. *Charles Darwin's Beagle Diary,* Richard D. Keynes, editor. Cambridge University Press. Cambridge, UK, 1988, p. 353.

5. Ibid.

6. Charles Robert Darwin. *Narrative of the surveying voyages of His Majesty's Ships Adventure and Beagle between the years 1826 and 1836, describing their examination of the southern shores of South America, and the Beagle's circumnavigation of the globe. Journal and remarks. 1832–1836.* Henry Colburn, London, 1839, p. 471.

7. Ibid., p. 475.

8. Nora Barlow, editor. 1963. Darwin's ornithological notes. *Bulletin of the British Museum (Natural History)*. Historical Series. 2(7): 201–278 (p. 261).

9. Richard D. Keynes, editor. *Charles Darwin's Zoology Notes & Specimen Lists from H.M.S. Beagle*. Cambridge University Press, Cambridge, UK, 2000, pp. 297–298.

10. Ibid.

11. Nora Barlow, editor. 1963. Darwin's ornithological notes. *Bulletin of the British Museum (Natural History)*. Historical Series. 2(7): 201–278 (p. 261).

12. Frank Sulloway. 1982. Darwin and His Finches: The Evolution of a Legend. *Journal of the History of Biology* 15(1):, 1–53.

13. Richard D. Keynes, editor. *Charles Darwin's Zoology Notes & Specimen Lists from H.M.S. Beagle*. Cambridge University Press, Cambridge, UK, 2000, pp. 297–298.

14. Nora Barlow, editor. 1963. Darwin's ornithological notes. *Bulletin of the British Museum (Natural History)*. Historical Series. 2(7): 201–278 (p. 261).

15. Ibid.

16. Frank Sulloway. 1982. Darwin and His Finches: The Evolution of a Legend. *Journal of the History of Biology* 15(1):, 1–53.

17. Charles Darwin. *Journal of Researches into the natural history and geology of the countries visited during the voyage of H.M.S. Beagle round the world, under the Command of Capt. Fitz Roy, R.N.*, 2nd ed. John Murray, London, 1845, p. 395.

18. Ibid.

19. Charles Darwin. Letter to John Stevens Henslow, 3 November 1838. In: *The Correspondence of Charles Darwin*, Vol. 13: *1865*, Frederick Burkhardt, Duncan M. Porter, Sheila Ann Dean, Samantha Evans, Shelley Innes, and Alison M. Pearn, editors. Cambridge University Press, Cambridge, UK, 2002, p. 350.

20. Charles Darwin. Letter to J. D. Hooker, 11–12 July 1845. In: *The Correspondence of Charles Darwin*, Vol. 3: *1844–1846*, Frederick Burkhardt and Sydney Smith, editors. Cambridge University Press, Cambridge, UK, 1987, p. 216.

21. Charles Darwin. *On the Origin of Species by Means of Natural Selection, or the Preservation of Favoured Races in the Struggle for Life*. John Murray, London, 1859, p. 400.

22. Ibid.

23. Ibid., p. 401.

24. Richard D. Keynes, editor. *Charles Darwin's Zoology Notes & Specimen Lists from H.M.S. Beagle*. Cambridge University Press, Cambridge, UK, 2000, p. 297.

25. Charles Darwin. *Journal of Researches into the natural history and geology of the countries visited during the voyage of H.M.S. Beagle round the world, under the Command of Capt. Fitz Roy, R.N.*, 2nd ed. John Murray, London, 1845, p. 395.

26. Peter R. Grant. *Ecology and Evolution of Darwin's Finches*. Princeton University Press, Princeton, NJ, 1986, p. 9.

27. Frank Sulloway. 1982. Darwin and His Finches: The Evolution of a Legend. *Journal of the History of Biology* 15(1): 41.

28. See Peter R. Grant. *Ecology and Evolution of Darwin's Finches*. Princeton University Press, Princeton, NJ, 1986.

29. P. R. Grant and B. R. Grant. 2006. Evolution of Character Displacement in Darwin's Finches. *Science* 313: 224–226.

30. Jonathan Weiner. *The Beak of the Finch*. Vintage Books, New York, 1994, p. 8.

31. Charles Darwin. Letter to Charles Kingsley, 10 June 1867. In: *The Correspondence of Charles Darwin*, Vol. 15: *1867*, Frederick Burkhardt, editor. Cambridge University Press, Cambridge, UK, 2006, pp. 297–299.

32. For additional reading on Darwin's finches see: P. R. Grant and B. R. Grant. *How and Why Species Multiply: The Radiation of Darwin's Finches*. Princeton University Press, Princeton, NJ, 2008.

CHAPTER 6: ALBEMARLE ISLAND (ISLA ISABELA)

1. James Colnett. *A Voyage to the South Atlantic and Round Cape Horn into the Pacific Ocean, for the Purpose of Extending the Spermaceti Whale Fisheries, and other objects of commerce, by ascertaining the ports, bays, harbours, and anchoring births, in certain islands and coasts in those seas at which the ships of the British merchants might be refitted*. W Bennett, London, 1798, p. 143.

2. Robert FitzRoy. Remarks on Galapagos Islands, the NE Coast of Tierra del Fuego, and Magellan Strait. H.M.S. Beagle 1835. In: *A list of Documents Sent to the Hydrographer of the Admiralty Sept 16, 1846 and April 14, 1845. Sailing Directions and Nautical Remarks referring to the Coasts of South America, and the Galapagos Islands, intended to be incorporated with the Directions and Remarks published by the Admiralty on the authority of Captain Phillip Parker King*, p. 5.

3. Charles Darwin. In: *Charles Darwin's Beagle Diary*, Richard D. Keynes, editor. Cambridge University Press, Cambridge, UK, 1988, p. 358.

4. Ibid.

5. Charles Darwin. *Notes on the geology of places visited on the voyage*. Darwin Archive, Cambridge University Library DAR 37.2 (1835) Folio 790(41).

6. Ibid.

7. Charles Darwin. *Geological Observations on the Volcanic Islands visited during the Voyage of H. M. S. Beagle, together with some brief notices of the Geology of Australia and the Cape of Good Hope. Being the second part of the Geology of the Voyage of the Beagle, under the Command of Capt. FitzRoy, R.N. during the years 1832 to 1836*. Smith Elder, London, 1844, pp. 35–36.

8. Ibid., p. 114.

9. Robert FitzRoy. *Narrative of the surveying voyages of His Majesty's Ships Adventure and Beagle between the years 1826 and 1836, describing their examination of the southern shores of South America, and the Beagle's circumnavigation of the globe: Proceedings of the second expedition, 1831–36, under the command of Captain Robert Fitz-Roy, R.N.* Henry Colburn, London, 1839, p. 494.

10. Robert FitzRoy. *Narrative of the surveying voyages of His Majesty's Ships Adventure and Beagle between the years 1826 and 1836, describing their examination of the southern shores of South America, and the Beagle's circumnavigation of the globe: Appendix to Volume II*. Henry Colburn, London, 1839, pp. 45–46.

11. Robert FitzRoy. *Narrative of the surveying voyages of His Majesty's Ships Adventure and Beagle between the years 1826 and 1836, describing their examination of*

the southern shores of South America, and the Beagle's circumnavigation of the globe: Proceedings of the second expedition, 1831–36, under the command of Captain Robert Fitz-Roy, R.N. Henry Colburn, London, 1839, p. 494.

12. Ibid.

13. Charles Darwin. *Notes on the geology of places visited on the voyage.* Darwin Archive, Cambridge University Library DAR 37.2 (1835) Folio 745(20).

14. Charles Darwin. In: *Charles Darwin's Beagle Diary,* Richard D. Keynes, editor. Cambridge University Press, Cambridge, UK, 1988, p. 357.

15. Ibid.

16. Charles Darwin. *Notes on the geology of places visited on the voyage.* Darwin Archive, Cambridge University Library DAR 37.2 (1835) Folio 759(13).

17. Ibid., Folio 744(19)–745 (20).

18. Ibid., Folio 759(13), footnote a.

19. Robert FitzRoy. *Narrative of the surveying voyages of His Majesty's Ships Adventure and Beagle between the years 1826 and 1836, describing their examination of the southern shores of South America, and the Beagle's circumnavigation of the globe: Proceedings of the second expedition, 1831–36, under the command of Captain Robert Fitz-Roy, R.N.* Henry Colburn, London, 1839, pp. 494–495.

20. Lord George Anson Byron. *Voyage of H.M.S. Blonde to the Sandwich Islands, in the years 1824–25,* Maria Graham, editor. John Murray, London, 1826.

21. Charles Darwin. *Notes on the geology of places visited on the voyage.* Darwin Archive, Cambridge University Library DAR 37.2 (1835) Folio 760(14).

22. A. F. Richards. Archipelago de Colon, Isla San Felix and Islas Juan Fernandez. In: *Catalog of Active Volcanoes of the World and Solfatara Fields. Rome:* IAVCEI, Rome, 1962, 14: 1–50. See also: T. Simkin. Geology of the Galápagos Islands, In: *Galápagos: Key Environments,* Roger Perry, editor. Pergamon Press, Oxford, UK, 1984, pp. 15–41.

23. Benjamin Morrel. *A Narrative of four voyages to the South Sea, North and South Pacific Ocean, Chinese Sea, Ethiopic and Southern Atlantic Ocean, Indian and Antarctic Ocean. From the Year 1822 to 1831.* J. & J. Harper, New York, 1832. Republished in 1970 by The Gregg Press, New Jersey. pp. 194–195.

24. Charles Darwin. In: *Charles Darwin's Beagle Diary,* Richard D. Keynes, editor. Cambridge University Press, Cambridge, UK, 1988, p. 363.

25. Charles Darwin. Letter to Charles Lyell, 11 February 1857. In: *The Correspondence of Charles Darwin,* Vol. 6: *1856–1857,* Frederick Burkhardt and Sydney Smith, editors. Cambridge University Press, Cambridge, UK 1990, p. 337.

26. Dennis Geist, Terry Naumann, and Karen Harpp. November 2005. *Report on Sierra Negra eruption October 2005.* Darwin Foundation Newsroom article. www.darwinfoundation.org

27. Footage of the eruption can be viewed on *Galápagos,* a three-part documentary series that was produced for release on the BBC Natural History Channel and the National Geographic Channel in 2006.

28. Lord George Anson Byron. *Voyage of H.M.S. Blonde to the Sandwich Islands, in the years 1824–25,* Maria Graham, editor. John Murray, London, 1826 (March 25, 1825).

29. Masters log of the Beagle. ADM 52/4002. October 01, 1835. Public Records Office, UK.

30. Lord George Anson Byron. *Voyage of H.M.S. Blonde to the Sandwich Islands, in the years 1824–25*, Maria Graham, editor. John Murray, London, 1826 (March 25, 1825).

31. David Porter. *Journal of a Cruise made to the Pacific Ocean in the United States Frigate Essex, in the years 1812, 1813, and 1814. Containing descriptions of the Cape Verd Islands, Coasts of Brazil, Patagonia, Chili, and Peru, and of the Gallapagos Islands.*Two volumes in one. Bradford & Inskeep, Philadelphia, 1815, Chapter VI

32. Charles Darwin. In: *Charles Darwin's Beagle Diary*, Richard D. Keynes, editor. Cambridge University Press, Cambridge, UK, 1988, p. 360.

33. Browne, Janet. *Charles Darwin: Voyaging.* AA Knopf, 1995, p. 300. Also, Alan Moorehead. *Darwin and the Beagle.* Penguin Books, London, 1969, pp. 155–156.

34. Flamingo bones were used to make flutes and pipe stems in Haiti in the early 1800s . José A. Ottenwalder, Charles A. Woods, Galen B. Rathburn. and John B. 1990. Thorbjarnarson. Status of the Greater Flamingo in Haiti. *Colonial Waterbirds* 13(2): 115–123.

35. Richard D. Keynes, editor. *Charles Darwin's Zoology Notes & Specimen Lists from H.M.S. Beagle.* Cambridge University Press, Cambridge, UK, 2000, p. 300.

36. Ibid.

37. Charles Darwin. *Journal of Researches into the natural history and geology of the countries visited during the voyage of H.M.S. Beagle round the world, under the Command of Capt. Fitz Roy, R.N.*, 2nd ed. John Murray, London, 1845, p. 380.

38. Charles Robert Darwin, editor. *Birds Part 3 No. 5 of The zoology of the voyage of H.M.S. Beagle, by John Gould. Edited and superintended by Charles Darwin.* Smith Elder, London, 1841, p. 142.

39. Richard D. Keynes, editor. *Charles Darwin's Zoology Notes & Specimen Lists from H.M.S. Beagle.* Cambridge University Press, Cambridge, UK, 2000, p. 300.

40. Charles Darwin. In: *Charles Darwin's Beagle Diary*, Richard D. Keynes, editor. Cambridge University Press, Cambridge, UK, 1988, p. 347.

41. Richard D. Keynes, editor. *Charles Darwin's Zoology Notes & Specimen Lists from H.M.S. Beagle.* Cambridge University Press, Cambridge, UK, 2000, p. 213

42. Ibid., p. 396.

43. Charles Darwin. In: *Charles Darwin's Beagle Diary*, Richard D. Keynes, editor. Cambridge University Press, Cambridge, UK, 1988, p. 278.

44. Ibid., p. 82.

45. Charles Robert Darwin, editor. *Mammalia Part 2 No. 2 of The zoology of the voyage of H.M.S. Beagle. By George R. Waterhouse. Edited and superintended by Charles Darwin.* Smith Elder, London, 1838, pp. 25–26.

46. Stephen Leatherwood and Randall R. Reeves. *The Sierra Club Handbook of Whales and Dolphins.* Sierra Club Books, San Francisco, 1883, pp. 200–203.

47. Charles Darwin. *Notes on the geology of places visited on the voyage.* Darwin Archive, Cambridge University Library DAR 37.2 (1835) Folio 767(21).

48. Robert FitzRoy. *Narrative of the surveying voyages of His Majesty's Ships Adventure and Beagle between the years 1826 and 1836, describing their examination of the southern shores of South America, and the Beagle's circumnavigation of the globe: Proceedings of the second expedition, 1831–36, under the command of Captain Robert Fitz-Roy, R.N.* Henry Colburn, London, 1839, p. 495.

49. Charles Darwin. In: *Charles Darwin's Beagle Diary*, Richard D. Keynes, editor. Cambridge University Press, Cambridge, UK, 1988, p. 359.

50. Charles Darwin. *Notes on the geology of places visited on the voyage*. Darwin Archive, Cambridge University Library DAR 37.2 (1835) Folio 759(13), footnote a.

51. Ibid., Folio 734(9).

52. Charles Darwin. Ibid., Folio 678 [*sic*, 768] (22).

53. Ibid., Folio 769(23).

54. Ibid., Folio 678 [*sic*, 768] (22).

55. Charles Darwin. *Geological Observations on the Volcanic Islands visited during the Voyage of H. M. S. Beagle, together with some brief notices of the Geology of Australia and the Cape of Good Hope. Being the second part of the Geology of the Voyage of the Beagle, under the Command of Capt. FitzRoy, R.N., during the years 1832 to 1836*. Smith Elder, London, 1844, p. 106.

56. Charles Darwin. *Notes on the geology of places visited on the voyage*. Darwin Archive, Cambridge University Library DAR 37.2 (1835) Folio 762(16).

57. Dennis Geist, personal communication, 2006.

58. Charles Darwin. *Notes on the geology of places visited on the voyage*. Darwin Archive, Cambridge University Library DAR 37.2 (1835) Folio 762(16).

59. Ibid., Folio 763(17).

60. Ibid.

61. Ibid., Folio 737(12).

62. Ibid., Folio 763(17).

63. Ibid., Folio 762(16).

64. Ibid., Folio 761(15).

65. Ibid., Folio 763(17).

66. Ibid.

67. Charles Darwin. In: *Charles Darwin's Beagle Diary*, Richard D. Keynes, editor. Cambridge University Press, Cambridge, UK, 1988, p. 359.

68. Charles Darwin. *Notes on the geology of places visited on the voyage*. Darwin Archive, Cambridge University Library DAR 37.2 (1835) Folio 761(15).

69. Charles Darwin. *Journal of Researches into the natural history and geology of the countries visited during the voyage of H.M.S. Beagle round the world, under the Command of Capt. Fitz Roy, R.N.*, 2d ed. John Murray, London, 1845, p. 377. Darwin used these same words in describing the ecology of a brine lake near Bahía Blanca (p. 67).

70. Charles Darwin. *Journal of Researches into the natural history and geology of the countries visited during the voyage of H.M.S. Beagle round the world, under the Command of Capt. Fitz Roy, R.N.*, 2d ed. John Murray, London, 1845, p. 377.

71. Robert FitzRoy. *Narrative of the surveying voyages of His Majesty's Ships Adventure and Beagle between the years 1826 and 1836, describing their examination of the southern shores of South America, and the Beagle's circumnavigation of the globe: Proceedings of the second expedition, 1831–36, under the command of Captain Robert Fitz-Roy, R.N.* Henry Colburn, London, 1839, p. 500.

72. Lord George Anson Byron. *Voyage of H.M.S. Blonde to the Sandwich Islands, in the years 1824–25*, Maria Graham, editor. John Murray, London, 1826 (March 25, 1825).

73. Ibid.

74. Ibid.

75. Charles Darwin. *Notes on the geology of places visited on the voyage.* Darwin Archive, Cambridge University Library DAR 37.2 (1835) Folio 678 [*sic*, 768] (22).

76. Charles Darwin. In: *Charles Darwin's Beagle Diary*, Richard D. Keynes, editor. Cambridge University Press, Cambridge, UK, 1988, p. 359.

77. Richard D. Keynes, editor. *Charles Darwin's Zoology Notes & Specimen Lists from H.M.S. Beagle.* Cambridge University Press, Cambridge, UK, 2000, p. 295.

78. Patrick Syme. *Werner's nomenclature of colours with additions, arranged so as to render it highly useful to the arts and science* . . . 2d (edition) Edinburgh, 1821.

79. Charles Robert Darwin, editor. *Fish Part 4 No. 4 of The zoology of the voyage of H.M.S. Beagle, by Leonard Jenyns. Edited and superintended by Charles Darwin.* Smith Elder, London, 1842, p. x.

80. Charles Robert Darwin. *Narrative of the surveying voyages of His Majesty's Ships Adventure and Beagle between the years 1826 and 1836, describing their examination of the southern shores of South America, and the Beagle's circumnavigation of the globe. Journal and remarks. 1832–1836.* Henry Colburn, London, 1839, p. 471.

81. Richard D. Keynes, editor. *Charles Darwin's Zoology Notes & Specimen Lists from H.M.S. Beagle.* Cambridge University Press, Cambridge, UK, 2000, p. 295.

82. Julian Fitter, Daniel Fitter, and David Hosking. *Wildlife of the Galápagos.* Safari Guides. HarperCollins, London, 2000, p. 94.

83. Charles Robert Darwin, editor. *Reptiles Part 5 No. 2 of The zoology of the voyage of H.M.S. Beagle. By Thomas Bell. Edited and superintended by Charles Darwin.* Smith Elder, London, 1843, p. 24.

84. Richard D. Keynes, editor. *Charles Darwin's Zoology Notes & Specimen Lists from H.M.S. Beagle.* Cambridge University Press, Cambridge, UK, 2000, p. 360.

85. Charles Darwin. Letter to Albrecht Günther, 6 March 1860. In: *The Correspondence of Charles Darwin,* Vol. 8: *1860,* Frederick Burkhardt, Janet Browne, Duncan M. Porter, and Marsha Richmond, editors. Cambridge University Press, Cambridge, UK, 1993, p.121. See also Charles Robert Darwin. 1839. *Narrative of the surveying voyages of His Majesty's Ships Adventure and Beagle between the years 1826 and 1836, describing their examination of the southern shores of South America, and the Beagle's circumnavigation of the globe. Journal and remarks. 1832–1836.* Henry Colburn, London, p. 462.

86. Charles Darwin. *Journal of Researches into the natural history and geology of the countries visited during the voyage of H.M.S. Beagle round the world, under the Command of Capt. Fitz Roy, R.N.,* 2d ed. John Murray, London, 1845, p. 381.

87. Charles Darwin. Letter to Albrecht Günther, 6 March 1860. In: *The Correspondence of Charles Darwin,* Vol. 8: *1860,* Frederick Burkhardt, Janet Browne, Duncan M. Porter, and Marsha Richmond, editors. Cambridge University Press, Cambridge, UK, 1993, p. 121.

88. Julian Fitter, Daniel Fitter, and David Hosking. *Wildlife of the Galápagos.* Safari Guides. HarperCollins, London, 2000, p. 98.

89. Ibid.

90. Ibid.

91. Charles Robert Darwin. *Narrative of the surveying voyages of His Majesty's Ships Adventure and Beagle between the years 1826 and 1836, describing their examination of the southern shores of South America, and the Beagle's circumnavigation of the globe. Journal and remarks. 1832–1836.* Henry Colburn, London, 1839, p. 462.

92. Ibid., p. 473.

93. Charles Darwin. Letter to Alexander von Humboldt, 1 November 1839. In: *The Correspondence of Charles Darwin,* Vol. 2: *1837–1843,* Frederick Burkhardt and Sydney Smith, editors. Cambridge University Press, Cambridge, UK, 1986, pp. 239–240.

94. Charles Darwin. *Notes on the geology of places visited on the voyage.* Darwin Archive, Cambridge University Library DAR 37.2 (1835) Folio 792(43).

95. Ibid.

96. Charles Lyell. *Principals of Geology,* Vol. 1. republished by The University of Chicago Press, Chicago, 1990, pp. 211–212.

97. Charles Darwin. *The structure and distribution of coral reefs. Being the first part of the geology of the voyage of the Beagle, under the command of Capt. Fitzroy, R.N. during the years 1832 to 1836.* Smith Elder, London, 1842, pp. 60–61.

98. Cleveland P. Hickman, Jr. *A Field Guide to Corals and other Radiates of Galápagos.* Sugar Spring Press, Lexington, Virginia, 2008, pp 1-4. See also: Gerard Wellington. Marine Environment and Protection. In: *Galápagos: Key Environments,* Roger Perry, editor. Pergamon Press, Oxford, UK, 1984, pp. 250–254.

99. It is not known when it was decided that Darwin would remain on James Island while FitzRoy returned to Chatham to water the ship.

100. Charles Robert Darwin. *Narrative of the surveying voyages of His Majesty's Ships Adventure and Beagle between the years 1826 and 1836, describing their examination of the southern shores of South America, and the Beagle's circumnavigation of the globe. Journal and remarks. 1832–1836.* Henry Colburn, London, 1839, p. 458.

101. Ibid., p. 602.

102. Charles Darwin. In: *Charles Darwin's Beagle Diary,* Richard D. Keynes, editor. Cambridge University Press, Cambridge, UK, 1988, p. 360.

103. Ibid., p. 359.

CHAPTER 7: JAMES ISLAND (ISLA SANTIAGO)

1. James Colnett. *A Voyage to the South Atlantic and Round Cape Horn into the Pacific Ocean, for the Purpose of Extending the Spermaceti Whale Fisheries, and other objects of commerce, by ascertaining the ports, bays, harbours, and anchoring births, in certain islands and coasts in those seas at which the ships of the British merchants might be refitted.* London: W Bennett, London, 1798, p. 156.

2. Robert FitzRoy *Narrative of the surveying voyages of His Majesty's Ships Adventure and Beagle between the years 1826 and 1836, describing their examination of the southern shores of South America, and the Beagle's circumnavigation of the globe: Proceedings of the second expedition, 1831–36, under the command of Captain Robert Fitz-Roy, R.N.* Henry Colburn, London, 1839, p. 497.

3. Charles Darwin (Richard D. Keynes, editor) *Charles Darwin's Beagle Diary.* Cambridge University Press, Cambridge, UK, 1988, p. 362.

4. Octavio Latorre. *El Hombre en las Islas Encantadas: La Historia Humana de Galápagos. (Man in the Enchanted Isles: The Human History of Galápagos).* Producción Gráfica (printer), Quito, Ecuador, 1999, pp. 88–89.

5. Robert FitzRoy *Narrative of the surveying voyages of His Majesty's Ships Adventure and Beagle between the years 1826 and 1836, describing their examination of the southern shores of South America, and the Beagle's circumnavigation of the globe: Proceedings of the second expedition, 1831–36, under the command of Captain Robert Fitz-Roy, R.N.* Henry Colburn, London, 1839, p. 497.

6. Charles Darwin. In: *Charles Darwin's Beagle Diary*, Richard D. Keynes, editor. Cambridge University Press, Cambridge, UK, 1988, p. 361.

7. Ibid., p. 445.

8. R. FitzRoy 1836. Sketch of the Surveying Voyages of his Majesty's Ships Adventure and Beagle, 1825-1836. Commanded by Captains P. P. King, P. Stokes, and R. Fitz-Roy, Royal Navy. *Journal of the Geological Society of London* 6: 311–343 (p. 333).

9. Charles Darwin. In: *Charles Darwin's Beagle Diary*, Richard D. Keynes, editor. Cambridge University Press, Cambridge, UK, 1988, p. 361.

10. Ibid.

11. K. Thalia Grant and Greg Estes. *Writings on Walls.* 2004. www.galapagos.to.

12. Charles Robert Darwin. *Narrative of the surveying voyages of His Majesty's Ships Adventure and Beagle between the years 1826 and 1836, describing their examination of the southern shores of South America, and the Beagle's circumnavigation of the globe: Journal and remarks, 1832–1836.* Henry Colburn, London, 1839, p. 469.

13. Ibid., p. 470.

14. Richard D. Keynes, editor. *Charles Darwin's Zoology Notes & Specimen Lists from H.M.S. Beagle.* Cambridge University Press, Cambridge, UK, 2000, p. 295.

15. Charles Darwin. *Beagle* field notebook *Galapagos Otaheite Lima* (1835). Darwin Archive, Cambridge University Library Microfilm EH1.17, p. 49b.

16. Richard D. Keynes, editor. *Charles Darwin's Zoology Notes & Specimen Lists from H.M.S. Beagle.* Cambridge University Press, Cambridge, UK, 2000, p. 295.

17. Charles Darwin. *Beagle* field notebook *Galapagos Otaheite Lima* (1835). Darwin Archive, Cambridge University Library Microfilm EH1.17, p. 43b.

18. Ibid.

19. Richard D. Keynes, editor. *Charles Darwin's Zoology Notes & Specimen Lists from H.M.S. Beagle.* Cambridge University Press, Cambridge, UK, 2000, p. 296.

20. Charles Robert Darwin. *Narrative of the surveying voyages of His Majesty's Ships Adventure and Beagle between the years 1826 and 1836, describing their examination of the southern shores of South America, and the Beagle's circumnavigation of the globe: Journal and remarks, 1832–1836.* Henry Colburn, London, 1839, p. 470.

21. Richard D. Keynes, editor. *Charles Darwin's Zoology Notes & Specimen Lists from H.M.S. Beagle.* Cambridge University Press, Cambridge, UK, 2000, p. 295.

22. Charles Robert Darwin. *Narrative of the surveying voyages of His Majesty's Ships Adventure and Beagle between the years 1826 and 1836, describing their examination of the southern shores of South America, and the Beagle's circumnavigation of the globe: Journal and remarks, 1832–1836.* Henry Colburn, London, 1839, p. 470.

23. Richard D. Keynes, editor. *Charles Darwin's Zoology Notes & Specimen Lists from H.M.S. Beagle.* Cambridge University Press, Cambridge, UK, 2000, p. 362.

24. Elizabeth Cabot Agassiz. 1873. A Cruise through the Galapagos. *Atlantic Monthly* 31(187, May 1873): 597–584. See also: Notes of the Hassler Expedition by a correspondent of the *New York Tribune* in *New York Tribune* July 22, 1872, p. 235.

25. Ibid.

26. Charles Robert Darwin. *Narrative of the surveying voyages of His Majesty's Ships Adventure and Beagle between the years 1826 and 1836, describing their examination of the southern shores of South America, and the Beagle's circumnavigation of the globe: Journal and remarks, 1832–1836.* Henry Colburn, London, 1839, p. 471.

27. Richard D. Keynes, editor. *Charles Darwin's Zoology Notes & Specimen Lists from H.M.S. Beagle.* Cambridge University Press, Cambridge, UK, 2000, p. 296.

28. Gabriele Gentile, Anna Fabiani, Cruz Marquez, Howard L. Snell, Heidi M. Snell, Washington Tapia, and Valerio Sbordoni. 2009. An Overlooked Pink Species of Land Iguana in the Galápagos. *Proceedings of the National Academy of Sciences* 106(2):507–511.

29. Charles Darwin. In: *Charles Darwin's Beagle Diary*, Richard D. Keynes, editor. Cambridge University Press, Cambridge, UK, 1988, p. 355.

30. Ibid., p. 361.

31. Charles Robert Darwin. *Narrative of the surveying voyages of His Majesty's Ships Adventure and Beagle between the years 1826 and 1836, describing their examination of the southern shores of South America, and the Beagle's circumnavigation of the globe: Journal and remarks, 1832–1836.* Henry Colburn, London, 1839, p. 458.

32. Charles Darwin. In: *Charles Darwin's Beagle Diary*, Richard D. Keynes, editor. Cambridge University Press, Cambridge, UK, 1988, p. 361.

33. Ibid., p. 363.

34. Ibid., p. 361.

35. Ibid., p. 363.

36. Richard D. Keynes, editor. *Charles Darwin's Zoology Notes & Specimen Lists from H.M.S. Beagle.* Cambridge University Press, Cambridge, UK, 2000, p. 296.

37. Charles Robert Darwin. *Narrative of the surveying voyages of His Majesty's Ships Adventure and Beagle between the years 1826 and 1836, describing their examination of the southern shores of South America, and the Beagle's circumnavigation of the globe: Journal and remarks, 1832–1836.* Henry Colburn, London, 1839, p. 471.

38. Richard D. Keynes, editor. *Charles Darwin's Zoology Notes & Specimen Lists from H.M.S. Beagle.* Cambridge University Press, Cambridge, UK, 2000, p. 296.

39. Charles Robert Darwin. *Narrative of the surveying voyages of His Majesty's Ships Adventure and Beagle between the years 1826 and 1836, describing their examination of the southern shores of South America, and the Beagle's circumnavigation of the globe: Journal and remarks, 1832–1836.* Henry Colburn, London, 1839, p. 474.

40. Ibid.

41. Charles Darwin. Letter to J. D. Hooker, 11 March 1844. In: *The Correspondence of Charles Darwin,* Vol. 3: *1844–1846,* Frederick Burkhardt and Sydney Smith, editors. Cambridge University Press, Cambridge, UK, 1987, p. 20.

42. Richard D. Keynes, editor. *Charles Darwin's Zoology Notes & Specimen Lists from H.M.S. Beagle.* Cambridge University Press, Cambridge, UK, 2000, pp. 291–292.

43. Ibid., p. 292.

44. D. M. Porter. 1987. Darwin's notes on *Beagle* plants. *Bulletin of the British Museum (Natural History) Historical Series* 14(2): 145–233 (p. 185).

45. Ibid.

46. Charles Darwin. *Journal of Researches into the natural history and geology of the countries visited during the voyage of H.M.S. Beagle round the world, under the Command of Capt. Fitz Roy, R.N.,* 2nd ed. John Murray, London, 1845, p. 377.

47. D. M. Porter. 1987. Darwin's notes on *Beagle* plants. *Bulletin of the British Museum (Natural History) Historical Series* 14(2): 145–233 (p. 185).

48. Joseph Hooker. Letter to Charles Darwin, 29 January 1844. In: *The Correspondence of Charles Darwin,* Vol. 3: *1844–1846,* Frederick Burkhardt and Sydney Smith, editors. Cambridge University Press, Cambridge, UK, 1987, p. 6.

49. Frank Bungartz, personal communication.

50. Charles Darwin. 1851. In: J. D. Hooker, An enumeration of the plants of the Galapagos Archipelago; with descriptions of those which are new. [Read 4 March, 6 May, and 16 December 1845.] *Transactions of the Linnean Society of London* 20: 164. See also: Richard D. Keynes, editor. *Charles Darwin's Zoology Notes & Specimen Lists from H.M.S. Beagle.* Cambridge University Press, Cambridge, UK, 2000, p. 292.

51. Joseph Hooker. Letter to Charles Darwin, 29 January 1844. In: *The Correspondence of Charles Darwin,* Vol. 3: *1844–1846,* Frederick Burkhardt and Sydney Smith, editors. Cambridge University Press, Cambridge, UK, 1987, p. 6.

52. Charles Darwin. Letter to Charles Lyell, 1 September 1860. In: *The Correspondence of Charles Darwin,* Vol. 8: *1860,* Frederick Burkhardt, Janet Browne, Duncan M. Porter, and Marsha Richmond, editors. Cambridge University Press, Cambridge, UK 1993, pp. 339–340.

53. J. D. Hooker. 1847. An enumeration of the plants of the Galapagos Archipelago; with descriptions of those which are new. *Transactions of the Linnean Society of London* 20: 163–233.

54. W. A. Weber. Lichenology and bryology in the Galápagos Islands, with check lists of the lichens and bryophytes thus far reported. In: *The Galápagos,* R. I. Bowman, editor. University of California Press, Berkeley 1966, pp. 190–200.

55. Kornelia Rassman. 1997. Evolutionary Age of the Galápagos Iguanas Predates the Age of the Present Galápagos Islands. *Molecular Phylogenetics and Evolution* 7(2): 158–172.

56. Roderick Mackie, University of Illinois, personal communication.

57. Charles Darwin. In: *Charles Darwin's Beagle Diary,* Richard D. Keynes, editor. Cambridge University Press, Cambridge, UK, 1988, p. 361.

58. Duncan M. Porter. 1980. The vascular plants of Joseph Dalton Hooker's An enumeration of the plants of the Galapagos Archipelago; with descriptions of those which are new. *Journal of the Linnean Society* 81: 79–134 (p. 100).

59. Charles Darwin. In: *Charles Darwin's Beagle Diary,* Richard D. Keynes, editor. Cambridge University Press, Cambridge, UK, 1988, p. 361.

60. Charles Darwin. *Journal of Researches into the natural history and geology of the countries visited during the voyage of H.M.S. Beagle round the world, under the Command of Capt. Fitz Roy, R.N.,* 2nd ed. John Murray, London, 1845, p. 376.

61. Richard D. Keynes, editor. *Charles Darwin's Zoology Notes & Specimen Lists from H.M.S. Beagle.* Cambridge University Press, Cambridge, UK 2000, p. 299.

62. Ibid.

63. Charles Robert Darwin. *Narrative of the surveying voyages of His Majesty's Ships Adventure and Beagle between the years 1826 and 1836, describing their examination of the southern shores of South America, and the Beagle's circumnavigation of the globe: Journal and remarks, 1832–1836.* Henry Colburn, London, 1839, p. 463.

64. Hartmut Wolfgang Baitis. *Geology, Petrology, and Petrology of Pinzon and Santiago Islands, Galápagos Archipelago.* PhD dissertation, University of Oregon, 1976.

65. Charles Darwin. In: *Charles Darwin's Beagle Diary,* Richard D. Keynes, editor. Cambridge University Press, Cambridge, UK, 1988, p. 361.

66. Charles Darwin. *Notes on the geology of places visited on the voyage.* Darwin Archive, Cambridge University Library DAR 37.2 (1835) Folio 770(24). See also Charles Darwin. *Beagle* field notebook *Galapagos Otaheite Lima* (1835). Darwin Archive, Cambridge University Library Microfilm EH1.17, p. 46b.

67. Charles Darwin. *Notes on the geology of places visited on the voyage.* Darwin Archive, Cambridge University Library DAR 37.2 (1835) Folio 770(24).

68. Ibid.

69. Charles Darwin. In: *Charles Darwin's Beagle Diary,* Richard D. Keynes, editor. Cambridge University Press, Cambridge, UK, 1988, p. 361.

70. Charles Darwin. *Notes on the geology of places visited on the voyage.* Darwin Archive, Cambridge University Library DAR 37.2 (1835) Folio 770(24).

71. Ibid., Folio 723(7).

72. Darwin wrote about the absence of tree ferns in Galápagos in his diary (Charles Darwin. In: *Charles Darwin's Beagle Diary,* Richard D. Keynes, editor. Cambridge University Press, Cambridge, UK, 1988, pp. 361–362) and again in his *Journal* (Charles Robert Darwin. *Narrative of the surveying voyages of His Majesty's Ships Adventure and Beagle between the years 1826 and 1836, describing their examination of the southern shores of South America, and the Beagle's circumnavigation of the globe: Journal and remarks, 1832–1836.* Henry Colburn, London, 1839, p. 460). In a letter to J. D. Hooker (Charles Darwin. Letter to J. D. Hooker, 11 March 1844. In: *The Correspondence of Charles Darwin,* Vol. 3: *1844–1846,* Frederick Burkhardt and Sydney Smith, editors. Cambridge University Press, Cambridge, UK, 1987, p. 20) he clarified his definition of tree ferns as meaning any "large Ferns or Palms or Palmettos."

73. Charles Darwin. *Beagle* field notebook *Galapagos Otaheite Lima* (1835). Darwin Archive, Cambridge University Library Microfilm EH1.17, p. 46b.

74. Charles Darwin. *Notes on the geology of places visited on the voyage.* Darwin Archive, Cambridge University Library DAR 37.2 (1835) Folio 781(35).

75. For a discussion of the problems surrounding this specimen see Sandra Herbert. *Charles Darwin, Geologist.* Cornell University Press, Ithaca, NY, 2005, pp. 121–123.

76. Sally Gibson, David Norman, Andrew Thurman, and Andrew Miles.

77. Dennis Geist.

78. Sandra Herbert, Sally Gibson, David Norman, Dennis Geist, Greg Estes, Thalia Grant, and Andrew Miles. 2009. Into the Field Again: Re-examining Charles

Darwin's 1825 Geological Work on Isa Santiago (James Island) in the Galápagos Archipelago. *Earth Sciences History* 28(1): 1–31.

79. Gregory Estes, K. Thalia Grant, and Peter R. Grant. 2000. Darwin in Galápagos: His Footsteps through the Archipelago. *Notes and Records of the Royal Society of London* 54(3): 343–368.

80. Uno Eliasson. Native Climax Forests. In: *Galápagos: Key Environments,* Roger Perry, editor. Pergamon Press, Oxford, UK, 1984, p. 111.

81. Ole Hamann. Changes and Threats to the Vegetation. In: *Galápagos: Key Environments,* Roger Perry, editor. Pergamon Press, Oxford, UK, 1984, pp. 115–131.

82. Charles Darwin. Letter to Caroline Darwin, 29 April 1836. In: *The Correspondence of Charles Darwin,* Vol. 1: *1821–1836.* Frederick Burkhardt, Sydney Smith, David Kohn, and William Montgomery, editors. Cambridge University Press, Cambridge, UK, 1985, p. 495.

83. Charles Darwin. *Notes on the geology of places visited on the voyage.* Darwin Archive, Cambridge University Library DAR 37.2 (1835) Folio 716(1).

84. Ibid., Folio 777(31).

85. Ibid., Folio 778(32).

86. Ibid., Folio 723(7).

87. Ibid., Folio 770(24).

88. Ibid., Folio 72(7).

89. Sandra Herbert. Charles Darwin, *Geologist.* Cornell University Press, Ithaca, NY, 2005, pp. 116–123.

90. Charles Darwin. *Notes on the geology of places visited on the voyage.* Darwin Archive, Cambridge University Library DAR 37.2 (1835) Folio 776(30).

91. Ibid.

92. Paul N. Pearson. 1996. Charles Darwin on the Origin and Diversity of Igneous Rocks. *Earth Sciences History* 15; 49–67.

93. Charles Darwin. *Notes on the geology of places visited on the voyage.* Darwin Archive, Cambridge University Library DAR 37.2 (1835) Folio 717(2).

94. Ibid., Folio 720(5).

95. Gregory Estes, K. Thalia Grant, and Peter R. Grant. 2000. Darwin in Galápagos: His Footsteps through the Archipelago. *Notes and Records of the Royal Society of London* 54(3): 357.

96. Charles Darwin. *Notes on the geology of places visited on the voyage.* Darwin Archive, Cambridge University Library DAR 37.2 (1835) Folio 718(3).

97. Ibid.

98. Ibid., Folio 775(29).

99. Charles Darwin. *Geological observations on the volcanic islands visited during the voyage of H.M.S. Beagle, together with some brief notices of the geology of Australia and the Cape of Good Hope. Being the second part of the geology of the voyage of the Beagle, under the command of Capt. Fitzroy, R.N., during the years 1832 to 1836.* Smith Elder, London, 1844, p. 109.

100. Charles Darwin. *Notes on the geology of places visited on the voyage.* Darwin Archive, Cambridge University Library DAR 37.2 (1835) Folio 718(3).

101. Ibid.

102. Charles Darwin. In: *Charles Darwin's Beagle Diary*, Richard D. Keynes, editor. Cambridge University Press, Cambridge, UK, 1988, p. 362.

103. Charles Darwin. *Notes on the geology of places visited on the voyage.* Darwin Archive, Cambridge University Library DAR 37.2 (1835) Folio 779(33).

104. Ibid.

105. Ibid., Folio 723(7).

106. Ibid., Folio 781(35).

107. Charles Robert Darwin. *Narrative of the surveying voyages of His Majesty's Ships Adventure and Beagle between the years 1826 and 1836, describing their examination of the southern shores of South America, and the Beagle's circumnavigation of the globe: Journal and remarks, 1832–1836.* Henry Colburn, London, 1839, p. 459.

108. Charles Darwin. *Notes on the geology of places visited on the voyage.* Darwin Archive, Cambridge University Library DAR 37.2 (1835) Folio 780(34).

109. Ibid., Folio 722(8).

110. Ibid.

111. Ibid., Folio 780(34).

112. Ibid., reverse side of Folio 723(8).

113. Charles Darwin. Letter to Ernst Haeckel. After 10 August 10–8 October 1864. In: *The Correspondence of Charles Darwin*, Vol. 12: *1864,* Frederick Burkhardt, Duncan M. Porter, Sheila Ann Dean, Paul S. White, and Sarah Wilmot, editors. Cambridge University Press, Cambridge, UK, 2001, p. 302.

114. Charles Darwin. *Notes on the geology of places visited on the voyage.* Darwin Archive, Cambridge University Library DAR 37.2 (1835) Folio 780(34).

115. Ibid., Folio 723(8).

116. Ibid., Folio 780(34).

117. Tom Simkin. Geology of Galápagos Islands. In: *Galápagos: Key Environments,* Roger Perry, editor. Pergamon Press, Oxford, UK, 1984, p. 18.

118. Charles Darwin. *Notes on the geology of places visited on the voyage.* Darwin Archive, Cambridge University Library DAR 37.2 (1835) Folio 724(9).

119. Ibid., Folio 781(35).

120. Charles Darwin. In: *Charles Darwin's Beagle Diary*, Richard D. Keynes, editor. Cambridge University Press, Cambridge, UK, 1988, p. 363.

121. Charles Darwin. Letter to J. D. Hooker, 10 February 1846. In: *The Correspondence of Charles Darwin*, Vol. 3: *1844–1846,* Frederick Burkhardt and Sydney Smith, editors. Cambridge University Press, Cambridge, UK, 1987, p. 288.

122. Ira L. Wiggins and Duncan M. Porter. *Flora of the Galápagos Islands.* Stanford University Press, 1971, pp. 537–546.

123. Charles Darwin. In: *Charles Darwin's Beagle Diary*, Richard D. Keynes, editor. Cambridge University Press, Cambridge, UK, 1988, p. 363.

124. J. S. Henslow. 1837. Description of two new species of *Opuntia* with remarks on the structure of the fruit of *Rhipsalis. Magazine of Zoology and Botany* 1: 466–469 (p. 468).

125. Ibid.

126. Michael H. Jackson. *Galápagos: A Natural History.* University of Calgary Press, Calgary, 1993, p. 74.

127. Ibid.

128. Charles Darwin. Letter to J. S. Henslow. 17 November 1845. In: *The Correspondence of Charles Darwin,* Vol. 3: *1844–1846,* Frederick Burkhardt and Sydney Smith, editors. Cambridge University Press, Cambridge, UK, 1987, p. 267.

129. Charles Darwin. *Notes on the geology of places visited on the voyage.* Darwin Archive, Cambridge University Library DAR 37.2 (1835) Folio 781(35)–782(36).

130. Ibid., Folio 78 (35).

131. Ibid., Folio 782(36).

132. Charles Darwin. In: *Charles Darwin's Beagle Diary*, Richard D. Keynes, editor. Cambridge University Press, Cambridge, UK, 1988, p. 363.

133. Charles Darwin. *Notes on the geology of places visited on the voyage.* Darwin Archive, Cambridge University Library DAR 37.2 (1835) Folio 783(37).

134. Charles Darwin. In: *Charles Darwin's Beagle Diary*, Richard D. Keynes, editor. Cambridge University Press, Cambridge, UK, 1988, p. 363.

135. Charles Darwin. *Notes on the geology of places visited on the voyage.* Darwin Archive, Cambridge University Library DAR 37.2 (1835) Folio 725(16).

136. Ibid., Folio 783(37).

137. Ibid.

138. Ibid.

139. Charles Darwin. In: *Charles Darwin's Beagle Diary*, Richard D. Keynes, editor. Cambridge University Press, Cambridge, UK, 1988, p. 363.

140. Charles Darwin. *Notes on the geology of places visited on the voyage.* Darwin Archive, Cambridge University Library DAR 37.2 (1835) Folio 783(37).

141. Ibid.

142. Ibid., Folio 784(38).

143. Ibid., Folio 783(37).

144. Charles Darwin. In: *Charles Darwin's Beagle Diary*, Richard D. Keynes, editor. Cambridge University Press, Cambridge, UK, 1988, p. 361.

145. Charles Robert Darwin. *Narrative of the surveying voyages of His Majesty's Ships Adventure and Beagle between the years 1826 and 1836, describing their examination of the southern shores of South America, and the Beagle's circumnavigation of the globe: Journal and remarks, 1832–1836.* Henry Colburn, London, 1839, p. 463.

146. Charles Darwin. In: *Charles Darwin's Beagle Diary*, Richard D. Keynes, editor. Cambridge University Press, Cambridge, UK, 1988, p. 361.

147. Charles Darwin. *Beagle* field notebook *Galapagos Otaheite Lima* (1835). Darwin Archive, Cambridge University Library Microfilm EH1.17, p. 37b.

148. Richard D. Keynes, editor. *Charles Darwin's Zoology Notes & Specimen Lists from H.M.S. Beagle.* Cambridge University Press, Cambridge, UK, 2000, p. 292.

149. Ibid.

150. Charles Darwin. *Beagle* field notebook *Galapagos Otaheite Lima* (1835). Darwin Archive, Cambridge University Library Microfilm EH1.17, p. 51b.

151. Richard D. Keynes, editor. *Charles Darwin's Zoology Notes & Specimen Lists from H.M.S. Beagle.* Cambridge University Press, Cambridge, UK, 2000, p. 292.

152. Ibid.

153. Charles Darwin. *Beagle* field notebook *Galapagos Otaheite Lima* (1835). Darwin Archive, Cambridge University Library Microfilm EH1.17, p. 37b.

154. Charles Darwin. In: *Charles Darwin's Beagle Diary*, Richard D. Keynes, editor. Cambridge University Press, Cambridge, UK, 1988, p. 362.

155. Charles Darwin. *Beagle* field notebook *Galapagos Otaheite Lima* (1835). Darwin Archive, Cambridge University Library Microfilm EH1.17, p. 37b.

156. Richard D. Keynes, editor. *Charles Darwin's Zoology Notes & Specimen Lists from H.M.S. Beagle*. Cambridge University Press, Cambridge, UK, 2000, p. 292.

157. Charles Robert Darwin. *Narrative of the surveying voyages of His Majesty's Ships Adventure and Beagle between the years 1826 and 1836, describing their examination of the southern shores of South America, and the Beagle's circumnavigation of the globe: Journal and remarks, 1832–1836*. Henry Colburn, London, 1839, p. 465.

158. Richard D. Keynes, editor. *Charles Darwin's Zoology Notes & Specimen Lists from H.M.S. Beagle*. Cambridge University Press, Cambridge, UK, 2000, p. 293.

159. Ibid.

160. Ibid., p. 292.

161. Ibid., pp. 292–293.

162. Ibid., p. 293.

163. Ibid., p. 292.

164. Darwin used several Spanish words and catch phrases in his writings. Some of them are "Quién sabe" (who knows), "Tampoco" (neither), "Cuidado" (careful), and "no hay remedio" (it can't be helped).

165. Richard D. Keynes, editor. *Charles Darwin's Zoology Notes & Specimen Lists from H.M.S. Beagle*. Cambridge University Press, Cambridge, UK, 2000, p. 292.

166. The rate at which the scutes wear down to a glossy sheen varies with respect to sex and vegetation density. The growth rings on a female's carapace will disappear faster than those on a male's carapace, presumably because of the rubbing effect of the male's plastron against the female's carapace during copulation. Personal communication from James Gibbs.

167. Peter C. H. Pritchard. *The Galápagos Tortoises: Nomenclatural and Survival Status*. Chelonian Research Monographs, No. 1, July 1996, p. 66.

168. In 1996 the authors also found the ancient remains of two tortoise eggs embedded in a ravine between Puerto Egas and the James Bay lava flow.

169. Charles Darwin. *Beagle* field notebook *Galapagos Otaheite Lima* (1835). Darwin Archive, Cambridge University Library Microfilm EH1.17, p. 41b.

170. Charles Darwin. In: *Charles Darwin's Beagle Diary*, Richard D. Keynes, editor. Cambridge University Press, Cambridge, UK, 1988, p. 363.

171. Ibid.

172. Ibid.

173. Ibid.

174. Robert FitzRoy. *Narrative of the surveying voyages of His Majesty's Ships Adventure and Beagle between the years 1826 and 1836, describing their examination of the southern shores of South America, and the Beagle's circumnavigation of the globe: Proceedings of the second expedition, 1831–36, under the command of Captain Robert Fitz-Roy, R.N.* Henry Colburn, London, 1839, p. 498.

175. Charles Darwin. *Beagle* field notebook *Galapagos Otaheite Lima* (1835). Darwin Archive, Cambridge University Library Microfilm EH1.17, p. 44b–45b.

176. Charles Robert Darwin. *Narrative of the surveying voyages of His Majesty's Ships Adventure and Beagle between the years 1826 and 1836, describing their examination of the southern shores of South America, and the Beagle's circumnavigation of the globe: Journal and remarks, 1832–1836.* Henry Colburn, London, 1839, p. 459.

177. Ibid.

178. Robert FitzRoy. *Narrative of the surveying voyages of His Majesty's Ships Adventure and Beagle between the years 1826 and 1836, describing their examination of the southern shores of South America, and the Beagle's circumnavigation of the globe: Appendix to Volume II.* Henry Colburn, London, 1839, pp. 45–47.

179. Robert B. Dunbar, Gerard Wellington, Mitchell W. Colgan, and Peter W. Glynn. April 1994. Eastern Pacific Sea Surface Temperature Since 1600 A.D.: The δ18O Record of Climate Variability in Galápagos Corals. *Paleoceanography* 9(2): 291–315.

180. William H. Quinn and Victor T. Neal. December 15, 1987. El Niño Occurrences over the Past Four and a Half Centuries. *Journal of Geophysical Research* 92(C13): 14,449–14,461.

181. Robert B. Dunbar, Gerard Wellington, Mitchell W. Colgan, and Peter W. Glynn. April 1994. Eastern Pacific Sea Surface Temperature Since 1600 A.D.: The δ18O Record of Climate Variability in Galápagos Corals. *Paleoceanography* 9(2): 291–315.

182. Charles Darwin. *Beagle* field notebook *Galapagos Otaheite Lima* (1835). Darwin Archive, Cambridge University Library Microfilm EH1.17, p. 42b.

183. Susan Darwin. Letter to Charles Darwin, 12 February 1836. In: *The Correspondence of Charles Darwin*, Vol. 1: *1821–1836*. Frederick Burkhardt, Sydney Smith, David Kohn, and William Montgomery, editors. Cambridge University Press, Cambridge, UK, 1985, p. 489.

184. Charles Robert Darwin, editor. *Birds Part 3 No. 2 of The zoology of the voyage of H.M.S. Beagle, by John Gould. Edited and superintended by Charles Darwin.* Smith Elder, London, 1839, p. 25.

185. Richard D. Keynes, editor. *Charles Darwin's Zoology Notes & Specimen Lists from H.M.S. Beagle.* Cambridge University Press, Cambridge, UK, 2000, p. 299.

186. Charles Robert Darwin. *Narrative of the surveying voyages of His Majesty's Ships Adventure and Beagle between the years 1826 and 1836, describing their examination of the southern shores of South America, and the Beagle's circumnavigation of the globe: Journal and remarks, 1832–1836.* Henry Colburn, London, 1839, p. 461.

187. Ibid., p. 474.

188. Jennifer L. Bollmer, Rebecca T. Kimball, Noah Kerness Whiteman, José Hernán Sarasola, and Patricia G. Parker. 2006. Phylogeography of the Galápagos Hawk (*Buteo galapagoensis*): A Recent Arrival to the Galápagos Islands. *Molecular Phylogenetics and Evolution* 39: 237–247.

189. Roger McDonald. *Darwin's Shooter.* Atlantic Monthly Press, 1999, 364 pages.

190. David Attenborough. *The Life of Birds.* Princeton University Press, Princeton, NJ, 1998, p. 67.

191. Agassiz, Elizabeth Cabot. 1873. A Cruise through the Galapagos. *Atlantic Monthly* 31(187): 597–584.

192. Richard D. Keynes, editor. *Charles Darwin's Zoology Notes & Specimen Lists from H.M.S. Beagle.* Cambridge University Press, Cambridge, UK, 2000, pp. 299 and 416.

193. Agassiz, Elizabeth Cabot. 1873. A Cruise through the Galapagos. *Atlantic Monthly* 31(187): 597–584.

194. Charles Robert Darwin, editor. *Mammalia Part 2 No. 3 of The zoology of the voyage of H.M.S. Beagle. By George R. Waterhouse. Edited and superintended by Charles Darwin.* Smith Elder, London, 1838, p. 35.

195. Ibid.

195. Richard D. Keynes, editor. *Charles Darwin's Zoology Notes & Specimen Lists from H.M.S. Beagle.* Cambridge University Press, Cambridge, UK, 2000, p. 416.

197. Charles Robert Darwin, editor. *Mammalia Part 2 No. 3 of The zoology of the voyage of H.M.S. Beagle, by George R. Waterhouse. Edited and superintended by Charles Darwin.* Smith Elder, London, 1838, p. 35.

198. Another example of change through adaptation to local circumstances that Darwin used was the fearlessness of the birds of Galápagos. Darwin was convinced, from reading Dampier and Cowley's accounts of their visit to Galápagos in 1684, that the doves and finches had been even more fearless then than when Darwin visited the islands 150 years later. He concluded that the increase of human presence in the islands between 1684 and 1835 had caused the birds to change their behavior. He wrote, "[T]he wildness of birds with regard to man, is a particular instinct directed against him, and not dependent on any . . . other sources of danger . . . [I]t is not acquired by them in a short time . . . but . . . in the course of successive generations it becomes hereditary." (Charles Robert Darwin. *Narrative of the surveying voyages of His Majesty's Ships Adventure and Beagle between the years 1826 and 1836, describing their examination of the southern shores of South America, and the Beagle's circumnavigation of the globe: Journal and remarks, 1832–1836.* Henry Colburn, London, 1839, pp. 475–478.)

199. James L. Patton, Suh Y. Yang, and Philip Myers. 1975. Genetic and Morphologic Divergence among Introduced Rat Populations (*Rattus rattus*) of the Galápagos Archipelago, Ecuador. *Systematic Zoology* 24 (3): 296–310.

200. Richard D. Keynes, editor. *Charles Darwin's Zoology Notes & Specimen Lists from H.M.S. Beagle.* Cambridge University Press, Cambridge, UK, 2000, p. 362.

201. Mary K. Wicksten. Caridean and Stenopodid Shrimp of the Galápagos Islands. In: *Galápagos Marine Invertebrates: Taxonomy, Biogeography, and Evolution in Darwin's Islands,* Matt James, editor. Plenum Press, 1991, pp. 147–156. See also Cleveland P. Hickman, Jr. and Todd L. Zimmerman. *A Field Guide to Crustaceans of Galápagos.* Sugar Spring Press, Lexington, VA, 2000, p. 23.

202. Charles Darwin. Letter to Caroline Darwin, 27 December 1835. In: *The Correspondence of Charles Darwin,* Vol. 1: *1821–1836.* Frederick Burkhardt, Sydney Smith, David Kohn, and William Montgomery, editors. Cambridge University Press, Cambridge, UK 1985, pp. 471–472.

203. Caroline Darwin. Letter to Charles Darwin, 30 March 1835. In: *The Correspondence of Charles Darwin,* Vol. 1: *1821–1836.* Frederick Burkhardt, Sydney Smith, David Kohn, and William Montgomery, editors. Cambridge University Press, Cambridge, UK, 1985, p. 438.

204. Catherine Darwin. Letter to Charles Darwin, 28 January 1835. In: *The Correspondence of Charles Darwin*, Vol. 1: *1821–1836*. Frederick Burkhardt, Sydney Smith, David Kohn, and William Montgomery, editors. Cambridge University Press, Cambridge, UK, 1985, p. 424.

205. Robert FitzRoy. *Narrative of the surveying voyages of His Majesty's Ships Adventure and Beagle between the years 1826 and 1836, describing their examination of the southern shores of South America, and the Beagle's circumnavigation of the globe: Appendix to Volume II.* Henry Colburn, London, 1839, p. 300.

206. Robert FitzRoy. *Narrative of the surveying voyages of His Majesty's Ships Adventure and Beagle between the years 1826 and 1836, describing their examination of the southern shores of South America, and the Beagle's circumnavigation of the globe: Proceedings of the second expedition, 1831–36, under the command of Captain Robert Fitz-Roy, R.N.* Henry Colburn, London, 1839, pp. 498–500.

207. Ship's log. HMS *Beagle*. ADM 53/236. Public Records Office, London, October 18, 1835.

208. Charles Darwin. In: *Charles Darwin's Beagle Diary*, Richard D. Keynes, editor. Cambridge University Press, Cambridge, UK, 1988, p. 362.

BOX: DARWIN'S SNAILS

1. Charles Darwin. Letter to William Henry Benson, 7 December 1855. In: *The Correspondence of Charles Darwin*, Vol. 13: *1865*, Frederick Burkhardt, Duncan M. Porter, Sheila Ann Dean, Samantha Evans, Shelley Innes, and Alison M. Pearn, editors. Cambridge University Press, Cambridge, UK, 2002..

2. Christine E. Parent and Bernard J. Crespi. 2006. Sequential Colonization and Diversification of Galápagos Endemic Land Snail Genus *Bulimulus* (Gastropoda, Stylommatophora). *Evolution* 60 (11):2311–2328.

3. L. Pfeiffer. 1846. Descriptions of thirty new species of *Helicea*, belonging to the collection of H. Cuming, Esq. *Proceedings of the Zoological Society of London* 28–34.

4. Richard D. Keynes, editor. *Charles Darwin's Zoology Notes & Specimen Lists from H.M.S. Beagle.* Cambridge University Press, Cambridge, UK, 2000, p. 416.

5. W. H. Dall, 1896. Insular landshell faunas, especially as illustrated by the data obtained by Dr. G. Baur in the Galapagos Islands. *Proceedings of the Academy of Natural Sciences of Philadelphia* 395–460.

6. Ibid.

7. Charles Darwin. *Journal of Researches into the natural history and geology of the countries visited during the voyage of H.M.S. Beagle round the world, under the Command of Capt. Fitz Roy, R.N., 2nd ed.* John Murray, London, 1845, p. 390.

8. Abel du Petit-Thouars. *Voyage Autour Du Monde Sur La Frégate Vénus Pendant Les Années 1836–1839 (Voyage Around the World on the Frigate La Vénus, during the Years 1836–1839).* Gide, Paris, 1841.

9. Christine E. Parent, personal communication.

10. Christine E. Parent and Bernard J. Crespi. 2006. Sequential Colonization and Diversification of Galápagos Endemic Land Snail Genus *Bulimulus* (Gastropoda, Stylommatophora). *Evolution* 60 (11):2311–2328.

11. Charles Darwin. Letter to Asa Gray, 5 September 1857. In: *The Correspondence of Charles Darwin*, Vol. 6: *1856–1857*, Frederick Burkhardt and Sydney Smith, editors. Cambridge University Press, Cambridge, UK, 1990, pp. 448–449.

12. Charles Darwin. Letter to Karl Semper, 26 November 1878. In: *The life and letters of Charles Darwin, including an autobiographical chapter*, Vol. 3. Francis Darwin, editor. John Murray, London, 1887, p. 160.

13. Ibid.

14. Ibid.

15. Charles Darwin. *The Autobiography of Charles Darwin, 1809–1882*, Nora Barlow, editor. WW Norton, New York, 1958, pp. 120–121.

16. Charles Darwin. Letter to Moritz Wagner, 13 October 1876. In: *The life and letters of Charles Darwin, including an autobiographical chapter*, Vol. 3. Francis Darwin, editor. John Murray, London, 1887, p. 159.

17. Charles Darwin. *The Autobiography of Charles Darwin, 1809–1882*, Nora Barlow, editor. WW Norton, New York, 1958, pp. 120–121.

18. Ibid.

19. Charles Darwin. Letter to P. H. Gosse., 28 September 1856. In: *The Correspondence of Charles Darwin*, Vol. 6: *1856–1857*, Frederick Burkhardt and Sydney Smith, editors. Cambridge University Press, Cambridge, UK, 1990, p. 232.

20. Charles Darwin. *On the Origin of Species by Means of Natural Selection, or the Preservation of Favoured Races in the Struggle for Life*. John Murray, London, 1859, p. 353.

21. Christine E. Parent and Bernard J. Crespi. 2006. Sequential Colonization and Diversification of Galápagos Endemic Land Snail Genus *Bulimulus* (Gastropoda, Stylommatophora). *Evolution* 60 (11):2311–2328.

BOX: HALLEY'S COMET

1. Erasmus Darwin. *The Botanic Garden, A Poem, in two parts; containing The Economy of Vegetation, and The Loves of Plants. With Philosophical Notes.* Jones and Company, London, 1825, p.12.

2. Charles Darwin. *Charles Darwin's Beagle Diary*, Richard D. Keynes, editor. Cambridge University Press, Cambridge, UK, 1988, p. 363.

3. Ibid., p. 41.

4. That Halley's comet could have been visible from Buccaneer Cove in mid October 1835 was determined from data on the comet's altitude, azimuth, and brightness provided by Dr. Robin Catchpole of the Institute of Astronomy, Cambridge, UK, as well as from measurements taken by the authors, in the field where Darwin was camped.

5. Charles Darwin. *Beagle* field notebook *Galapagos Otaheite Lima* (1835). Darwin Archive, Cambridge University Library Microfilm EH1.17, p. 50b.

6. Charles Darwin. Letter to Susan Darwin, 9 September 1831. In: *The Correspondence of Charles Darwin*, Vol 1: *1821–1836*, Frederick Burkhardt, Sydney Smith, David Kohn, and William Montgomery, editors. Cambridge University Press, Cambridge, UK, 1985, p. 147.

7. Charles Darwin. Letter to J. S. Henslow, 11 April 1833. In: *The Correspondence of Charles Darwin,* Vol. 1: *1821–1836,* Frederick Burkhardt, Sydney Smith, David Kohn, and William Montgomery, editors. Cambridge University Press, Cambridge, UK, 1985, p. 308.

8. J. S. Henslow. Letter to Charles Darwin, 31 August 1833. In: *The Correspondence of Charles Darwin,* Vol 1: *1821–1836,* Frederick Burkhardt, Sydney Smith, David Kohn, and William Montgomery, editors. Cambridge University Press, Cambridge, UK, 1985, p. 327.

9. Charles Darwin. Letter to J. S. Henslow, 9 July 1836. In: *The Correspondence of Charles Darwin,* Vol 1: *1821–1836,* Frederick Burkhardt, Sydney Smith, David Kohn, and William Montgomery, editors. Cambridge University Press, Cambridge, UK 1985, p. 500.

10. Catherine Darwin. Letter to Charles Darwin, 30 October 1835. In: *The Correspondence of Charles Darwin,* Vol 1: *1821–1836,* Frederick Burkhardt, Sydney Smith, David Kohn, and William Montgomery, editors. Cambridge University Press, Cambridge, UK 1985, p. 468.

11. Robert FitzRoy. *Narrative of the surveying voyages of His Majesty's Ships Adventure and Beagle between the years 1826 and 1836, describing their examination of the southern shores of South America, and the Beagle's circumnavigation of the globe. Proceedings of the second expedition, 1831–36, under the command of Captain Robert Fitz-Roy, R.N.* Henry Colburn, London, 1839, p. 36.

12. Robert FitzRoy 1839. *Narrative of the surveying voyages of His Majesty's Ships Adventure and Beagle between the years 1826 and 1836, describing their examination of the southern shores of South America, and the Beagle's circumnavigation of the globe: Appendix to Volume II.* Henry Colburn, London, 1839, pp. 46–47.

CHAPTER 8: HOMEWARD BOUND

1. Charles Robert Darwin. *Narrative of the surveying voyages of His Majesty's Ships Adventure and Beagle between the years 1826 and 1836, describing their examination of the southern shores of South America, and the Beagle's circumnavigation of the globe: Journal and remarks, 1832–1836.* Henry Colburn, London, 1839, p. 474.

1. Richard D. Keynes, editor. *Charles Darwin's Zoology Notes & Specimen Lists from H.M.S. Beagle.* Cambridge University Press, Cambridge, UK, 2000, p. 288.

2. In: *The Correspondence of Charles Darwin,* Vol. 1: *1821–1836.* Frederick Burkhardt, Sydney Smith, David Kohn, and William Montgomery, editors. Cambridge University Press, Cambridge, UK, 1985, p. 485.

3. Charles Darwin. In: *Charles Darwin's Beagle Diary,* Richard D. Keynes, editor. Cambridge University Press, Cambridge, UK, 1988, p. 364.

4. Charles Darwin. 1836 Red Notebook 55e. In: *Charles Darwin's Notebooks 1836–1844,* Paul H. Barrett, Peter J. Gautrey, Sandra Herbert, David Kohn, and Sydney Smith, editors. Cornell University Press, Ithaca, NY, 1987.

5. Charles Darwin. Letter to Caroline Darwin, 27 December 1835. In: *The Correspondence of Charles Darwin,* Vol. 1: *1821–1836.* Frederick Burkhardt, Sydney Smith,

David Kohn, and William Montgomery, editors. Cambridge University Press, Cambridge, UK, 1985, p. 471.

6. Charles Darwin. *Notes on the geology of places visited on the voyage.* Darwin Archive, Cambridge University Library DAR 37.2 (1835) Folio 786(40).

7. Richard D. Keynes, editor. *Charles Darwin's Zoology Notes & Specimen Lists from H.M.S. Beagle.* Cambridge University Press, Cambridge, UK, 2000, p. 298.

8. James Colnett. *A Voyage to the South Atlantic and Round Cape Horn into the Pacific Ocean, for the Purpose of Extending the Spermaceti Whale Fisheries, and other objects of commerce, by ascertaining the ports, bays, harbours, and anchoring births, in certain islands and coasts in those seas at which the ships of the British merchants might be refitted.* W Bennett, London, 1798, pp. 52 and 156–157.

9. William Dampier. *A New Voyage Round the World. Describing particularly, The Isthmus of America, several Coasts and Islands in the West Indies, the Isles of Cape Verd, the Passage by Terra del Fuego, the South Sea Coasts of Chili, Peru, and Mexico; the Isle of Guam one of the Ladrones, Mindanao, and other Philippine and East-India Islands near Cambodia, China, Formosa, Luconia, Celebes, &c. New Holland, Sumatra, Nicobar Isles; the Cape of Good Hope, and Santa Hellena. THEIR Soil, Rivers, Harbours, Plants, Fruits, Animals, and Inhabitants. THEIR Customs, Religion, Government, Trade, &c. Illustrated with Particular Maps and Draughts.* James Knapton, London, 1697, p. 102.

10. Lord George Anson Byron. *Voyage of H.M.S. Blonde to the Sandwich Islands, in the years 1824–25,* Maria Graham, editor. John Murray, London, 1826 (March 25, 1825).

11. Charles Darwin. *Beagle* field notebook *Galapagos Otaheite Lima* (1835). Darwin Archive, Cambridge University Library Microfilm EH1.17, p. 30b.

12. Charles Darwin. In: *Charles Darwin's Beagle Diary,* Richard D. Keynes, editor. Cambridge University Press, Cambridge, UK, 1988, p. 356.

13. Charles Darwin. Charles Darwin. *Journal of Researches into the natural history and geology of the countries visited during the voyage of H.M.S. Beagle round the world, under the Command of Capt. Fitz Roy, R.N.,* 2nd ed. John Murray, London, 1845, p. 393.

14. Ibid., pp. 393–394.

15. Ibid., p. 394.

16. Ibid., p. 393.

17. Charles Darwin. Letter to Catherine Darwin, 3 June 1836. In: *The Correspondence of Charles Darwin,* Vol. 1: *1821–1836.* Frederick Burkhardt, Sydney Smith, David Kohn, and William Montgomery, editors. Cambridge University Press, Cambridge, UK, 1985, p. 498.

18. Charles Robert Darwin. *Narrative of the surveying voyages of His Majesty's Ships Adventure and Beagle between the years 1826 and 1836, describing their examination of the southern shores of South America, and the Beagle's circumnavigation of the globe: Journal and remarks, 1832–1836.* Henry Colburn, London, 1839, p. 629 (addenda to p. 465).

19. Loren Eiseley. *Darwin's Century. Evolution and the Men Who Discovered It.* Anchor Books, Doubleday, New York, 1961, pp. 167–168.

20. Charles Darwin. *Geological observations on the volcanic islands visited during the voyage of H.M.S. Beagle, together with some brief notices of the geology of Austra-*

lia and the Cape of Good Hope. Being the second part of the geology of the voyage of the Beagle, under the command of Capt. Fitzroy, R.N., during the years 1832 to 1836. Smith Elder, London, 1844, p. 114.

21. Charles Darwin. In: *Charles Darwin's Beagle Diary*, Richard D. Keynes, editor. Cambridge University Press, Cambridge, UK, 1988, p. 443.

22. Charles Darwin. 1837. On certain areas of elevation and subsidence in the Pacific and Indian oceans, as deduced from the study of coral formations. *Proceedings of the Geological Society of London* 2 (Read May 31): 552–554.

23. Charles Darwin. *The structure and distribution of coral reefs. Being the first part of the geology of the voyage of the Beagle, under the command of Capt. Fitzroy, R.N., during the years 1832 to 1836.* Smith Elder, London, 1842.

24. Nora Barlow, editor. 1963. Darwin's Ornithological Notes. *Bulletin of the British Museum (Natural History).* Historical Series 2(7): 201–278 (p. 262).

25. Charles Lyell. *Principles of Geology, Being An Attempt To Explain The Former Changes Of The Earth's Surface, By Reference To Causes Now In Operation. Volume II.* John Murray, London, 1832. Republished by The University of Chicago Press, Chicago, 1991, p. 64.

26. D. Kohn, G. Murrell, J. Parker, and M. Whitehorn. 2005. What Henslow Taught Darwin. *Nature* 436: 643–645.

27. Charles Darwin. *The Autobiography of Charles Darwin 1809–1882,* Nora Barlow, editor. WW Norton, New York, 1958, p. 119.

28. Charles Darwin. *The Autobiography of Charles Darwin 1809–1882,* Nora Barlow, editor. WW Norton, New York, 1958, p. 119.

29. Charles Darwin. In: *Charles Darwin's Beagle Diary*, Richard D. Keynes, editor. Cambridge University Press, Cambridge, UK, 1988, p. 447.

30. Charles Darwin. Letter to Robert FitzRoy, 6 October 1836. In: *The Correspondence of Charles Darwin,* Vol. 1: *1821–1836.* Frederick Burkhardt, Sydney Smith, David Kohn, and William Montgomery, editors. Cambridge University Press, Cambridge, UK, 1985, p. 506.

CHAPTER 9: A NEW VOYAGE

1. Francis Darwin, editor. *The life and letters of Charles Darwin, including an autobiographical chapter.* John Murray, London, Vol. 1. p. 298.

2. Charles Darwin. Letter to W. D. Fox, 6 November 1836. In: *The Correspondence of Charles Darwin,* Vol. 1: *1821–1836.* Frederick Burkhardt, Sydney Smith, David Kohn, and William Montgomery, editors. Cambridge University Press, Cambridge, UK, 1985, p. 516.

3. Niles Eldredge. *Darwin: Discovering the Tree of Life.* WW Norton, New York, 2005, pp. 116–121.

4. Charles Darwin. *Journal of Researches into the natural history and geology of the countries visited during the voyage of H.M.S. Beagle round the world, under the Command of Capt. Fitz Roy, R.N.,* 2nd ed. John Murray, London, 1845, p. 380.

5. Charles Darwin. *On the Origin of Species by Means of Natural Selection, or the Preservation of Favoured Races in the Struggle for Life.* John Murray, London, 1859, p. 481.

6. Charles Darwin. Letter to J. D. Hooker, 10 September 1845. In: *The Correspondence of Charles Darwin*, Vol. 3: *1844–1846*, Frederick Burkhardt and Sydney Smith, editors. Cambridge University Press, Cambridge, UK, 1987, p. 253.

7. Charles Darwin. *The Autobiography of Charles Darwin 1809–1882*, Nora Barlow, editor. WW Norton, New York, 1958, pp. 117–124.

8. Charles Darwin. Journal 1837–1843 (DAR 158 in the Cambridge University Library), Appendix II. In: *The Correspondence of Charles Darwin*, Vol. 2: *1837–1843*, Frederick Burkhardt and Sydney Smith, editors. Cambridge University Press, Cambridge, UK, 1986, p. 431.

9. Charles Robert Darwin. 1837. Observations of proofs of recent elevation on the coast of Chili, made during the survey of His Majesty's Ship Beagle commanded by Capt. Fitzroy R.N. *Proceedings of the Geological Society of London* 2: 446–449.

10. Alexander von Humboldt. Letter to Charles Darwin, 18 September 1839. In: *The Correspondence of Charles Darwin*, Vol. 2: *1837–1843*, Frederick Burkhardt and Sydney Smith, editors. Cambridge University Press, Cambridge, UK, 1986, pp. 218–222.

11. Charles Darwin. *The Autobiography of Charles Darwin 1809–1882*, Nora Barlow, editor. WW Norton, New York, 1958, p. 81.

12. R. B. Freeman. *The Works of Charles Darwin: An Annotated Bibliographical Handlist*. 2nd ed. Dawsons, Folkstone, UK, 1977, p. 57,

13. Charles Darwin. *The Autobiography of Charles Darwin 1809–1882*, Nora Barlow, editor. WW Norton, New York, 1958, p. 119.

14. Ibid.

15. Ibid., p. 120.

16. Thomas Malthus. *An Essay on the Principle of Population, as it affects the future improvement of society with remarks on the speculations of Mr. Godwin, M. Condorcet, and other writers*. Printed for J. Johnson in St. Paul's Church-Yard, London, 1798.

17. Charles Darwin. Letter to Ernst Haeckle, after 10 August–8 October 1864. In: *The Correspondence of Charles Darwin*, Vol. 12: *1864*, Frederick Burkhardt, Duncan M. Porter, Sheila Ann Dean, Paul S. White, and Sarah Wilmot, editors. Cambridge University Press, Cambridge, UK 2001, p. 302.

18. Charles Darwin. *On the Origin of Species by Means of Natural Selection, or the Preservation of Favoured Races in the Struggle for Life*. John Murray, London, 1859.

19. Charles Darwin. *The Autobiography of Charles Darwin 1809–1882*, Nora Barlow, editor. WW Norton, New York, 1958, p. 120.

20. Ibid.

21. Francis Darwin, editor. *The foundations of The origin of species. Two essays written in 1842 and 1844*. Cambridge University Press, Cambridge, UK, 1909, p. 7.

22. Charles Darwin. Charles Darwin. *On the Origin of Species by Means of Natural Selection, or the Preservation of Favoured Races in the Struggle for Life*. John Murray, London, 1859, p. 48.

23. Charles Darwin. *Journal of Researches into the natural history and geology of the countries visited during the voyage of H.M.S. Beagle round the world, under the Command of Capt. Fitz Roy, R.N.*, 2nd ed. John Murray, London, 1845, p. 397.

24. Charles Darwin 1872. *The origin of species by means of natural selection, or the preservation of favoured races in the struggle for life*, 6th ed., with additions and corrections. John Murray, London, 1872, pp. 41–42.

25. Charles Darwin. Letter to George Robert Waterhouse, after 22 May 1845. In: *The Correspondence of Charles Darwin*, Vol. 13: *1865*, Frederick Burkhardt, Duncan M. Porter, Sheila Ann Dean, Samantha Evans, Shelley Innes, and Alison M. Pearn, editors. Cambridge University Press, Cambridge, UK, 2002, p. 363.

26. Charles Darwin. Letter to J. D. Hooker, 11–12 July 1845. In: *The Correspondence of Charles Darwin*, Vol. 3: *1844–1846*, Frederick Burkhardt and Sydney Smith, editors. Cambridge University Press, Cambridge, UK, 1987, p. 218.

27. Matthew J. James, editor. *Galápagos Marine Invertebrates Taxonomy, Biogeography, and Evolution in Darwin's Islands*. Plenum, New York, 1991, p. 333.

28. Charles Darwin. Letter to J.D. Hooker, 31 January 1846. In: *The Correspondence of Charles Darwin*, Vol. 3: *1844–1846*, Frederick Burkhardt and Sydney Smith, editors. Cambridge University Press, Cambridge, UK, 1987, p. 279.

29. Joseph Hooker. Letter to Charles Darwin, 28 April 1845. In: *The Correspondence of Charles Darwin*, Vol. 3: *1844–1846*, Frederick Burkhardt and Sydney Smith, editors. Cambridge University Press, Cambridge, UK, 1987, p. 183.

30. Charles Darwin. *Journal of Researches into the natural history and geology of the countries visited during the voyage of H.M.S. Beagle round the world, under the Command of Capt. Fitz Roy, R.N.*, 2nd ed. John Murray, London, 1845, p. 396.

31. J. D. Hooker. Letter to Charles Darwin, after 12 July 1845. In: *The Correspondence of Charles Darwin*, Vol. 3: *1844–1846*, Frederick Burkhardt and Sydney Smith, editors. Cambridge University Press, Cambridge, UK, 1987, p. 221.

32. Duncan Porter. 1980. The Vascular Plants of Joseph Dalton Hooker's *An enumeration of the plants of the Galapagos Archipelago; with descriptions of those which are new. Botanical Journal of the Linnean Society* 81: 79–134 (p. 84).

33. J. D. Hooker. Letter to Charles Darwin, after 12 July 1845. In: *The Correspondence of Charles Darwin*, Vol. 3: *1844–1846*, Frederick Burkhardt and Sydney Smith, editors. Cambridge University Press, Cambridge, UK, 1987, p. 220.

34. Ibid.

35. Duncan Porter. 1980. The Vascular Plants of Joseph Dalton Hooker's *An enumeration of the plants of the Galapagos Archipelago; with descriptions of those which are new. Botanical Journal of the Linnean Society* 81: 79–134.

36. Charles Darwin. Letter to J. D. Hooker, May 1846. In: *The Correspondence of Charles Darwin*, Vol. 3: *1844–1846*, Frederick Burkhardt and Sydney Smith, editors. Cambridge University Press, Cambridge, UK, 1987, p. 317.

37. Charles Darwin. Letter to J. D. Hooker, 8 or 15 July 1846. In: *The Correspondence of Charles Darwin*, Vol. 3: *1844–1846*, Frederick Burkhardt and Sydney Smith, editors. Cambridge University Press, Cambridge, UK, 1987, p. 328.

38. Charles Darwin. Letter to J. D. Hooker, 11 January 1844. In: *The Correspondence of Charles Darwin*, Vol. 3: *1844–1846*, Frederick Burkhardt and Sydney Smith, editors. Cambridge University Press, Cambridge, UK, 1987, p. 2.

39. Ibid.

40. Charles Darwin. *Journal of Researches into the natural history and geology of the countries visited during the voyage of H.M.S. Beagle round the world, under the Command of Capt. Fitz Roy, R.N.,* 2nd ed. John Murray, London, 1845, pp. 378–397.

41. J. D. Hooker. Letter to Charles Darwin, 10 September 1845. In: *The Correspondence of Charles Darwin,* Vol. 3: *1844–1846,* Frederick Burkhardt and Sydney Smith, editors. Cambridge University Press, Cambridge, UK, 1987, p. 253.

42. Ibid.

43. Ibid.

44. Frederick Burkhardt and Sydney Smith. Darwin's Study of the Cirripedia. Appendix II In: *The Correspondence of Charles Darwin,* Vol. 4: *1847–1850,* Frederick Burkhardt and Sydney Smith editors. Cambridge University Press, Cambridge, UK, 1989, p. 390.

45. Charles Darwin. *A monograph on the sub-class Cirripedia, with figures of all the species.* Vol. 1: *The Lepadidæ; or, pedunculated cirripedes.* and Vol. 2: *The Balanidæ, (or sessile cirripedes); the Verrucidæ, etc. etc. etc.* The Ray Society, London, 1851–1854.

46. Twenty-one barnacle species have since been recorded in Galápagos. See: Cleveland P. Hickman, Jr. and Todd L. Zimmerman. *A Field Guide to Crustaceans of Galapagos; An Illustrated Guidebook to the Common Barnacles, Shrimps, Lobsters, and Crabs of the Galapagos Islands.* Sugar Spring Press, Lexington, VA, 2000, pp. 2–4.

47. Janet Browne. *Charles Darwin: Voyaging.* AA Knopf, New York, 1995, p. 473.

48. Charles Darwin. In: *Charles Darwin's Beagle Diary,* Richard D. Keynes, editor. Cambridge University Press, Cambridge, UK, 1988, p. 113.

49. Charles Darwin. *On the Origin of Species by Means of Natural Selection, or the Preservation of Favoured Races in the Struggle for Life.* John Murray, London, 1859, p. 386.

50. Charles Darwin. Letter to John Stevens Henslow, 28 March, 1837. In: *The Correspondence of Charles Darwin,* Vol. 2: *1837–1843,* Frederick Burkhardt and Sydney Smith, editors. Cambridge University Press, Cambridge, UK, 1986 p. 14.

51. Charles Darwin. Letter to J. D. Hooker, May 1846. In: *The Correspondence of Charles Darwin,* Vol. 3: *1844–1846,* Frederick Burkhardt and Sydney Smith, editors. Cambridge University Press, Cambridge, UK, 1987, p. 316.

52. Ibid.

53. Charles Darwin. Letter to W. D. Fox, 22 February 1857. In: *The Correspondence of Charles Darwin,* Vol. 6: *1856–1857,* Frederick Burkhardt and Sydney Smith, editors. Cambridge University Press, Cambridge, UK, 1990, pp. 346–347.

54. Charles Darwin. Letter to James Lamont, 5 March, 1860. In: *The Correspondence of Charles Darwin,* Vol. 8: *1860,* Frederick Burkhardt, Janet Browne, Duncan M. Porter, and Marsha Richmond, editors. Cambridge University Press, Cambridge, UK, 1993, p. 120.

55. Charles Darwin. Letter to William Darwin, 22 September 1858. In: *The Correspondence of Charles Darwin,* Vol. 7: *1858–1859,* Frederick Burkhardt and Sydney Smith, editors. Cambridge University Press, Cambridge, UK, 1992, p. 158.

56. Charles Darwin. *The Autobiography of Charles Darwin 1809–1882,* Nora Barlow, editor. WW Norton, New York, 1958, p. 118.

57. A transcription of Darwin's book on natural selection, commonly referred to as "the long version of *The Origin of Species*," was first published in 1975 by Robert Clinton Stauffer.

58. Charles Darwin. Letter to Charles Lyell, 18 June 1858. In: *The Correspondence of Charles Darwin*, Vol. 7: *1858–1859*, Frederick Burkhardt and Sydney Smith, editors. Cambridge University Press, Cambridge, UK, 1992, p. 107.

59. Robert Chambers. *Vestiges of the natural history of creation*. John Churchill, London, 1844.

60. Charles Darwin. Letter to Charles Lyell. 8 October 1845. In: *The Correspondence of Charles Darwin*, Vol. 3: *1844–1846*, Frederick Burkhardt and Sydney Smith, editors. Cambridge University Press, Cambridge, UK, 1987, p. 258.

61. Robert Chambers. *Vestiges of the natural history of creation*. John Churchill, London, 1844, p. 196.

62. Charles Darwin. Letter to J. D. Hooker, 10 September 1845. In: *The Correspondence of Charles Darwin*, Vol. 3: *1844–1846*, Frederick Burkhardt and Sydney Smith, editors. Cambridge University Press, Cambridge, UK, 1987, p. 253.

63. Charles Darwin. Letter to Charles Lyell, 18 June 1858. In: *The Correspondence of Charles Darwin*, Vol. 7: *1858–1859*, Frederick Burkhardt and Sydney Smith, editors. Cambridge University Press, Cambridge, UK, 1992, p. 107.

64. Ernst Mayr. *Introduction to On the Origin of Species: A Facsimile of the First Edition, by C. Darwin*. Harvard University Press, Cambridge, MA, 1964. See also Frank J. Sulloway, 1984. Darwin and the Galápagos. *Biological Journal of the Linnean Society* 21: 52.

65. Charles Darwin. *On the Origin of Species by Means of Natural Selection, or the Preservation of Favoured Races in the Struggle for Life*. John Murray, London, 1859, pp. 397–400.

66. Ibid., p. 402.

67. Charles Darwin. Letter to Ernst Haeckel, after 10 August–8 October 1864. In: *The Correspondence of Charles Darwin*, Vol. 12: *1864*, Frederick Burkhardt, Duncan M. Porter, Sheila Ann Dean, Paul S. White, and Sarah Wilmot, editors. Cambridge University Press, Cambridge, UK, 2001, p. 302.

68. Charles Darwin. Letter to A. R. Wallace, 6 April 1859. In: *The Correspondence of Charles Darwin*, Vol. 7: *1858–1859*, Frederick Burkhardt and Sydney Smith, editors. Cambridge University Press, Cambridge, UK, 1992, p. 279.

69. Charles Darwin. Letter to Charles Lyell, 27 December 1859. In: *The Correspondence of Charles Darwin*, Vol. 7: *1858–1859*, Frederick Burkhardt and Sydney Smith, editors. Cambridge University Press, Cambridge, UK, 1992, p. 456.

70. Charles Darwin. Letter to Ernst Haeckel, after 10 August–8 October 1864. In: *The Correspondence of Charles Darwin*, Vol. 12: *1864*, Frederick Burkhardt, Duncan M. Porter, Sheila Ann Dean, Paul S. White, and Sarah Wilmot, editors. Cambridge University Press, Cambridge, UK, 2001, p. 302.

71. Charles Darwin. Letter to J. D. Hooker, 8 or 15 July 1846. In: *The Correspondence of Charles Darwin*, Vol. 3: *1844–1846*, Frederick Burkhardt and Sydney Smith, editors. Cambridge University Press, Cambridge, UK, 1987, p. 327.

72. Charles Darwin. Letter to Osbert Salvin, 11 May 1863. In: *The Correspondence of Charles Darwin*, Vol. 11: *1863*, Frederick Burkhardt, Duncan M. Porter, Sheila

Ann Dean, Jonathan R. Topham, and Sarah Wilmot, editors. Cambridge University Press, Cambridge, UK, 1999, pp. 404–405.

73. J. D. Hooker. Letter to Charles Darwin, 24 May 1863. In: *The Correspondence of Charles Darwin*, Vol. 11: *1863*, Frederick Burkhardt, Duncan M. Porter, Sheila Ann Dean, Jonathan R. Topham, and Sarah Wilmot, editors. Cambridge University Press, Cambridge, UK, 1999, p. 443.

74. Osbert Salvin. May 1876. On the Avifauna of the Galápagos Archipelago. *Transactions of the London Zoological Society* 9(9): 509.

75. Charles Robert Darwin. *On the various contrivances by which British and foreign orchids are fertilised by insects, and on the good effects of intercrossing.* John Murray London, 1862.

76. Charles Robert Darwin. *The movements and habits of climbing plants*, 2d ed. John Murray, London, 1875.

77. Charles Robert Darwin. *Insectivorous Plants*. John Murray, London, 1875.

78. Charles Robert Darwin. *The effects of cross and self fertilisation in the vegetable kingdom*. John Murray, London, 1876.

79. Charles Robert Darwin. *The different forms of flowers on plants of the same species*. John Murray, London, 1877.

80. Charles Robert Darwin. *The power of movement in plants*. John Murray, London, 1880.

81. Charles Robert Darwin. *The formation of vegetable mould, through the action of worms, with observations on their habits*. John Murray, London, 1881.

82. Charles Robert Darwin. *The variation of animals and plants under domestication*. John Murray, London, 1868.

83. Charles Robert Darwin. *The descent of man, and selection in relation to sex*, 1st ed. John Murray, London, 1871.

84. Charles Robert Darwin. *The expression of the emotions in man and animals*. John Murray, London, 1872.

85. Charles Robert Darwin. Preliminary notice. In: Krause, E., *Erasmus Darwin. Translated from the German by W. S. Dallas, with a preliminary notice by Charles Darwin*. John Murray, London, 1879.

86. Charles Darwin. Letter to J. D. Hooker, February 3 1868. In: *More letters of Charles Darwin. A record of his work in a series of hitherto unpublished letters*, Vol. 2, Francis Darwin and A. C. Seward, editors. John Murray, London, 1903, p. 9.

87. H. C. Watson. Letter to Charles Darwin, 21 November, 1859. In: *The Correspondence of Charles Darwin*, Vol. 7: *1858–1859*, Frederick Burkhardt and Sydney Smith, editors. Cambridge University Press, Cambridge, UK, 1992, p. 385.

88. Charles Darwin. *The Autobiography of Charles Darwin 1809–1882*, Nora Barlow, editor. WW Norton, New York, 1958, p. 80.

APPENDIX 1: ISLAND AND SITE NAMES IN GALÁPAGOS

1. John Woram. *Human and Cartographic History of the Galápagos Islands*. www .galapagos.to

2. Robert FitzRoy. *Narrative of the surveying voyages of His Majesty's Ships Adventure and Beagle between the years 1826 and 1836, describing their examination of the southern shores of South America, and the Beagle's circumnavigation of the globe: Appendix to Volume II.* Henry Colburn, London, 1839.

3. Charles Darwin. *Notes on the geology of places visited on the voyage.* Darwin Archive, Cambridge University Library DAR 37.2 (1835).

4. *Ship's Log of the Beagle 1831–1836.* ADM53/236 Part 2. Public Records Office, Kew. Transcribed by Thalia Grant.

5. Robert FitzRoy. *Narrative of the surveying voyages of His Majesty's Ships Adventure and Beagle between the years 1826 and 1836, describing their examination of the southern shores of South America, and the Beagle's circumnavigation of the globe: Proceedings of the second expedition, 1831–36, under the command of Captain Robert Fitz-Roy, R.N.* Henry Colburn, London, 1839.

6. Charles Darwin. In: *Charles Darwin's Beagle Diary,* Richard D. Keynes, editor. Cambridge University Press. Cambridge, UK, 1988.

7. British Admiralty Charts L945, L946, L947, L948, L949, L950 printed in 1837 and Chart 1375 printed in 1841. *Galapagos Islands Surveyed by Capt. Robt. Fitz Roy R. N. and the Officers of H. M. S. Beagle, 1836.* Hydrographic Office of Taunton, Taunton, UK.

APPENDIX 3: HMS *BEAGLE*'S COMPLEMENT

1. Persons on board the Beagle, Appendix III. In: *The Correspondence of Charles Darwin,* Vol. 1: *1821–1836,* Frederick Burkhardt, Sydney Smith, David Kohn, and William Montgomery, editors. Cambridge University Press, Cambridge, UK, 1985, pp. 549–551.

ACKNOWLEDGMENTS

1. Robert FitzRoy. Letter to Charles Darwin, 16 November 1837. In: *The Correspondence of Charles Darwin,* Vol. 2: *1837–1843,* Frederick Burkhardt and Sydney Smith, editors. Cambridge University Press, Cambridge, UK, 1986, pp. 57–59.

2. Charles Darwin. *Journal and Remarks.* Volume III of *Narrative of the Surveying Voyages of his Majesty's Ships Adventure and Beagle Between the Years 1826 and 1836, describing their Examination of the Southern Shores of South America and The Beagle's Circumnavigation of the Globe.* Henry Colburn, London, 1839. Reprinted by AMS Press, New York, 1966, pp. vii–viii.

Further Reading

For further reading on Charles Darwin, the voyage of the Beagle, and the Galápagos Islands we suggest the following books, articles and websites.

Janet Browne. *Charles Darwin: Voyaging.* AA Knopf, New York, 1995.

Charles Darwin Foundation (www.darwinfoundation.org).

Charles Darwin. *The Autobiography of Charles Darwin 1809–1882,* Nora Barlow, editor. WW Norton, New York, 1958.

Charles Darwin. In: *Charles Darwin's Beagle Diary,* Richard D. Keynes, editor. Cambridge University Press, Cambridge, UK, 1988.

Charles Darwin. *Voyage of the Beagle,* Janet Browne and Michael Neve, editors. Penguin Books, London, 1989.

Charles Darwin. *The Origin of Species,* Gillian Beer, editor. Oxford University Press. Oxford, UK, and New York, 1996.

Darwin Correspondence Project (www.darwinproject.ac.uk).

Darwin Digital Library (darwinlibrary.amnh.org).

Darwin Online (www.darwin-online.org.uk).

Adrian Desmond and James Moore. *Darwin.* Penguin Books, 1992.

Niles Eldredge. *Darwin, Discovering the Tree of Life.* WW Norton, New York and London, 2005.

Gregory Estes, K. Thalia Grant, and Peter R. Grant. 2000. Darwin in Galápagos: His Footsteps through the Archipelago. *Notes and Records of the Royal Society of London* 54(3): 343–368.

Robert FitzRoy, *Narrative of the Surveying Voyages of his Majesty's Ships Adventure and Beagle Between the Years 1826 and 1836, describing their Examination of the Southern Shores of South America and The Beagle's Circumnavigation of the Globe. Volume II.* Henry Colburn, London, 1839. Reprinted by AMS Press, New York, 1966.

P. R. Grant and B. R. Grant. *How and Why Species Multiply: The Radiation of Darwin's Finches.* Princeton University Press, Princeton, NJ, 2008.

Sandra Herbert. *Darwin, Geologist.* Cornell University Press, Ithaca, NY, and London, 2005.

Sandra Herbert, Sally Gibson, David Norman, Dennis Geist, Greg Estes, Thalia Grant, and Andrew Miles. 2009. Into the Field Again: Re-examining

Charles Darwin's 1825 Geological Work on Isa Santiago (James Island) in the Galápagos Archipelago. *Earth Sciences History* 28(1): 1–31.

Michael H. Jackson. *Galápagos: A Natural History.* University of Calgary Press, Calgary, Alberta, 2004.

Richard Keynes. *Fossils, Finches and Fuegians: Charles Darwin's Adventures and Discoveries on the Beagle, 1832–1836.* HarperCollins, London, 2003.

Edward J. Larson. *Evolution's Workshop: God and Science on the Galapagos Islands.* Basic Books, New York, 2001.

Roger Perry, editor. *Galápagos: Key Environments.* Pergamon Press, Oxford, UK, 1984.

Paul Stewart. *Galápagos: The Islands that Changed the World.* BBC Books, London, 2006.

Jonathan Weiner. *The Beak of the Finch.* Vintage Books, New York, 1994.

John Woram. *Human and Cartographic History of the Galápagos Islands* (www .galapagos.to).

Index

Abingdon Island, 70, 79, 86, 176, 213, 245, 301n120
Acacia, 179. *See also* mimosa
acquired characteristics, 4, 29
adaptation, 16, 26, 28, 55, 60, 89, 95, 148, 159, 192, 193, 322n198
adaptive radiation, 120, 143, 191-193
Admiralty, British, 39-41, 54, 60, 61, 76-78, 100, 211, b/w plate 5B; and instructions for the voyage of the *Beagle*, 39-41, 78, 79, 211
Adventure, schooner, 60, 62
Africa, 35, 52, 55, 211, 219, 222, 237
Agassiz, Louis, 178
albatross, waved (*Phoebastria irrorata*), 159-161, 178, color plate 29A
Albemarle Island, 11, 13, 69, 78, 79, 85-87, 89, 90, 119, 121, 125, 135, 139, 140, 142, 144, 155-173, 176, 178, 183, 187, 188, 196-201, 205, 210, 212, 213, 234, 245, 246, 248, 250, 298n77, color plates 14A, 14B, 15D, 25D
Alcedo volcano, 157
Alsophis spp. *See* snakes
Alternanthera filifolia, 93
Amblyrhynchus cristatus. *See* iguanas, marine
America, whaling ship, 86
Anaximander, 4
Andes, 44, 53, 62, 66, 218, 223
Antillophis spp. *See* snakes
Argentina, 53, 56, 57, 59, 65, 66, 88, 207. *See also* Patagonia
armadillo, 56, 57, 64, 227, 241, b/w plates 10A, 10B
arthropods, 12, 14, 34, 48, 53, 62, 96, 106, 108, 109, 131, 132, 135, 140, 148, 170, 183, 217, 227, 232, 249

Ascension Island, 34, 156
Asilo de Paz, 112, 113, 115, 116, 141, 195, 246, 295n30, 298n82, b/w plates 22A, 22B
astronomy, 30, 209, 210, 324n4. *See also* stars
Australia, 49, 52, 130, 131, 171, 222

Bahía Blanca, 56, 310n69
Bahía de Tortuga Agua Dulce (Chatham Island). *See* Terrapin Road
Bahia, Brazil, 56
Baitis, Harmut Wolfgang, 185
Baltra Island, 87, 134, 156, 178, 303n2
barnacles, 8, 236, 238, 330n46
Barrington Island, 83, 107, 140, 178, 245, 298n76
Bartholomew Island, 78, 245
Bartolomé. *See* Bartholomew Island
basalt, 86, 91, 97, 101, 102, 164, 166, 183, 189, 190, 200, 218, 282n56
bat, Galápagos red (*Lasiurus brachyotis*), 133
Beagle channel, 58
Beagle crater, 78, 162-166, 171, 173, 188, 200, 246, 248, color plates 15D, 25D
Beagle, HMS, b/w plates 6A, 14A, 14B, 16B, 22C, 28A, color plate 4C ; boats of, 44, 50, 65, 75-78, 82, 83, 92, 100-103, 112, 130, 156, 176, 213, 220, 282n56, 290n6, b/w plate 28A; books on board, 35, 44, 55, 56, 68-71, 80, 82, 104, 105, 100, 267n26; Darwin's quarters on, 43, 44, 92, 223, 267n26, b/w plates 6B, 7; design of, 43, 44, b/w plates 6B, 7; first voyage of, 43, 58; officers and crew of, 42, 44, 45, 47-49, 60, 77-79, 96, 97, 99, 101, 130, 140, 166, 167, 171, 206, 212, 213, 251-253, 289n154; route of, 52, 53, color plates 3, 6A; second voyage of, 6, 7, 11, 12, 18-20, 37, 38, 39-72, 78,

Vertebrate Species Collected in Galápagos, on the Voyage of the Beagle*

Modern common name	Modern scientific name	Scientific name in Zoology of the Beagle
Land Birds		
Galápagos hawk	*Buteo galapagoensis*	*Craxirex galapagoensis*
Short eared owl	*Asio flammeus galapagoensis*	*Otus galapagosensis*
Barn owl	*Tyto alba punctatissima*	*Strix piunctatissima*
Galápagos martin	*Progne modesta*	*Progne modesta*
Vermilion flycatcher	*Pyrocephalus rubinus nanus*	*Pyrocepalus nanus*
Vermilion flycatcher (Chatham Island)	*Pyrocephalus rubinus dubius*	*Pyrocepalus dubius*
Galápagos flycatcher	*Myiarchus magnirostris*	*Myiobius magnirostris*
Chatham Island mockingbird	*Mimus melanotis*	*Mimus melanotis*
Charles Island mockingbird	*Mimus trifasciatus*	*Mimus trifasciatus*
Galápagos mockingbird	*Mimus parvulus*	*Mimus parvulus*
Yellow warbler	*Dendroica petechia aureola*	*Sylvicola aureola*
Large ground finch	*Geospiza magnirostris*	*Geospiza magnirostris* *Geospiza strenua*
Medium ground finch	*Geospiza fortis*	*Geospiza fortis* *Geospiza dubia* *Geospiza dentirostris*
Small ground finch	*Geospiza fuliginosa*	*Geospiza fuliginosa*
Sharp-beaked ground finch	*Geospiza difficilis*	*Geospiza nebulosa*
Cactus finch	*Geospiza scandens*	*Cactornis scandens* *Cactornis assimilis*
Small tree finch	*Camarhynchus parvulus*	*Geospiza parvula*
Large tree finch	*Camarhynchus psittacula*	*Camarhynchus psittaculus*
Vegetarian tree finch	*Platyspiza crassirostris*	*Camarhynchus crassirostris*

Modern common name	Modern scientific name	Scientific name in Zoology of the Beagle
LAND BIRDS (*continued*)		
Warbler finch	*Certhidia olivacea*	*Certidea olivacea*
Bobolink	*Dolichonyx oryzivorus*	*Dolichonyx oryzivorus*
Galápagos dove	*Zenaida galapagoenis*	*Zenaida galapagoenis*
SHORE BIRDS/WATER BIRDS		
Wandering tattler	*Tringa incana*	*Totanus fuliginosus*
Least sandpiper	*Calidris mutilla*	*Pelidna minutilla*
Ruddy turnstone	*Arenaria interpres*	*Strepsilas interpres*
Galápagos rail	*Laterallus spilonotus*	*Zapornia spilonota*
White-cheeked pintail duck	*Anas bahamensis galapagensis*	*Pœcilonitta bahamensis*
Lava gull	*Leucophaeus fuliginosus*	*Larus fuliginosus*
Brown noddy	*Anous stolidus galapagensis*	*Megalopterus stolidus*
Greater flamingo (stomach contents only)	*Phoenicopterus ruber*	
Semi-palmated plover	*Charadrius semipalmatus*	*Hiaticula semipalmata*
Great blue heron	*Ardea herodias*	*Ardea herodias*
Yellow-crowned night heron	*Nyctanassa violacea*	*Nycticorax violaceus*
REPTILES		
Charles Island lava lizard	*Microlophus grayi*	*Leiocephalus grayii*
Chatham Island lava lizard	*Microlophus bivittatus*	*Leiocephalus grayii*
Marine iguana	*Conolophus subcristatus*	*Amblyrynchus cristatus*
Land iguana	*Amblyrhynchus cristatus*	*Amblyrynchus demarlii*
Striped Galápagos snake	*Antillophis steindachneri*	
Eastern Galápagos racer snake	*Alsophis biseralis biseralis*	
James Island tortoise (live young)	*Geochelone nigra darwini*	
Hood Island tortoise (live young)	*Geochelone nigra hoodensis*	
Charles Island tortoise (live young)	*Geochelone nigra galapagoensis*	

Modern common name	Modern scientific name	Scientific name in Zoology of the Beagle
MAMMALS		
Black rat (from James Island)	*Rattus rattus*	*Mus jacobiae*
Chatham Island rice rat	*Oryzomys galapagoensis*	*Mus galapagoensis*
FISH		
Bullseye puffer	*Sphoeroides annulatus*	*Tetrodon annulatus*
Lentil Moray	*Muraena lentiginosa*	*Muræna lentiginosa*
Pacific red sheephead	*Semicossyphus darwini*	*Cossyphus Darwini*
Southern frillfin	*Bathygobius lineatus*	*Gobius lineatus*
Galápagos porgy	*Calamus taurinus*	*Chrysophrys taurina*
Ocean whitefish	*Caulolatilus princeps*	*Latilus princeps*
Misty grouper	*Epinephelus mystacinus*	*Prionodes fasciatus*
Bandin scorpionfish	*Scorpaena histrio*	*Scorpæna Histrio*
Bacalao	*Mycteroperca olfax*	*Serranus olfax*
Flag cabrilla grouper	*Epinephelus labriformis*	*Serranus labriformis*
Orangethroat searobin	*Prionotus miles*	*Prionotus miles*
Red clingfish	*Arcos poecilophthalmus*	*Gobiesox pœcilophthalmos*
Concave puffer	*Sphoeroides angusticeps*	*Tetrodon angusticeps*
Sheephead grunt	*Orthopristis cantharinus*	*Pristipoma cantharinum*
Whitespotted sand bass	*Paralabrax albomaculatus*	*Serranus albo-maculatus*

* Darwin also observed but did not collect: frigatebird spp. (*Fregata minor* and/or *F. magnificens*), blue-footed boobies (*Sula nebouxii*), Nazca boobies (*Sula granti*), brown pelicans (*Pelecanus occidentalis urinator*), "mother carys chickens" (unspecified storm petrel spp.), Procellaria (most likely Audubon's shearwater, *Puffinus lherminieri subalaris*), Galápagos lava lizards (*Microlophus albemarlensis*) from Albemarle and James Islands, and Pacific green sea turtles (*Chelonia mydas agassisi*). He apparently saw the striated heron (*Butorides striata*) because in addition to the great blue heron, he recorded "two kinds of Bittern" but collected only one (the yellow-crowned night heron). Captain FitzRoy recorded sealions (*Zalophus wollebaeki*) on Barrington Island. The crew of the *Beagle* collected tortoises (*Geochelone nigra chathamensis*) from Chatham Island for food.

Illustration Credits

~~~~~~

BLACK AND WHITE PLATES

1A. Charles Darwin, age 31. Sketch for a portrait by George Richmond. From: *Charles Darwin's diary of the voyage of H.M.S.* Beagle. Nora Barlow, editor. Cambridge University Press, Cambridge, UK, 1933. (Authors' collection)

2A. "Charles Darwin (aged six) and Catherine," by Sharples, circa 1816. (Courtesy of Darwin Heirlooms Trust)

2B. The Mount, Shrewsbury. From: *The Sphere*, July 18, 1903. (Authors' collection)

3A. Royal Grammar School, Shrewsbury. From: *The Illustrated London News*, December 14, 1861. (Authors' collection)

3B–D. Beetles. From: *The British* Coleoptera *Delineated, consisting of Figures of All the Genera of British Beetles, Drawn in Outline by W. Spry*. W. E. Shuckard, editor. W. Crofts, Chancery Lane, London, 1840.

3E. Dr. Robert Darwin. A silhouette, circa 1826. From: *The autobiography of Charles Darwin 1809–1882. With the original omissions restored*. Nora Barlow, editor. Collins, London, 1958. (Authors' collection)

4A. Adam Sedgwick. From a portrait painted by Thomas Phillips, R. A., 1832. In: John Willis Clark and Thomas McKenny Hughes. *The Life and Letters of The Reverend Adam Sedgwick*. Cambridge University Press, Cambridge, UK 1890.

4B. John Stevens Henslow. Illustration by T. H. Maguire, 1849. In: *Darwin and Henslow: The growth of an idea*. Nora Barlow, editor. Bentham-Moxon Trust, John Murray, London, 1967.

4C. Charles Lyell. In: Francis Darwin. *More Letters of Charles Darwin, Volume II.* John Murray, London, 1903.

4D. Joseph Dalton Hooker. From a portrait by George Richmond. 1855. In: Leonard Huxley. *Life and Letters of Sir Joseph Dalton Hooker, Volume 1.* John Murray, London, 1918.

5A. Captain Robert FitzRoy, Commander of HMS *Beagle*, circa 1836. In: *Darwin and Henslow: The growth of an idea*. Nora Barlow, editor. Bentham-Moxon Trust, John Murray, London, 1967.

5B. "The Admiralty-Board Room. Meeting of the Lords of the Admiralty." By Thomas Rowlandson. Engraved by Henry Melville. Published for the Proprietors by J. Mead, 10 Gough Square, Fleet Street, London, 1843. (Authors' collection)

6A. HMS *Beagle* in the straits of Magellan. From: Charles Darwin. *Journal of Researches into the natural history and geology of the various countries visited by H.M.S. Beagle etc.* John Murray, London, 1890. (Authors' collection)

6B. Diagram of HMS *Beagle*. From: Charles Darwin. *Journal of Researches into the natural history and geology of the various countries visited by H.M.S. Beagle etc.* John Murray, London, 1890. (Authors' collection)

7A. Diagram of HMS *Beagle*. From: *Charles Darwin's diary of the voyage of H.M.S. Beagle.* Nora Barlow, editor. Cambridge University Press, Cambridge, UK, 1933. (Authors' collection)

8A. "The Breakwater, From Mount Edgcumbe, Plymouth." In: *Devonshire and Cornwall Illustrated, with Original Drawings by Thomas Allom, W. H. Bartlett etc. With historical and topographical descriptions by J. Britton and E. W. Brayley.* H. Fisher, R. Fisher, and P. Jackson, Newgate St., London, 1832. (Authors' collection)

8B. "Falmouth Harbour." (Authors' collection)

9A. "Waldscenerie aus dem Küftengebirge Venezuela. Nach der Natur gezeichnet von U. Göring." In: *Meisterwerke der Holzschneidekunst,* J. J. Weber, Leipzig, 1882. (Authors' collection)

10A. "Extinct gigantic Armadillo (*Glyptodon clavipes*)." From: Richard Owen. *Palaeontology or a Systematic Summary of Extinct Animals and their Geological Relations.* Adam and Charles Black, Edinburgh, 1860. (Courtesy of Paul D. Stewart picture collection)

10B. "Hairy Armadillo." Lascelles & Co. Ltd, Willesden, 1902. (Authors' collection)

10C. "The Hunterian Museum, at the Royal College of Surgeons." *Illustrated London News,* October 4, 1845. (Courtesy of Paul D. Stewart picture collection)

11A. Fuegia Basket, Jemmy Button, and York Minster. From Robert FitzRoy. *Proceedings of the second expedition, 1831–36, under the command of Captain Robert Fitz-Roy, R.N.* Henry Colburn, London, 1839.

11B. "Natives of Tierra del Fuego." *Saturday Magazine,* 1841. From a drawing by Robert FitzRoy in *Proceedings of the second expedition, 1831–36, under the command of Captain Robert Fitz-Roy, R.N.* Henry Colburn, London, 1839. (Authors' collection)

12A. "Glacier of Mount Sarmiento," c. 1890. From an illustration by Conrad Martens. (Authors' collection)

12B. "Berkeley Sound, Falkland Islands." From: Charles Darwin. *Journal of researches into the natural history and geology of the various countries visited by H.M.S. Beagle etc.* John Murray, London, 1890. (Authors' collection)

12C. "Settlement at Port Louis." From: Robert FitzRoy. *Proceedings of the second expedition, 1831–36, under the command of Captain Robert Fitz-Roy, R.N.* Henry Colburn, London, 1839.

13A. "The Inca's Bridge (Natural Formation), Pass of Uspallata, South America." From: *The Illustrated London News,* 1868. (Authors' collection)

13B. "Travelling in the Andes." From: Charles Darwin. *Journal of Researches into the natural history and geology of the various countries visited by H.M.S. Beagle etc.* Thomas Nelson, London, 1890. (Authors' collection)

14A. "Coquimbo, Chile." From: Charles Darwin. *Journal of researches into the natural history and geology of the various countries visited by H.M.S. Beagle etc.* John Murray, London, 1890. (Authors' collection)

14B. "Homeward bound." From: Charles Darwin. *Journal of researches into the natural history and geology of the various countries visited by H.M.S. Beagle etc.* John Murray, London, 1890. (Authors' collection)

15A. Map of Galápagos drawn by Emanuel Bowen 1744, after an earlier map produced by Herman Moll in 1699. From: John Harris. *Navigantium atque Itinerantium Bibliotheca, Volume 1.* T. Woodward et al., London. (Authors' collection)

15B. "Chart of the Galapagos, Surveyed in the Merchant-Ship Rattler, and Drawn by Capt James Colnett, of the Royal Navy, in 1793 1794." Engraved by T. Foot, Weston Place, St. Pancras, London. Published 1st January 1798 by A. Arrowsmith, Rathbone Place. Courtesy of Library of Congress. (Authors' collection)

16A. Fresh Water Bay, Chatham Island. Engraving by E. De Bérard and Sargent, from a sketch by P. G. King engraved by S. Bull and published in: Robert FitzRoy. *Proceedings of the second expedition, 1831–36, under the command of Captain Robert Fitz-Roy, R.N.* Henry Colburn, London, 1839. In: Voyages d'un naturaliste: l'Archipel Galapagos et les attolls de coraux. *Le Tour du Monde*, 1860. Edited by Édouard Charton. (Authors' collection)

16B. Leaving Chatham Island. Engraving by E. De Bérard and Sargent, from a sketch by P.G. King engraved by S. Bull and published in: Robert FitzRoy. *Proceedings of the second expedition, 1831-36, under the command of Captain Robert Fitz-Roy, R.N.* Henry Colburn, London, 1839. In: Voyages d'un naturaliste: l'Archipel Galapagos et les attolls de coraux. *Le Tour du Monde*, 1860. Edited by Édouard Charton. (Authors' collection)

17A. Giant tortoises. In: Joseph B. Holder. *Animate Creation (A Popular edition of Our Living World), Volume III.* Selmar Hess, New York, 1885. (Authors' collection)

17B. "Testudo Abingdonii." From: Charles Darwin. *Journal of Researches into the natural history and geology of the various countries visited by H.M.S. Beagle etc.* John Murray, London, 1890. (Authors' collection)

17C. Turning a tortoise with a boat hook. *Illustrated London News*, July 13, 1850. (Authors' collection)

18A. Kicker Rock. From: Charles Darwin. *Geological observations on the volcanic islands visited during the voyage of H.M.S. Beagle, together with some brief notices of the geology of Australia and the Cape of Good Hope. Being the second part of the geology of the voyage of the Beagle, under the command of Capt. Fitzroy, R.N. during the years 1832 to 1836.* Smith Elder, London, 1844. (Authors' collection)

18B. Kicker Rock. Photograph by the authors, 1996.

19A. "Scorpæna Histrio." From: Charles Darwin, editor. *Fish Part 4 No. 1 of The zoology of the voyage of H.M.S. Beagle. By Leonard Jenyns. Edited and superintended by Charles Darwin.* Smith Elder, London, 1841.

19B. "Cossyphus Darwini." From: Charles Darwin, editor. *Fish Part 4 No. 3 of The zoology of the voyage of H.M.S. Beagle. By Leonard Jenyns. Edited and superintended by Charles Darwin.* Smith Elder, London, 1841.

19C. *"Tetrodon Angusticeps."* From: Charles Darwin, editor. *Fish Part 4 No. 4 of The zoology of the voyage of H.M.S. Beagle.* By Leonard Jenyns. Edited and superintended by Charles Darwin. Smith Elder, London, 1841.

20A. One of the authors (GBE) in the Craterized District, Chatham Island. Photograph by the authors, 1996.

20B. View of the Chaine des Puys near Clermont. In: George Poulett Scrope. *Considerations on Volcanos.* W. Phillips, George Yard, Lombard Street, London, 1825.

20C. View of "The Phlegræan Fields." In: Charles Lyell. *Principles of Geology, Volume 1.* John Murray, Albemarle Street, London, 1830.

20D. Finger Hill (Cerro Brujo). Photograph by the authors, 1996.

21A. Folio 753 of Darwin's notes on the geology of Galápagos (DAR 37.2). Reproduced by kind permission of the Syndics of Cambridge University Library.

21B–E. Sketches from Darwin's notes on the geology of Galápagos (DAR 37.2), redrawn by K. Thalia Grant.

22A. Charles Island. Engraving by S. Bull after P. G. King. In: Robert FitzRoy. *Proceedings of the second expedition, 1831–36, under the command of Captain Robert Fitz-Roy, R.N.* Henry Colburn, London, 1839.

22B. Charles Island. Engraving by E. De Bérard and Sargent, from a sketch by P. G. King engraved by S. Bull and published in: Robert FitzRoy. *Proceedings of the second expedition, 1831–36, under the command of Captain Robert Fitz-Roy, R.N.* Henry Colburn, London, 1839. In: Voyages d'un naturaliste: l'Archipel Galapagos et les attolls de coraux. *Le Tour du Monde,* 1860. Edited by Édouard Charton. (Authors' collection)

22C. Leaving Charles Island. Engraving by E. De Bérard and Sargent, from a sketch by P. G. King engraved by S. Bull and published in: Robert FitzRoy. *Proceedings of the second expedition, 1831–36, under the command of Captain Robert Fitz-Roy, R.N.* Henry Colburn, London, 1839. In: Voyages d'un naturaliste: l'Archipel Galapagos et les attolls de coraux. Le Tour du Monde, 1860. Edited by Édouard Charton. (Authors' collection)

23A–C. Details from paintings by Thalia Grant. (Authors' collection)

23D–F. Radiographs by Christine Parent. Courtesy of Christine Parent.

24A. Engraving by E. De Bérard and Huyot. In: Voyages d'un naturaliste: l'Archipel Galapagos et les attolls de coraux. *Le Tour du Monde,* 1860. Edited by Édouard Charton. (Authors' collection)

24B. "The Whale Fishery 'Laying On.' " Lithograph by Nathaniel Currier, 1852. (Authors' collection)

25A. Land iguana (*Conolophus subcristatus*) and small ground finch (*Geospiza fuliginosa*). Illustration by Thalia Grant. Reprinted with permission, from: Jonathan Weiner. *The Beak of the Finch.* AA Knopf, New York, 1994.

25B. Marine iguana (*Amblyrhynchus cristatus*). From: Charles Darwin. *Journal of Researches into the natural history and geology of the various countries visited by H.M.S. Beagle etc.* John Murray, London, 1890. (Authors' collection)

25C. Marine iguana (*Amblyrhynchus cristatus*). Engraving by Rouyer and Huyot. In: Voyages d'un naturaliste: l'Archipel Galapagos et les attolls de coraux. *Le Tour du Monde,* 1860. Edited by Édouard Charton. (Authors' collection)

26A. "Landpadda." From: Carl Skogman and C. A. Virgin. *Fregatten Eugenies Resa Omkring Jorden.* Adolf Bonnier, Stockholm, 1855. (Authors' collection)

26B. Galápagos tortoise (*Geochelone nigra*) and small ground finch (*Geospiza fulginosa*). Illustration by Thalia Grant. Reprinted with permission, from: Jonathan Weiner. *The Beak of the Finch.* AA Knopf, New York, 1994.

27A. Engraving by Rouyer and Huyot. In: Voyages d'un naturaliste: l'Archipel Galapagos et les attolls de coraux. *Le Tour du Monde,* 1860. Edited by Édouard Charton. (Authors' collection)

27B. Cactus finch (*Geospiza scandens*). Illustration by Thalia Grant. Reprinted with permission, from: Jonathan Weiner. *The Beak of the Finch.* AA Knopf, New York, 1994.

28A. "Darwin's shore party returning to Beagle." Ink drawing by John Chancellor. (By courtesy of the trustees of John Chancellor)

28B. Illustrations by Thalia Grant. Reprinted with permission, from: Jonathan Weiner. *The Beak of the Finch.* AA Knopf, New York, 1994.

28C. Darwin's finches. From: Charles Darwin. *Journal of Researches into the natural history and geology of the various countries visited by H.M.S. Beagle etc.* John Murray, London, 1890. (Authors' collection)

29A. "British Museum—Zoological Gallery." Drawn by L. Jewitt. Engraved by Radclyffe. Published for the Proprietor by Joseph Meade, 10 Gough Square, Fleet Street, London, 1844. (Authors' collection)

29B. "Darwin's Theory Illustrated by Pigeons." From: *The Illustrated London News,* December 10, 1887. (Authors' collection)

30A. "Darwin's House at Down, near Beckenham, Kent." From: *The Illustrated London News,* December 10, 1887. (Authors' collection)

30B. "Darwin's Green-House." From: *The Illustrated London News,* December 10, 1887. (Authors' collection)

30C. Darwin's study. Copper engraving by Axel H. Haig. (Courtesy of Paul D. Stewart picture collection)

30D. "Darwin's Usual Walk." From: *The Illustrated London News,* December 10, 1887. (Authors' collection)

31A. Alfred Russel Wallace, 1869. (Courtesy of Paul D. Stewart picture collection)

31B. "View taken in a forest near Kupang, Timor," circa 1892, engraved by F. Meaulle. From: *The Universal Geography: Earth and its Inhabitants, Volume 14: Australasia.* Elisée Recluse, editor. JS Virtue, London. (Authors' collection)

32A. Charles Robert Darwin. Rotary photograph 2308, circa 1878. (Authors' collection)

COLOR PLATES
Photographs by the authors, unless otherwise indicated.

Cover. HMS *Beagle* in the Galápagos, 17 October 1835, 2.15 p.m. By John Chancellor. (By courtesy of the trustees of John Chancellor)
Charles Darwin, 1840. By George Richmond. (Courtesy of Darwin Heirlooms Trust)

1A. "English Bridge, Shrewsbury." Engraved by T. Radclyffe. Drawn by F. Calvert. From: *Picturesque Views and Description of Cities, Towns, Castles, Mansions, and other Objects of Interesting Features in Staffordshire and Shropshire.* W. Emans, Bromsgrove St., Birmingham, UK, 1830. (Authors' collection)
1B. "Wolverhampton, from the Penn Road." Engraved by T. Radclyffe. Drawn by F. Calvert. From: *Picturesque Views and Description of Cities, Towns, Castles, Mansions, and other Objects of Interesting Features in Staffordshire and Shropshire.* W. Emans, Bromsgrove St., Birmingham, UK, 1830. (Authors' collection)
1C. The Craterized District, Chatham Island, 1996.
2A. "The University, South Bridge Street, Edinburgh." Engraved by W. H. Lizars. Drawn by Thomas H. Shepherd. From: *Modern Athens! Displayed in a Series of Views, Or Edinburgh in the Nineteenth Century.* London, 1829–1831 (Authors' collection)
2B. "Christs College, 1838." Engraved by J. LeKeux after a picture by I.A. Bell. In: *Cambridge Described and Illustrated.* MacMillan, London, 1897. (Authors' collection)
3A. Route of HMS *Beagle*'s second surveying voyage. In: Charles Darwin. *Journal of Researches into the natural history and geology of the various countries visited by H.M.S. Beagle etc.* John Murray, London, 1890. (Authors' collection)
4A. "Der Pic Von Teneriffa," 1839. From the Art Foundation of the Bibliographic Institute, Hildburghausen. Published in Herman J. Meyer. *Universum 1833–1861.* (Authors' collection)
4B. "*Delphinus Fitz-Royi*" (Dusky dolphin, *Lagenorhynchus obscurus*). Charles Darwin, editor. *Mammalia Part 2 No. 1 of The zoology of the voyage of H.M.S. Beagle.* By George R. Waterhouse. Edited and superintended by Charles Darwin. Smith Elder, London, 1838.
4C. "Sorely Tried." HMS *Beagle* off Cape Horn, 13 January 1833 at 1:45 p.m. By John C. Chancellor. (By courtesy of the trustees of John Chancellor)
5A. "Squelette de Megatherium et Megatherium restauré" by Oudart and Fournier. Plate 5 in: Charles d'Orbigny. *Dictionnaire d'Histoire Naturelle: Mammiferes fossils,* 1849. (Authors' collection)
5B. South American rheas, by H. Grönvold. Plate 1 in: Wyndham Brabourne and Charles Chubb. *The Birds of South America,* Volume II. RH Porter, London, 1915–1917. (Authors' collection)
5C. "*Canis antarcticus*" (Falkland Island fox, *Dusicyon australis*). Charles Darwin, editor. *Mammalia Part 2 No. 1 of The zoology of the voyage of H.M.S. Beagle.* By George R. Waterhouse. Edited and superintended by Charles Darwin. Smith Elder, London, 1838.
6A. *Galápagos Islands, surveyed by Captain FitzRoy, R.N., and the Officers of H.M.S. Beagle with additions from later Admiralty Surveys.* Published at the Admiralty 9th August, 1886, under the Superintendence of Captain W. J. L. Wharton, R.N: F.R.S. Hydrographer. Courtesy of Library of Congress. (Route of HMS *Beagle* added by the authors)
6B. Sulivan Bay, James Bay. Detail from: *Galapagos Islands surveyed by Capt Robt Fitz Roy, R.N., and the Officers of H.M.S.* Beagle. London Published according to Act of Parliament at the Hydrographic Office of the Admiralty Jan 15th, 1841. (Courtesy of Library of Congress)

7A. Pan de Azucar, Chatham Island, 1996.

7B. *"Mimus melanotis"* (Chatham Island mockingbird). Charles Darwin, editor. *Birds Part 3 No. 2 of The zoology of the voyage of H.M.S. Beagle.* By John Gould. Edited and superintended by Charles Darwin. Smith Elder, London, 1839.

7C. Finger Hill (Cerro Brujo), Chatham Island.

7D. *"Mus Galapagoensis"* (Chatham Island rice rat, *Oryzomys galapagoensis*). Charles Darwin, editor. *Mammalia Part 2 No. 3 of The zoology of the voyage of H.M.S. Beagle.* By George R. Waterhouse. Edited and superintended by Charles Darwin. Smith Elder, London, 1838.

8A. Marine iguana (*Amblyrhynchus cristatus*) feeding underwater.

8B. Marine iguana head (*Amblyrhynchus cristatus*).

8C. Pacific green turtle (*Chelonia mydas agassisi*).

9A. Chatham Island tortoise (*Geochelone nigra chatamensis*), La Galapaguera Natural, Chatham Island, 1996.

9B. Striped Galápagos snake (*Antillophis steindachneri*), James Island.

9C. Sally Lightfoot crab (*Grapsus grapsus*).

10A. Darwin's cotton (*Gossypium darwinii*).

10B. Thin-leafed Darwin's shrub (*Darwiniothamnus tenuifolius*).

10C. Macrea (*Macrea laricifolia*).

10D. Wing-fruited lecocarpus (*Lecocarpus pinnatifidus*).

10E. Galápagos carpetweed (*Sesuvium edmonstonei*).

11A. Round Hill (*Cerro Pajas*), Charles Island.

11B. Snail (*Bulimuls ochsneri*) from Indefatigable Island. (Photograph courtesy of Christine Parent)

11C. The authors' young daughter mailing a letter at the post office barrel at Post Office Bay, Charles Island.

11D. Hood Island tortoise (*Geochelone nigra hoodensis*).

11E. *"Mimus trifasciatus"* (Charles Island mockingbird). From: Charles Darwin, editor. *Birds Part 3 No. 2 of The zoology of the voyage of H.M.S. Beagle.* By John Gould. Edited and superintended by Charles Darwin. Smith Elder, London, 1839.

11F. *"Zapornia spilonota"* (Galápagos rail, *Laterallus spilonotus*). From: Charles Darwin, editor. *Birds Part 3 No. 5 of The zoology of the voyage of H.M.S. Beagle.* By John Gould. Edited and superintended by Charles Darwin. Smith Elder, London, 1841.

12A. Chatham Island mockingbird (*Mimus melanotis*), Chatham Island.

12B. Charles Island mockingbird (*Mimus trifasciatus*), Champion Islet off Charles Island.

12C. Galápagos mockingbird (*Mimus parvulus*), Tagus Cove, Albemarle Island.

12D. Galápagos mockingbird (*Mimus parvulus*), Playa Espumilla, James Island.

13A. Small ground finch (*Geospiza fulginosa*), Tagus Cove, Albemarle Island.

13B. Medium ground finch (*Geospiza fortis*), Daphne Island. (Photograph courtesy of Rosemary Grant)

13C. Warbler finch (*Certhidia olivacea*), Tower Island.

13D. Large ground finch (*Geospiza magnirostris*), Daphne Island. (Photograph courtesy of Rosemary Grant)

13E. Sharp-beaked ground finch (*Geospiza difficilis*), Tower Island.

14A. Eruption of Sierra Negra, Albemarle Island, October 2005.

14B. Craterized District, near Point Christopher, Albemarle Island, 1996.

14C. Ropey (pahoehoe) lava, Sulivan Bay, James Island.

15A. Tagus Cove and Darwin Lake during an El Niño year (1983), looking much greener than when Darwin visited the island.

15B. "*Cactornis scandens*" (Cactus finch, *Geospiza scandens*). Female (brown) and male (black). From: Charles Darwin, editor. 1841. *Birds Part 3 No. 5 of The zoology of the voyage of H.M.S.* Beagle, by John Gould. Edited and superintended by Charles Darwin. Smith Elder, London, 1841.

15C. "*Mimus parvulus*" (Galápagos mockingbird). From: Charles Darwin, editor. *Birds Part 3 No. 2 of The zoology of the voyage of H.M.S.* Beagle. By John Gould. Edited and superintended by Charles Darwin. Smith Elder, London, 1839.

15D. The lake in Beagle Crater, Albemarle Island, 1996.

16A. Female Chatham Island lava lizard (*Microlophus bivittatus*).

16B. Male Charles Island lava lizard (*Microlophus grayi*).

16C. Vermilion flycatcher (*Pyrocephalus rubinus nanus*) on a Galápagos tortoise (*Geochelone nigra porter*), highlands of Indefatigable Island.

16D. Land iguana (*Conolophus subcristatus*), Urvina Bay, Albemarle Island.

17A. Freshwater cove of the Buccaniers (Buccaneer Cove), James Island, from the general area of Darwin's campsite, 2007.

17B. "*Graxirex Galapagoesnis*" (Galápagos hawk, *Buteo galapagoensis*). Charles Darwin, editor. *Birds Part 3 No. 1 of The zoology of the voyage of H.M.S.* Beagle. By John Gould. Edited and superintended by Charles Darwin. Smith Elder, London, 1838.

17C. "Darwin's layer cake," Freshwater cove of the Buccaniers (Buccaneer Cove), James Island, 2007.

17D. Pinnacle in Freshwater Cove of the Buccaniers (Buccaneer Cove), James Island, 2007.

18A. Prickly pear tree (*Opuntia galapageia*), Buccaneer Cove, James Island, 2007.

18B. Drawing of *Opuntia galapageia*, after a sketch published by John Stevens Henslow in his 1837 article: Description of Two New Species of Opuntia; with Remarks on the Structure of the Fruit of Rhipsalis. *Magazine of Zoology and Botany 1:466–469*. From: *Charles Darwin. Journal of Researches into the natural history and geology of the various countries visited by H.M.S. Beagle etc.* John Murray, London, 1890. (Authors' collection)

18C. Flower of *Opuntia echios*, South Plaza Island.

18D. Tree scalesia (*Scalesia pedunculata*) flower, James Island.

18E. One of Darwin's specimens of tree scalesia (*Scalesia pedunculata*) from James Island. Housed in the University Herbarium of the University of Cambridge.

19A. Bulimulus snail (*B. reibischi*), Indefatigable Island. (Photograph courtesy of Christine Parent)

19B. Lichen (*Ramelina usnea*), James Island, 2007.

19C. Guayabillo (*Psidium galapageium*) berries, James Island.

19D. James Island tortoise (*Geochelone nigra darwini*), Highlands of James Island, 2007.

19E. Guayabillo (*Psidium galapageium*) tree, James Island, 2007.

20A. Cerro Pelado, highlands of James Island, 2007.

20B. *Scalesia pedunculata* trees, highlands of James Island, 2007.

20C. Feral goat (*Capra hircus*) skull, Playa Espumilla, James Island, 2007.

20D. Feral goats (*Capra hircus*), Urvina Bay, Albemarle Island.

21A. James Bay lava flow and Pan de Azucar, James Island, 2007.

22B. James Bay lava flow and highlands, James Island, 2007.

22C. Looking down toward Pan de Azucar from highlands of James Island, 2007.

22A. Cerro Cowan, James Island, 1996.

22B. "Volcanic sandstone" (tuff) specimen collected by Darwin from Pan de Azucar, Chatham Island. (Housed in The Sedgwick Museum of Earth Sciences, University of Cambridge)

22C. The Salina (Salt Mine) with Pan de Azucar in the distance, 2007.

23A. Galápagos hawk (*Buteo galapagoensis*), James Island.

23B. Barn owl (*Tyto alba punctatissima*), Indefatigable Island.

23C. Galápagos dove (*Zenaida galapagoensis*), James Island.

23D. Vermilion flycatcher (*Pyrocephalus rubinus nanus*), Indefatigable Island, 2004.

23E. Short-eared owl (*Asio flammeus galapagoensis*).

23F. Yellow warbler (*Dendroica petechia aureola*), Chatham Island.

24A. Male magnificent frigatebird (*Fregata magnificens*).

24B. Yellow-crowned night heron (*Nyctanassa violacea*), Tower Island.

24C. Brown pelican (*Pelecanus occidentalis urinator*).

24D. Striated (lava) heron (*Butorides striata*).

24E. Great blue heron (*Ardea herodias*). Painting by Thalia Grant.

24F. Lava gull (*Leucophaeus fuliginosus*). Painting by Thalia Grant.

25A. One of Darwin's large ground finch (*Geospiza magnirostris*) specimens. (Housed in the University Museum of Zoology, Cambridge)

25B. Darwin's specimen of an endemic fern (*Ctenitis pleiosoros*) from James Island. (Housed in the University Herbarium of the University of Cambridge)

25C. One of Darwin's fish specimens from the Voyage of the Beagle. (Housed in the University Museum of Zoology, Cambridge)

25D. One of Darwin's rock specimens, collected from the lava flow behind Beagle Crater, Albemarle Island. (Housed in the Sedgwick Museum of Earth Sciences, University of Cambridge)

25E. Darwin's geological specimens notebook. (The Sedgwick Museum of Earth Sciences, University of Cambridge)

26A. Castela (*Castela galapageia*), Indefatigable Island.

26B. Matazarno (*Piscidia carthagenensis*), Chatham Island.

26C. Hairy Galápagos tomato (*Lycopersicon cheesmanii* var. *minor*), James Island.

26D. Yellow cordia (*Cordia lutea*).

26E. Radiate-headed scalesia (*Scalesia affinis*), Sierra Negra, Albemarle Island.

27A. Hood Island tortoise (*Geochelone nigra hoodensis*).

27B. Chatham Island tortoise (*Geochelone nigra chathamensis*).

27C. James Island tortoise (*Geochelone nigra darwini*).

27D. Indefatigable Island tortoise (*Geochelone nigra porteri*).

28A. Galápagos Islands surveyed by Captain Robert FitzRoy, R.N., and the Officers of H.M.S. *Beagle*. London Published according to Act of Parliament at the Hydrographic Office of the Admiralty Jan 15th, 1841. (Courtesy of Library of Congress)

29A. Waved albatross (*Phoebastria irrorata*), Hood Island.

29B. Sea lion (*Zalophus wollebaeki*), Tower Island.

29C. Galápagos penguin (*Spheniscus mendiculus*), Albemarle Island.

29D. Flightless cormorant (*Phalacrocorax harrisi*), Narborough Island.

29E. Blue-footed booby (*Sula nebouxii*), Daphne Island.

29F. Great frigatebird (*Fregata minor*), Tower Island.

30A. Wedgwood pottery. (Authors' collection)

30B. Charles Darwin, age 31, in 1840. Portrait by George Richmond. (Courtesy of Darwin Heirlooms Trust)

30C. Down House, Kent, 2005.

30D. The Royal Society's Darwin Medal. (Permission for photograph courtesy of Peter and Rosemary Grant)

31A. Tourist yachts in Academy Bay, Indefatigable Island, 2008.

31B. Introduced wasps (*Polistes versicolor*), James Island, 2007.

31C. Sign for the Galápagos National Park Service, Indefatigable Island.

31D. Charles Darwin Avenue, Puerto Ayora, Indefatigable Island, 2008.

32A. Street signs, "Avenida Charles Darwin" and "Avenida 12 de Febrero," Puerto Ayora, Indefatigable Island.

32B. Bust of Charles Darwin, Puerto Baquerizo Moreno, Chatham Island.

32C. Sign for the Charles Darwin Research Station, Indefatigable Island.

32D. Plaque commemorating Charles Darwin on statue at Puerto Baquerizo Moreno, Chatham Island.